# Advanced Information and Knowledge Processing

**Series Editors**

Professor Lakhmi Jain
Lakhmi.jain@unisa.edu.au

Professor Xindong Wu
xwu@cs.uvm.edu

For other titles published in this series, go to
http://www.springer.com/series/4738

Y. Narahari · Dinesh Garg · Ramasuri Narayanam
Hastagiri Prakash

# Game Theoretic Problems in Network Economics and Mechanism Design Solutions

 Springer

Y. Narahari
Indian Institute of Science
Department of Computer Science &
  Automation
Bangalore-560012
India
hari@csa.iisc.ernet.in

Ramasuri Narayanam
Indian Institute of Science
Department of Computer Science &
  Automation
Bangalore-560012
India
nrsuri@csa.iisc.ernet.in

Dinesh Garg
IBM India Research Lab
Embassy Golf Links Business Park
Bangalore-560071
India
gargdinesh@gmail.com

Hastagiri Prakash
Indian Institute of Science
Department of Computer Science &
  Automation
Bangalore-560012
India
hastagiri@gmail.com

AI&KP ISSN: 1610-3947
ISBN: 978-1-84996-807-2          e-ISBN: 978-1-84800-938-7
DOI: 10.1007/978-1-84800-938-7

British Library Cataloguing in Publication Data
A catalogue record for this book is available from the British Library

Springer Science+Business Media
springer.com

**Dedicated to**

**Our Beloved Parents**

*for giving us this wonderful life,*
*for teaching us the fundamentals of the game of life,*
*and for continuously inspiring us in this life*
*through their exemplary mechanisms*

*Yadati Narahari*
*Dinesh Garg*
*Ramasuri Narayanam*
*Hastagiri Prakash*

# Preface

The project of writing this monograph was conceived in August 2006. It is a matter of delight and satisfaction that this monograph would be published during the centenary year (May 27, 2008 – May 26, 2009) of our dear alma mater, the Indian Institute of Science, which is truly a magnificent temple and an eternal source of inspiration, with a splendid ambiance for research.

Studying the rational behavior of entities interacting with each other in organized or ad-hoc marketplaces has been the bread and butter of our research group here at the Electronic Commerce Laboratory, Department of Computer Science and Automation, Indian Institute of Science. Specifically, the application of game theoretic modeling and mechanism design principles to the area of network economics was an area of special interest to the authors. In fact, the dissertations of the second, third, and fourth authors (Dinesh Garg, Ramasuri Narayanam, and Hastagiri Prakash) were all in this area. Dinesh Garg's Doctoral Thesis, which later won the Best Dissertation Award at the Department of Computer Science and Automation, Indian Institute of Science for the academic year 2006-07, included an interesting chapter on applying the brilliant work of Roger Myerson (Nobel laureate in Economic Sciences in 2007) to the topical problem of sponsored search auctions on the web. Ramasuri's Master's work applied mechanism design to develop robust broadcast protocols in wireless ad hoc networks while Hastagiri's Master's work developed resource allocation mechanisms for computational grids. The moment we realized, in these three strands of work, the common thread of applying mechanism design to solve important current problems in network economics, the monograph suggested itself, and we immediately embarked on this journey of putting together this monograph.

The last decade (1999-2008), as ingeniously observed by Christos Papadimitriou in his sparkling foreword to the edited volume *Algorithmic Game Theory* (Cambridge University Press, 2007), has seen the convergence of two intellectual currents, *Game Theory* and *Computer Science*, created more than six decades ago by the legendary John von Neumann. In particular, during this decade, the application of game theory and mechanism design to problem solving in electronic commerce, supply chain management, protocol design in networks, etc., has seen a dramatic rise. This phenomenon certainly inspired our foray into this area in the year 2000. Concurrently, there were other developments that helped us to get locked into this area during the past eight years. Intel India, Bangalore, funded a collaborative project in 2000 that required the development of a multi-attribute combinatorial procurement auction for their indirect materials procurement. General Motors R & D, Warren, Michigan, next collaborated with our group to develop procurement auction mechanisms for their direct materials procurement during 2002-2004. Following this, General Motors R & D again funded us to design game theory and mechanism design based algorithms for procurement network formation during 2005-2007. Meanwhile, Infosys Technologies, Bangalore, collaborated with us in 2006-07 on applying game theory and mechanism design to the web services composition problem. More recently, the Office of Naval Research, Arlington, Virginia, USA, has funded an ongoing project on applying game theory to problem solving in complex networks. These projects helped us to investigate deep practical problems, providing an ideal complement to our theoretical work in the area. We have also been very fortunate to be working in this area during an eventful period when game theorists and mechanism designers have been awarded the Nobel Prize in Economic Sciences. We were excited when Professors Robert Aumann and Thomas Schelling were awarded the Nobel Prize 2005. In fact, we had an illuminating visit by Robert Aumann in January 2007 to the Indian Institute of Science. Our excitement reached a crescendo when, just two years later, Professors Leonid Hurwicz, Eric Maskin, and Roger Myerson were awarded the Nobel Prize in Economic Sciences in 2007 for their fundamental contributions to mechanism design theory.

Set in the above backdrop, the monograph is organized into two logical parts. The first part comprises a long chapter (Chapter 2) that contains an overview of foundational concepts and key results in mechanism design. This chapter is intended as a self-sufficient introduction to mechanism design theory with the help of carefully crafted, stylized network economics examples. The chapter also includes interesting biographical notes on legendary researchers who have made key contributions to game theory and mechanism design. The second part of the monograph contains an exposition of representative game theoretic problems in three different network economics situations and a systematic exploration of mechanism design solutions to these problems. This part has three chapters: Chapter 3 deals with the sponsored search auction problem, Chapter 4 with the resource allocation problem in computational grids, and Chapter 5 with robust broadcast protocol design problem in ad hoc networks. We conclude the monograph with Chapter 6 where we provide several pointers to the relevant literature to facilitate a deeper and broader investigation of problem solving with mechanism design.

The monograph has been structured with the objective of providing a sound foundation of relevant concepts and theory to help apply mechanism design to problem solving in a rigorous way. At the end of a serious reading of this monograph, the readers should be able to model real-world situations using game theory, analyze the situations using game theoretic concepts, and design correct and robust solutions (mechanisms, algorithms, protocols) that would work for agents that are rational and intelligent.

All the authors of this monograph have learned game theory by reading their favorite book *Game Theory: Analysis of Conflict* by Roger Myerson, and they have learned mechanism design by reading the amazingly comprehensive book *Microeconomic Theory* by Mas-Colell, Whinston, and Green. Therefore the readers would find many discussions in this monograph inspired in a quite striking way by these two classics. This is an involuntary response to the indelible impact and awesome influence these two books have had on all of us.

We wish to draw the attention of our readers regarding the use of certain words and phrases. We use the words *players* and *agents* interchangeably throughout the text. The words *bidders*, *buyers*, and *sellers* are often used to refer to players in an auction or a market. The words *he* and *his* are used symbolically to refer to both the genders. This has been done to avoid frequent usage of phrases such as *he/she* or *his/her*. Also, we have occasionally used the words *it* and *its* while referring to players or agents.

The monograph is targeted at several categories of audience. The first target includes first year graduate students and final year Master's students in the departments of computer science, electrical engineering, communications, systems engineering, industrial engineering and operations research, management science, and economics, and business schools. The next target includes researchers from both academic institutions and industries working in the areas of Network and Internet Economics, Internet Analytics, Electronic Commerce, Supply Chain Management, Wireless Networks, Communication Networks, Social Networks, etc. The monograph would serve as a suitable supplementary reference for courses such as Game Theory, Mechanism Design, Microeconomic Theory, Electronic Commerce, Network Economics, and Supply Chain management. The first author has successfully used the material from this monograph in two courses: Game Theory and Electronic Commerce. It is our sincere hope that the monograph will whet the appetite of the intended audience and create an intense interest in this exciting subject.

# Acknowledgments

It is our pleasant duty to recall the exemplary support we have received from numerous individuals and organizations. First, we thank General Motors R & D, Warren, Michigan, and the General Motors India Science Lab, Bangalore for their wonderful support during the past eight years. We also thank, for their splendid support, Intel India, Bangalore, during 2000-2003; Infosys Technologies, Bangalore, during 2006-07; and the Office of Naval Research, Arlington, Virgina, during 2007-08. The first author would like to thank the Homi Bhabha Fellowships Council, Mumbai, for awarding him a fellowship during 2006-07 to undertake this work.

The first author would like to express special thanks to Professor N. Viswanadham, Indian School of Business, for his phenomenal inspiration and encouragement at all times. The first author would also like to thank Professors V.V.S. Sarma (IISc, Bangalore), U.R. Prasad, V. Rajaraman (IISc, Bangalore), Vivek Borkar (TIFR, Mumbai), Vijay Chandru (Strand Life Sciences), Peter Luh (University of Connecticut, Storrs), Krishna Pattipati (University of Connecticut, Storrs), Richard Volz (Texas A & M University), Sanjoy Mitter (Massachusetts Institute of Technology), P. R. Kumar (University of Illinois, Urbana), K.S. Trivedi (Duke University), Ram Sriram (National Institute of Standards and Technology, Gaithersberg), Debasis Mitra (Bell Laboratories), Karmeshu (Jawaharlal University, New Delhi), and T.L. Johnson (General Electric R & D) for their constant support. He also wishes to thank his research collaborators: Professor N. Viswanadham, Professor N. Hemachandra (IIT Bombay), Dr. Jeffrey D. Tew (General Motors R & D), Dr. Datta Kulkarni (General Motors R & D), and Dr. K. Ravi Kumar (General Motors Indian Science Lab).

It is a true privilege and pleasure to be associated with the Indian Institute of Science. As already stated, this pleasure is made even more intense by the fact that the institute is currently celebrating its Centenary Year. We would like to thank our Director Professor P. Balaram and our Associate Director Professor N. Balakrishnan for their wonderful support and encouragement. Similarly the Department of Computer Science and Automation is like a paradise for researchers. We would like to remember the support and encouragement of Professor M. Narasimha Murty, chairman, and other colleagues at the Department of Computer Science and Automation.

The Latexing of Chapter 2 of this document was done flawlessly by Mrs. Asha-lata, who also created many of the pictures appearing in the monograph. Doctoral candidate Sujit Gujar should take a major credit for contributing in an invisible way to several sections and several pictures in this monograph. Sujit also has gone through carefully several draft versions of this monograph. In fact, the students at the Electronic Commerce Laboratory have provided excellent input on the various drafts.

All members of the Electronic Commerce Laboratory have been directly or in-directly involved with this monograph project over the years. We wish to thank all of them: Shantanu Biswas, Venkatapathi Raju, S. Kameshwaran, T.S. Chan-drashekar, Sujit Gujar, Pankaj Dayama, A. Rajagopal, K. Rajanikanth, Sourav Sen, Karthik Subbian, Ramakrishnan Kannan, Sunil Shelke, Nagaraj, Ashwin, Sriram, Prashanth, Raghav Gautam, Megha, Santosh Srinivas, Nikesh Srivastava, Kalyan Chakravarty, Radhanikanth, Siva Sankar, Soujanya, Durgesh, Mukti Jain, M.D. Srinivasulu, Kalyan, Madhuri, Chetan, Ravi Shankar, Maria Praveen Kumar, Shar-vani, Devansh, Chaitanya, Vadlamani Sastry, Sharat, and N. Hemanth.

We have received excellent support from the Springer, UK team and we thank Beverley Ford, Catherine Brett, Rebecca Mowat, and Frank Ganz for their incred-ible support. Our sincere thanks to Professor Lakshmi Jain, University of South Australia, and editor of the Springer AI & KP Series, for his excellent and efficient handling during the review phase of this monograph. We thank Professor Debasis Ghose, Department of Aerospace Engineering, IISc, for introducing us to Professor Lakshmi Jain.

For the first author, behind any effort of this kind, there are two immortal per-sonalities whose blessings form the inspirational force. They are his divine parents, Brahmasri Y. Simhadri Sastry and Matrusri Y. Nagavenamma. They are not here anymore in physical form but their magnificent personalities continue to be a bea-con and a driving force. This work is dedicated with great humility at their lotus feet. He would like to lovingly thank his better half, Padmashri, who has made numerous sacrifices during the past decade because of her husband's continuous struggle with his research. The same applies to his son Naganand, who put up with his father dur-ing very crucial and formative years. The first author also cannot forget the love and affection of his brothers, sisters-in-law, sister, and brother-in-law.

The second author is grateful to his loving parents, parents-in-law, and family members who have been a source of great strength and support. He would like to thank his better half, Richa, for her wonderful support.

The third author is thankful and obliged to his research supervisor Professor Y. Narahari for providing valuable guidance and suggestions in shaping up his research career and also in his personal life. He is deeply indebted to his affectionate par-ents Sri N. Hanumantha Rao and Srimati N. Nagamani. He would like to thank his brother Ranga Suri and his sister-in-law Mohana for their love and affection. He is always motivated by the suggestions and advices from his sister Satya Vani and his brother-in-law Kumar. He never forgets the inspiration and great support from his younger brother Krishna Suri. Interestingly, the writing of this monograph has coin-cided with his wedding to Lavanya. He would like thank his wife for her continuous support and encouragement. He strongly believes that all the efforts involved in successfully completing this monograph are due to the divine blessings of Lord Sri Venkateswara.

The fourth author wishes to dedicate this and everything in his life to his beloved parents, Srimathi Chitra and Shri Vanchi, without whom his very existence would not have come to be. He wishes to thank his fiancee, Manjari, and his sister, Haripriya, for being understanding and loving all the time. The four of them have motivated, counseled, loved, and cared the right amount. He owes a debt of gratitude to his mentor, guide, and teacher, Professor Narahari, for his keen insight, inspiring guidance, and kind nature. Professor Narahari has made the fourth author a better researcher and more importantly, a better person. The fourth author also thanks his friends and colleagues who have been wonderful supporters right through his life. He has journeyed from the motto: "Life is a Game: Play it" to the motto "Life is a Game: find its equilibrium and stay in it." He wishes to thank everyone who was and will be part of his game.

# Contents

# Acronyms

| | |
|---|---|
| NE | Nash Equilibrium |
| PSNE | Pure Strategy Nash Equilibrium |
| MSNE | Mixed Strategy Nash Equilibrium |
| SCF | Social Choice Function |
| IC | Incentive Compatible |
| DSIC | Dominant Strategy Incentive Compatible |
| BIC | Bayesian (Nash) Incentive Compatible |
| G-DSIC | Grid - Dominant Strategy Incentive Compatible Mechanism |
| DSIC-B | Dominant Strategy Incentive Compatible Broadcast |
| G-BIC | Grid - Bayesian Incentive Compatible Mechanism |
| BIC-B | Bayesian Incentive Compatible Broadcast |
| AE | Allocative Efficiency (Allocatively Efficient) |
| BB | (Strictly) Budget Balanced |
| WBB | Weakly Budget Balanced |
| EPE | Ex-Post Efficient |
| IR | Individually Rational |
| IIR | Interim Individually Rational |
| VCG | Vickrey-Clarke-Groves Mechanisms |
| GVA | Generalized Vickrey Auction |
| dAGVA | d'Aspremont and Gérard-Varet mechanisms |
| SSA | Sponsored Search Auction |
| GFP | Generalized First Price |
| GSP | Generalized Second Price |
| OPT | Optimal Mechanism |
| G-OPT | Grid - Optimal Mechanism |
| SRBT | Source Rooted Broadcast Tree |

# Chapter 1
# Introduction

With the advent of the Internet and other modern information and communication technologies, a magnificent opportunity has opened up for introducing new, innovative models of commerce, markets, and business. Creating these innovations calls for significant interdisciplinary interaction among researchers in computer science, communication networks, operations research, economics, mathematics, sociology, and management science. In the emerging era of new problems and challenges, one particular tool that has found widespread applications is game theory and mechanism design. Application areas where game theory and mechanism design have been extensively used in recent times include: Internet advertising, spectrum and bandwidth trading, electronic procurement, logistics, supply chain management, grid computing, wireless networks, peer-to-peer networks, social networks, and many other emerging Internet-based applications. In this monograph, we use the phrase *network economics* to describe the above application areas. Our focus in this monograph is to explore game theoretic modeling and mechanism design for problem solving in network economics. In this introductory chapter, we first present four examples of representative problems that motivate the use of game theory and mechanism design. Next, we provide a brief introduction to mechanism design. We conclude the chapter with a chapter-by-chapter outline of the monograph.

## 1.1 Motivating Problems in Network Economics

We present below four contemporary problems in the Internet and network economics area. These are representative of situations where game theory and mechanism design provide a natural modeling tool for problem solving.

### 1.1.1 Sponsored Web Search Auctions

Sponsored search is a fascinating example of a successful business model in network economics. When a web user searches a keyword (that is, a search phrase) on a search engine, the user gets back a page with results containing the links most

Y. Narahari et al., *Game Theoretic Problems in Network Economics and Mechanism Design Solutions*, Advanced Information and Knowledge Processing,
DOI: 10.1007/978-1-84800-938-7_1, © Springer-Verlag London Limited 2009

relevant to the keyword and also sponsored links that correspond to specific advertisers. See Figure 1.1.

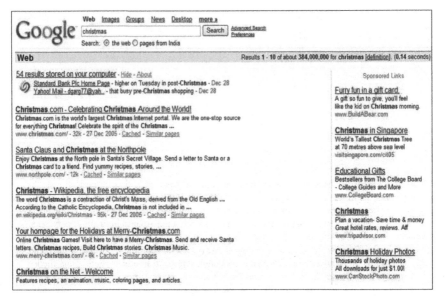

**Fig. 1.1** Result of a search performed on google.com

When a sponsored link is clicked, the user is directed to the corresponding advertiser's web page. The advertiser pays the search engine a certain amount of money for directing the user to its web page. Against every search performed by any user on any keyword, the search engine faces the problem of matching a set of advertisers to the (limited number of) sponsored slots. In addition, the search engine also needs to decide on a price that will be paid by the advertiser. Due to increasing demands for Internet advertising space, most search engines currently use an auction mechanism for this purpose. These are called sponsored search auctions. A significant percentage of the revenue of Internet giants such as Google, Yahoo!, MSN, etc., is contributed by sponsored search auctions.

In a typical sponsored search auction, advertisers are invited to specify their willingness to pay for their preferred keywords, that is, the maximum amount they would be willing to pay when an Internet user clicks on the respective sponsored slots. This willingness to pay is typically referred to as *cost-per-click*. Based on the bids submitted by the advertisers for a particular keyword, the search engine determines (1) a subset of advertisements (2) the order in which to display the selected advertisements (3) the payments to be made by the selected advertisers when their respective slots are clicked by a user. The actual price charged depends on the bids submitted by the advertisers. The decisions (1), (2), and (3) constitute the sponsored search auction mechanism.

The search engine would typically like to maximize its revenue whereas the advertisers would wish to achieve maximum payoffs within a given budget. This leads to a game situation where the search engine is the auctioneer and the advertisers are the players. The players are rational in the sense of trying to maximize their payoffs and this induces them to bid strategically after computing their best response bids. The problem of designing a sponsored search auction mechanism could be considered as a problem of designing a game involving the search engine and the advertisers. The rules of the game have to be designed in a way that a well defined set of criteria would be satisfied by the solution in an equilibrium of this game.

## 1.1.2 Resource Allocation in Grid Computing

A computational grid is a hardware and software infrastructure that provides dependable, consistent, pervasive, and inexpensive access to high-end computational capabilities. This access includes not only file exchange but also direct access to computers, software, data, and other resources, as is required by a range of collaborative problem solving in science, engineering, and the industry.

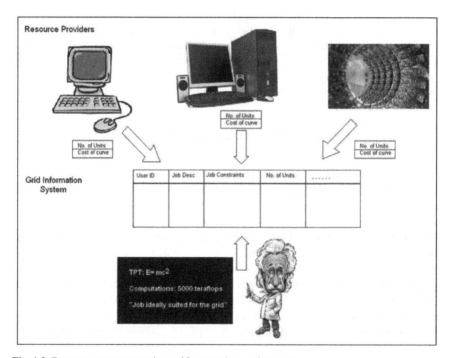

**Fig. 1.2** Resource procurement in a grid computing environment

The grid is an excellent example of innovative sharing of resources enabled by a high speed network. The stakeholders in a grid computing environment are the grid resource providers who are willing to make their resources available and the grid users who would like to consume these available resources. The sharing of resources is, necessarily, highly controlled, with grid resource providers and grid users defining clearly and carefully just what is shared, who is allowed to share, and the conditions under which sharing occurs. In this setting, *resource allocation* forms an important step in the overall facilitation. See Figure 1.2, which shows a grid user who requires to procure grid resources available with grid resource providers so as to execute a massive computing job. If the grid resource providers and the grid users are rational and intelligent, then there is likely to be an overall degradation of the total efficiency achieved by the computational grid when compared to what could be achieved if the participating resource providers and users were globally aligned under a single institution. This loss in efficiency might arise due to the unwillingness of a resource provider to either perform completely or perform to the fullest capability, the computational jobs of the grid users, or if the grid users are unwilling to reveal their true resource requirement. The grid resource providers and the grid users are potentially involved in a game situation with both conflict and cooperation possible in their interactions.

Given this setting, it is an interesting problem to study the behavior of the grid users and resource providers in the presence of economic incentives that would make them voluntarily participate in the grid functions and perform to their maximum capacity to fulfill the overall goals of the grid and increase its efficiency. Game theory and mechanism design have a central role to play in the design of these resource allocation protocols.

### 1.1.3 Protocol Design in Wireless Ad Hoc Networks

The wireless communications industry is currently one of the fastest growing industries in the world and is a primary source for network economics problems. The industry has several segments such as cellular telephony, satellite-based communication networks, wireless LANs, and ad hoc wireless networks. The current motivating example comes from *ad hoc wireless networks* that represent autonomous systems of nodes connected through wireless links. A wireless network does not have any fixed infrastructure such as base stations in cellular networks. The nodes in the network coordinate among themselves for communication. Hence, each node in the network, apart from being a source or destination, is also expected to route packets for other nodes in the network. Such networks find varied applications in real-life environments such as communication in battle fields, communication among rescue personnel in disaster affected areas, and wireless sensor networks. Figure 1.3 depicts typical elements in an ad hoc network.

The conventional protocols for routing, unicast, multicast, and broadcast in ad hoc wireless networks assume that nodes follow the prescribed protocol without

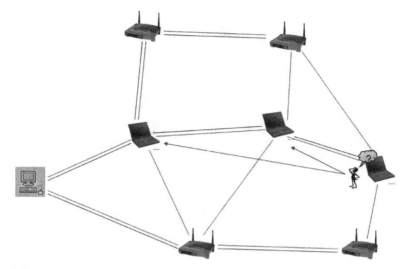

**Fig. 1.3** Communication in an ad hoc wireless network

any deviation and that the nodes cooperate with one another in performing network functions such as packet forwarding, etc. However, the nodes in a typical ad hoc network, if owned by independent and autonomous users, could be self-interested and may exhibit strategic behavior. Stimulating cooperation in these nodes could be achieved through incentive based mechanisms. The problem of designing communication and routing protocols that are robust to the strategic behavior of the nodes and that induce cooperative behavior by the nodes could be formulated as a mechanism design problem.

### 1.1.4 Supply Chain Network Formation

Consider a supply chain scenario where a supply chain planner is interested in forming an optimal network for delivering products/services. The supply chain would have multiple stages, and at each of these stages, the supply chain planner may have a choice of multiple service providers or supply chain partners. Figure 1.4 shows a multi-echelon supply chain with multiple service providers available at each echelon.

Each service provider has a service cost that is private information of the provider, and the service provider may not be willing to share this private information due to strategic reasons. The problem facing the supply chain planner here is to determine a network of service providers who will enable forming the supply chain at the least cost. Since the true private values of the service providers are not known, the supply chain planner is faced with an optimization problem with

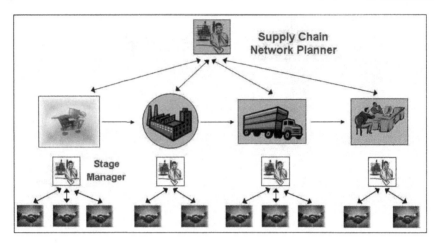

**Fig. 1.4** A supply chain formation scenario

incomplete information. Mechanism design offers a natural tool to be used in this setting.

A situation such as described above occurs in many other settings. The service providers could correspond to suppliers in which case we have a procurement network formation problem. The service providers could correspond to Internet/bandwidth service providers in which case we have a telecom service network formation problem between two cities.

## *1.1.5 Nature of the Problems*

In all of the problems above, problem solving involves implementing a *system-wide solution* that will satisfy certain *desirable properties*. In this process of problem solving, the following key common characteristics of these problems need to be taken into account:

- There is a set of decision makers or players who interact in a *strategic* way. The players have well defined payoff functions and are *rational* in the sense of having the sole objective of maximizing their own individual payoffs. The respective objectives of the individual players could be conflicting. Both conflict and cooperation could be possible during the interactions of these rational players.
- Each player holds certain information that is *private*, and only this player would know it deterministically; other players do not know this information deterministically. Thus the information in the system is decentralized, and each player only has incomplete information. Of course, there could be some information that all players know and all players know that all players know and so on. Such information is *common knowledge*.

- Each player has a choice of certain strategies that are available to him. The players have enough *intelligence* to determine their *best response strategies*.

Because of the above characteristics, the problems could be called *game theoretic problems*. Solving problems in such a setting would involve solving a *decision or optimization problem with incomplete information or incomplete specification*. Mechanism design offers a natural tool to solve such problems by providing an elegant way to do *reverse engineering* of games. Essentially, using techniques of mechanism design, we induce a game among the players in such a way that in an equilibrium of the induced game, the desired system-wide solution is implemented.

This monograph is all about how mechanism design could be used to solve game theoretic problems arising in Internet and network economics applications such as those described above.

## 1.2 Mechanism Design

In the second half of the twentieth century, game theory and mechanism design have found widespread use in a gamut of applications in engineering. In particular, game theory and mechanism design have emerged as an important tool to model, analyze, and solve *decentralized design problems* in engineering involving multiple autonomous agents that interact strategically in a *rational* and *intelligent* way. The field of mechanism design has been in intense limelight in recent times; the Nobel Prize in Economic Sciences for 2007 was jointly awarded to three economists, Leonid Hurwicz, Eric Maskin, and Roger Myerson *for having laid the foundations of mechanism design theory* [1]. Earlier, in 1996, William Vickrey, the inventor of the famous Vickrey auction, had been awarded the Nobel Prize in Economic Sciences.

The theory of *mechanism design* is concerned with settings where a policy maker (or social planner) faces the problem of aggregating the *announced preferences* of multiple agents into a collective (or social) decision when the *actual preferences* are not publicly known. Mechanism design theory uses the framework of *noncooperative games with incomplete information* and seeks to study how the privately held preference information can be elicited. The theory also clarifies the extent to which the preference elicitation problem constrains the way in which social decisions can respond to individual preferences. In fact, mechanism design can be viewed as *reverse engineering* of games or equivalently as the *art of designing the rules of a game to achieve a specific desired outcome*. The main focus of mechanism design is to design institutions or protocols that satisfy certain desired objectives, assuming that the individual agents, interacting through the institution, will act strategically and may hold private information that is relevant to the decision at hand.

## 1.2.1 Mechanisms: Some Simple Examples

Mechanisms have been used and practiced from times immemorial. Auctions provide a popular example of mechanisms; as is well known, auctions have been in vogue for a long time for selling, procuring, and exchanging goods and services.

Two simple, popular stories capture the idea behind mechanisms quite strikingly. The first story is that of a mother of two kids who has to design a mechanism to make her two kids share a cake equally. The mother is the social planner in this case, and the mechanism she designs is the following: (1) One of the kids would slice the cake into two pieces, and (2) the other kid would pick up one of the pieces, leaving the remaining piece to the kid who sliced the cake into two pieces. This mechanism implements the desirable outcome of the kids sharing the cake equally (of course, it would be interesting to see what a suitable mechanism would be if instead of two, there were more kids).

The second story is from ancient wisdom. This is attributed to several wise people. In India, it is attributed independently to Birbal, who was an adviser to Emperor Akbar in the late 1500s, and to Tenali Rama, who was a popular poet and adviser to the king in the court of the famous King Sri Krishna Devaraya of the Vijayanagara dynasty in the early 1500s. The story is also attributed to King Solomon in the Old Testament. In this fable, two women come to the king with a baby, each claiming to be the baby's mother, seeking justice. The clueless king turns to his adviser for advice. Birbal (Tenali Rama) (King Solomon) is supposed to have suggested that the baby be sliced into two pieces and the two pieces be equally shared by the two mothers. Upon which, one of the women (the real mother) immediately pleaded with the king not to resort to the cruelty. The king immediately ordered that the baby be handed over to that woman. This is an example of a truth elicitation mechanism.

Mechanisms such as above are ubiquitous in everyday life. The emergence of game theory during the 1940s and 1950s helped develop a formal theory of mechanism design starting from the 1960s.

## 1.2.2 Mechanism Design: A Brief History

Leonid Hurwicz (Nobel laureate in Economic Sciences in 2007) first introduced the notion of mechanisms with his work in 1960 [2]. He defined a mechanism as a communication system in which participants send messages to each other and perhaps to a *message center*, and a prespecified rule assigns an outcome (such as allocation of goods and payments to be made) for every collection of received messages. William Vickrey (Nobel laureate in Economic Sciences in 1996) wrote a classic paper in 1961 [3] that introduced the famous Vickrey auction (second price auction). To this day, the Vickrey auction continues to enjoy a special place in the annals of mechanism design. John Harsanyi (Nobel laureate in Economic Sciences in 1994 jointly with John Nash and Richard Selten) developed the theory of games with incomplete information, in particular Bayesian games, through a series of three seminal papers in 1967-68 [4, 5, 6]. Harsanyi's work later proved to be of foundational value to mechanism design. Hurwicz [7] introduced the key notion of incentive compatibility in 1972. This notion allowed mechanism design to incorporate the incentives of rational players and opened up mechanism design. Clarke [8] and Groves [9] came up with a generalization of the Vickrey mechanisms and helped define a broad class of dominant strategy incentive compatible mechanisms in the quasilinear environment.

There were two major advances in mechanism design in the 1970s. The first was the *revelation principle*, which essentially showed that direct mechanisms are the same as indirect mechanisms. This meant that mechanism theorists needed to worry only about direct mechanisms, leaving the development of real-world mechanisms (which are mostly indirect mechanisms) to mechanism designers and practitioners. Gibbard [10] formulated the revelation principle for dominant strategy incentive compatible mechanisms. This was later extended to Bayesian incentive compatible mechanisms through several independent efforts [1] — Maskin and Myerson (both Nobel laureates in Economic Sciences in 2007) had a leading role to play in this. In fact, Myerson developed the revelation principle in its greatest generality [1]. The second major advance in mechanism design in the 1970s was on *implementation theory*, which addresses the following problem: Can a mechanism be designed so that all its equilibria are optimal? Maskin [11] gave the first general solution to this problem.

Mechanism design has made phenomenal advances during 1980s, 1990s, and during the past few years. It has found widespread applicability in a variety of disciplines. These include: design of markets and trading institutions [1, 12, 13], regulation and auditing [1], social choice theory [1], and computer science [14]. The above list is by no means exhaustive. In this monograph, our focus is on applying mechanism design in the area of Internet and network economics.

## 1.3 Outline of the Monograph

The monograph has two logical parts. The first part comprises one chapter, namely Chapter 2, which contains an overview of foundational concepts and key results in mechanism design. This chapter is intended as a self-sufficient introduction to mechanism design theory with the help of stylized examples from network economics. The second part of the monograph contains an exposition of representative game theoretic problems in three different network economics situations and a systematic exploration of mechanism design solutions to these problems. This part has three chapters: Chapter 3 deals with the sponsored search auction problem, Chapter 4 with the resource allocation problem in computational grids, and Chapter 5 with the robust broadcast protocol design problem in ad hoc networks. We conclude the monograph with Chapter 6 where we provide several pointers to the relevant literature to facilitate a deeper and broader investigation of problem solving with mechanism design. We now provide a chapter-by-chapter outline.

## Chapter 2: Foundations of Mechanism Design

The chapter is organized into four parts. Part 1 sets the stage by describing essential aspects of game theory for understanding mechanism design, through five sections (Sections 2.1 to 2.5). The five sections deal with strategic form games, dominant strategy equilibria, pure strategy Nash equilibria, mixed strategy Nash equilibria, and Bayesian games. The results and the examples presented in these five sections provide a congenial backdrop for the rest of the chapter. In Part 2, which spans seven sections (Sections 2.6 to 2.12), fundamentals of mechanism design are brought out. The sections include a high level description of the mechanism design environment (Section 2.6); representative examples of social choice functions (Section 2.7); implementation of social choice functions by mechanisms (Section 2.8); incentive compatibility and revelation theorem (Section 2.9); properties of social choice functions (Section 2.10); the Gibbard–Satterthwaite impossibility theorem (Section 2.11); and the Arrow's impossibility theorem (Section 2.12). These seven sections set the stage for understanding the design of mechanisms, which constitutes the subject of Part 3. The sections in Part 3 include: The quasilinear environment (Section 2.13); Groves mechanisms (Section 2.14); Clarke mechanisms (Section 2.15); examples of VCG (Vickrey–Clarke–Groves) mechanisms (Section 2.16), the dAGVA mechanism (Section 2.17), Bayesian mechanisms in linear environment (Section 2.18), revenue equivalence theorem (Section 2.19), and optimal auctions (Section 2.20). Finally, in Part 4 (Section 2.21), we provide an expository introduction to further topics in mechanism design. A feature of this chapter is the inclusion of biographical notes, in appropriate contexts, of as many as 15 legendary researchers who have made significant contributions to game theory in general and to mechanism design in particular.

## Chapter 3: Mechanism Design for Sponsored Search Auctions

In Chapter 3, we study the sponsored web search problem as a compelling application of mechanism design. We first describe a framework to model this problem as a mechanism design problem under a reasonable set of assumptions. Using this framework, we describe three well known mechanisms for sponsored search auctions — *Generalized First Price (GFP), Generalized Second Price (GSP)*, and *Vickrey–Clarke–Groves (VCG)*. We then design an optimal auction mechanism by extending Myerson's optimal auction mechanism for a single indivisible good. For this, we impose the following well known requirements, which we feel are practical requirements for a sponsored search auction, namely — *revenue maximization, individual rationality*, and *Bayesian incentive compatibility* or *dominant strategy incentive compatibility*. We call this mechanism the *Optimal (OPT)* mechanism. We then make a comparative study of the GSP, VCG, and OPT mechanisms, along four different dimensions — *incentive compatibility, expected revenue earned by the search engine, individual rationality*, and *computational complexity*.

## Chapter 4: Resource Allocation in Computational Grids

In Chapter 4, we present a resource procurement problem in computational grids in the presence of strategic resource providers. The problem involves a grid user with a requirement to procure required resources for a computational job. The resources are to be procured from resource providers who might be strategic. The chapter offers three elegant mechanisms as solutions to the resource procurement problem. These mechanisms include:

- *G-DSIC* (Grid-Dominant Strategy Incentive Compatible) mechanism, which guarantees that truthful bidding is a best response for each resource provider, irrespective of what the other resource providers bid
- *G-BIC* (Grid-Bayesian Incentive Compatible) mechanism, which only guarantees that truthful bidding is a best response for each resource provider whenever all other resource providers also bid truthfully
- *G-OPT* (Grid-Optimal) mechanism, which minimizes the cost to the grid user, satisfying at the same time, (1) *Bayesian Incentive Compatibility* (which guarantees that truthful bidding is a best response for each resource provider whenever all other resource providers also bid truthfully) and (2) *Individual Rationality* (which guarantees that the resource providers have non-negative payoffs if they participate in the bidding process).

The mechanisms designed in this chapter are in the context of *parameter sweep* type of jobs, which consist of multiple homogeneous and independent tasks. However, the use of the mechanisms proposed transcends parameter sweep types of jobs, and in general, the proposed mechanisms could be extended to provide a robust way of procuring resources in any computational grid with strategically behaving resource providers.

## Chapter 5: Design of Incentive Compatible Broadcast Protocols in Ad Hoc Networks

Chapter 5 starts with an introduction to the broadcast problem in ad hoc networks with strategic nodes and brings out the need to use a game theoretic approach to solving the problem. We call this problem the incentive compatible broadcast problem (ICB). First we propose a dominant strategy incentive compatible mechanism, DSIC-B, that is built into the corresponding broadcast protocol, as a solution to the ICB problem. In the *DSIC-B* protocol, a node pays only for the node from which it has received the broadcast packet, and these payments are designed such that the mechanism is truthful in the dominant strategy sense, that is, true cost revelation is a best response for each node irrespective of what the other nodes report. There are, however, certain limitations of using the *DSIC-B* protocol. This provides the motivation to explore a Bayesian incentive compatible solution to the ICB problem. The proposed mechanism, which we call *BIC-B* (Bayesian incentive compatible broadcast) mechanism, satisfies several desirable economic properties and outperforms the DSIC-B protocol.

## *Chapter 6: To Probe Further*

This is a brief chapter where our main objective is to indicate the key advances in mechanism design in recent years and also suggest emerging application areas in Internet and network economics where mechanism design would be a useful tool to use. We also provide pointers to the current literature.

## References

1. The Nobel Foundation. The Sveriges Riksbank Prize in Economic Sciences in Memory of Alfred Nobel 2007: Scientific Background. Technical report, The Nobel Foundation, Stockholm, Sweden, December 2007.
2. L. Hurwicz. Optimality and informational efficiency in resource allocation processes. In K.J. Arrow, S. Karlin, and P. Suppes (eds.), *Mathematical Methods in the Social Sciences*, Stanford University Press, 1960.
3. W. Vickrey. Counterspeculation, auctions, and competitive sealed tenders. *Journal of Finance*, 16(1):8–37, 1961.
4. J.C. Harsanyi. Games with incomplete information played by Bayesian players. Part I: The basic model. *Management Science*, 14:159–182, 1967.
5. J.C. Harsanyi. Games with incomplete information played by Bayesian players. Part II: Bayesian equilibrium points. *Management Science*, 14:320–334, 1968.
6. J.C. Harsanyi. Games with incomplete information played by Bayesian players. Part III: The basic probability distribution of the game. *Management Science*, 14:486–502, 1968.
7. L. Hurwicz. On informationally decentralized systems. In *Decision and Organization. Radner and McGuire*. North-Holland, Amsterdam, 1972.
8. E. Clarke. Multi-part pricing of public goods. *Public Choice*, 11:17–23, 1971.
9. T. Groves. Incentives in teams. *Econometrica*, 41:617–631, 1973.
10. A. Gibbard. Manipulation of voting schemes. *Econometrica*, 41:587–601, 1973.
11. E. Maskin. Nash equilibrium and welfare optimality. *Review of Economic Studies*, 66:23–38, 1999.
12. A. Mas-Colell, M.D. Whinston, and J.R. Green. *Microeconomic Theory*. Oxford University Press, New York, 1995.
13. P. Milgrom. *Putting Auction Theory to Work*. Cambridge University Press, Cambridge, UK, 2004.
14. N. Nisan, T. Roughgarden, E. Tardos, and V.V. Vazirani. *Algorithmic Game Theory*. Cambridge University Press, New York, 2007.

# Chapter 2
# Foundations of Mechanism Design

This chapter forms the first part of the monograph and presents key concepts and results in mechanism design. The second part of the monograph explores application of mechanism design to contemporary problems in network economics. The chapter comprises 21 sections that can be logically partitioned into four groups. Sections 2.1 through 2.5 constitute Group 1, and they set the stage by describing essential aspects of game theory for understanding mechanism design. The five sections deal with strategic form games, dominant strategy equilibria, pure strategy Nash equilibria, mixed strategy Nash equilibria, and Bayesian games. Sections 2.6 through 2.12 constitute the next group of sections, and they deal with fundamental notions and results of mechanism design. The sections include a description of the mechanism design environment, social choice functions, implementation of social choice functions by mechanisms, incentive compatibility and revelation theorem, properties of social choice functions, the Gibbard–Satterthwaite impossibility theorem, and the Arrow's impossibility theorem. Following this, the sections in the third group (Sections 2.13 - 2.20) present useful mechanisms that provide the building blocks for solving mechanism design problems. The sections here include: The quasilinear environment, Groves mechanisms, Clarke mechanisms, examples of VCG mechanisms, the dAGVA mechanism, Bayesian mechanisms in linear environment, revenue equivalence theorem, and optimal auctions. Finally, in Section 2.21, we provide a sprinkling of further key topics in mechanism design. The chapter uses a fairly large number of stylized examples of network economics situations to illustrate the notions and the results.

## 2.1 Strategic Form Games

Game theory may be defined as the study of mathematical models of interaction between rational, intelligent decision makers [1]. The interaction may include both *conflict* and *cooperation*. The theory provides general mathematical techniques for analyzing situations in which two or more individuals (called players or agents) make decisions that influence one another's welfare. There are many categories of games that have been proposed and discussed in game theory. We introduce here a class of games called *strategic form games* or *normal form games*, which are most appropriate for the discussions in this monograph. We start with the definition of a strategic form game.

Y. Narahari et al., *Game Theoretic Problems in Network Economics and Mechanism Design Solutions*, Advanced Information and Knowledge Processing, DOI: 10.1007/978-1-84800-938-7_2, © Springer-Verlag London Limited 2009

**Definition 2.1 (Strategic Form Game).** A strategic form game $\Gamma$ is defined as a tuple $\langle N, (S_i)_{i \in N}, (u_i)_{i \in N} \rangle$, where $N = \{1, 2, \ldots, n\}$ is a finite set of players; $S_1, S_2, \ldots, S_n$ are the strategy sets of the players $1, \ldots, n$, respectively; and $u_i : S_1 \times S_2 \times \cdots \times S_n \to \mathbb{R}$ for $i = 1, 2, \ldots, n$ are mappings called the utility functions or payoff functions.

The strategies are also called *actions* or *pure strategies*. We denote by $S$, the Cartesian product $S_1 \times S_2 \times \cdots \times S_n$. The set $S$ is the collection of all strategy profiles of the players. Note that the utility of an agent depends not only on its own strategy but also on the strategies of the rest of the agents. Every profile of strategies induces an *outcome* in the game. A strategic form game is said to be *finite* if $N$ and all the strategy sets $S_1, \ldots, S_n$ are finite.

The idea behind a strategic form game is to capture each agent's decision problem of choosing a strategy that will counter the strategies adopted by the other agents. Each player is faced with this problem and therefore the players can be thought of as simultaneously choosing their strategies from the respective sets $S_1, S_2 \ldots, S_n$. We can view the play of a strategic game as follows: each player simultaneously writes down a chosen strategy on a piece of paper and hands it over to a referee who then computes the outcome and the utilities. Several examples of strategic form games will be presented in Section 2.1.2.

## 2.1.1 Key Notions

There are certain key notions underlying game theory. We discuss these notions and a few related issues.

### 2.1.1.1 Utilities

Utility theory enables the preferences of the players to be expressed in terms of payoffs in some utility scale. Utility theory is the science of assigning numbers to outcomes in a way that reflects the preferences of the players. The theory is an important contribution of von Neumann and Morgenstern, who stated and proved in [2] a crucial theorem called the *expected utility maximization theorem*. This theorem establishes for any rational decision maker that there must exist a way of assigning utility numbers to different outcomes in a way that the decision maker would always choose the option that maximizes his expected utility. This theorem holds under quite weak assumptions about how a rational decision maker should behave.

 **John von Neumann** (1903 - 1957) is regarded as one of the foremost mathematicians of the 20th century. In particular, he is regarded as the founding father of game theory. He was born in Budapest, Hungary on December 28, 1903. He was a mathematical genius from early childhood, but interestingly his first major degree was in chemical engineering from the Swiss Federal Institute of Technology in Zurich. In 1926, he earned a Doctorate in Mathematics from the University of Budapest, working with Professor Leopold Fezer. During 1926 to 1930, he taught in Berlin and Hamburg, and from 1930 to 1933, he taught at Princeton University. In 1933, he became the youngest of the six professors of the School of Mathematics at the Institute of Advanced Study in Princeton. Albert Einstein and Gödel were his colleagues at the center. During his glittering scientific career, von Neumann created several intellectual currents, two of the major ones being game theory and computer science. The fact that these two disciplines have converged during the 1990s and 2000s, almost sixty years after von Neumann brilliantly created them, is a testimony to his visionary genius. In addition to game theory and computer science, he made stunning contributions to a wide array of disciplines including set theory, functional analysis, quantum mechanics, ergodic theory, continuous geometry, numerical analysis, hydrodynamics, and statistics. He is best known for his minimax theorem, utility theory, von Neumann algebras, von Neumann architecture, and cellular automata.

In game theory, von Neumann's first significant contribution was the minimax theorem, which proves the existence of a randomized saddle point in two player zero sum games. His collaboration with Oskar Morgenstern at the Center for Advanced Study resulted in the classic book *The Theory of Games and Economic Behavior*, which to this day continues to be an excellent source for early game theory results. This, classic work develops many fundamental notions of game theory such as utilities, saddle points, coalitional games, bargaining sets, etc.

Von Neumann was associated with the development of the first electronic computer in the 1940s. He wrote a widely circulated paper entitled the First Draft of a Report on the EDVAC in which he described a computer architecture (which is now famously called the von Neumann architecture). He is also credited with the development of the notions of a computer algorithm and algorithm complexity.

### 2.1.1.2 Rationality

The first key assumption in game theory is that the players are rational. An agent is said to be rational if the agent makes decisions consistently in pursuit of its own objectives. It is assumed that each agent's objective is to maximize the expected value of its own payoff measured in some utility scale. The above notion of rationality (maximization of expected utility) was initially proposed by Bernoulli (1738) and later formalized by von Neumann and Morgenstern (1944) [2]. A key observation here would be that *self-interest* is essentially an implication of rationality.

It is important to note that maximizing expected utility is not necessarily the same as maximizing expected monetary returns. For example, a given amount of money may have significant utility to a person desperately in need of the money. The same amount of money may have much less utility to a person who is already rich. In general, utility and money are nonlinearly related.

When there are two or more players, it would be the case that the rational solution to each player's decision problem depends on the others' individual problems and vice-versa. None of the problems may be solvable without understanding the solutions of the other problems. When such rational decision makers interact, their decision problems must be analyzed together, like a system of simultaneous equations. Game theory, in a natural way, deals with such analysis.

### 2.1.1.3 Intelligence

Another key notion in game theory is that of intelligence of the players. This notion connotes that each player in the game knows everything about the game that a game theorist knows, and the player can make any inferences about the game that a game theorist can make. In particular, an intelligent player is *strategic*, that is, would fully take into account his knowledge or expectation of behavior of other agents in determining what his best response strategy should be. Each player is assumed to have enough computational resources to perform the required computations involved in determining a best response strategy.

Myerson [1] and several other authors provide the following convincing explanation to show that the assumptions of rationality and intelligence are indeed reasonable. The assumption that all individuals are rational and intelligent may not exactly be satisfied in a typical real-world situation. However, any theory that is not consistent with the assumptions of rationality and intelligence is fallible for the following reason: If a theory predicts that some individuals will be systematically deceived into making mistakes, then such a theory will lose validity when individuals learn through mistakes to understand the situations better. On the other hand, a theory based on rationality and intelligence assumptions would be credible.

### 2.1.1.4 Common Knowledge

The notion of common knowledge is an important implication of *intelligence*. Aumann (1976) [3] defined *common knowledge* as follows: A fact is common knowledge among the players if every player knows it, every player knows that every player knows it, and so on. That is, every statement of the form "every player knows that every player knows that ⋯ every player knows it" is true ad infinitum. If it happens that a fact is known to all the players, without the requirement of all players knowing that all players know it, etc., then such a fact is called *mutual knowledge*. In game theory, analysis often requires the assumption of common knowledge to be true; however, sometimes, the assumption of mutual knowledge suffices for the analysis. A player's *private information* is any information that the player has that is not common knowledge among all the players.

 **Robert Aumann** is a versatile game theorist who has stamped his authority with creative contributions in a wide range of topics in game theory such as repeated games, correlated equilibria, bargaining theory, cooperative game theory, etc. It was Aumann, who provided in 1976 [3], a convincing explanation of the notion of common knowledge in game theory, in a classic paper entitled *Agreeing to disagree* (which appeared in the Annals of Statistics). Aumann's work in the 1960s on repeated games clarified the difference between infinitely and finitely repeated games. With Bezalel Peleg in 1960, Aumann formalized the notion of a coalitional game without transferable utility (NTU), a significant result in cooperative game theory. With Michael Maschler (1963) he introduced the concept of a *bargaining set*, an important solution concept in cooperative game theory. In 1974, Aumann went on to define and formalize the notion of *correlated equilibrium* in Bayesian games. In 1975, Aumann proved a convergence theorem for the Shapley value. In 1976, in an unpublished paper with Lloyd Shapley, Aumann provided the perfect folk theorem using the limit of means criterion. All of these contributions have advanced game theory in significant ways. His books on *Values of Non-Atomic Games* (1984) co-authored with Lloyd Shapley and on *Repeated Games with Incomplete Information* (1995) co-authored with Michael Maschler are considered game theory classics.

Aumann was born in Frankfurt am Main, Germany on June 8, 1930. He earned an MSc Degree in Mathematics in 1952 from the Massachusetts Institute of Technology where he also received his Ph D Degree in 1955. His doctoral adviser at MIT was Professor George Whitehead Jr. and his doctoral thesis was on knot theory. He has been a professor at the Center for Rationality in the Hebrew University of Jerusalem, Israel, since 1956, and he also holds a visiting appointment with Stonybrook University, USA.

Robert Aumann and Thomas Schelling received the 2005 Nobel Prize in Economic Sciences for their contributions toward a clear understanding of conflict and cooperation through game theory analysis.

The intelligence assumption means that whatever a game theorist may know or understand about the game must be known or understood by the players of the game. Thus the model of the game is also known to the players. Since all the players know the model and they are intelligent, they also know that they all know the model. Thus the model is common knowledge.

In a *strategic form game with complete information*, the set $N$, the strategy sets $S_1, \ldots, S_n$, and the utility functions $u_1, \ldots, u_n$ are common knowledge, that is every player knows them, every player knows that every player knows them, etc. We will be studying strategic form games with complete information in this and the next three sections. We will study games with incomplete information in Section 2.5.

## 2.1.2 Examples of Strategic Form Games

We now provide several examples of game theoretic situations and formulate them as strategic form games.

*Example 2.1 (Matching Companies Game).* This example is developed on the lines of the famous matching pennies game, where there are two players who simultaneously put down a coin each, heads up or tails up. Each player is unaware of the move made by the other. If the two coins match, player 1 wins; otherwise, player 2 wins. In the version that we develop here, there are two companies, call them 1 and 2. Each company is capable of producing two products A and B, but at any given time, a company can only produce one product, owing to high setup and switchover costs. Company 1 is known to produce superior quality products but company 2 scores over company 1 in terms of marketing power and advertising innovations.

|   |   | 2 | |
|---|---|---|---|
| 1 | | A | B |
| A | | +1, −1 | −1, +1 |
| B | | −1, +1 | +1, −1 |

**Table 2.1** Payoff matrix for the matching companies game

If both the companies produce the same product (A or B), it turns out that company 1 makes all the profits and company 2 loses out, because of the superior quality of products produced by company 1. This is reflected in our model with a payoff of +1 for company 1 and a payoff of −1 for company 2, corresponding to the strategy profiles (A,A) and (B,B).

On the other hand, if one company produces product A and the other company produces product B, it turns out that because of the marketing skills of company 2 in differentiating the product offerings A and B, company 2 captures all the market, resulting in a payoff of +1 for company 2 and a payoff of −1 for company 1.

The two companies have to simultaneously decide (each one does not know the decision of the other) which product to produce. This is the strategic decision facing the two companies. This situation is captured by a strategic form game $\Gamma = \langle N, S_1, S_2, u_1, u_2 \rangle$, where $N = \{1,2\}$; $S_1 = S_2 = \{A,B\}$, and the utility functions are as described in Table 2.1.

*Example 2.2 (Battle of Companies Game).* This game is developed on the lines of the famous Battle of Sexes problem. Consider two companies, 1 and 2. As in Example 2.1, each company can produce two products A and B, but at any given time, a company can only produce one type of product, owing to high setup and switchover costs. The products A and B are competing products. Product A is a niche product of company 1 while product B is a niche product of company 2. If both the companies produce product A the consumers are compelled to buy product A and would naturally prefer to buy it from company 1 rather than from 2. Assume that company 1 will capture two thirds of the market. We will reflect this fact by saying that the payoff to company 1 is twice as much as for company 2. If both the companies produce product B, the reverse situation will prevail and company 2 will make twice as much payoff as company 1.

On the other hand, if the two companies decide to produce different products, then the market gets segmented, and each company tries to outwit the other through increased spending on advertising. In fact, their competition may actually benefit a third company, and effectively, neither of the original companies 1 or 2 makes any payoff. Table 2.2 depicts the payoff structure for this game.

|   | 2 | |
|---|---|---|
| 1 | A | B |
| A | 2, 1 | 0,0 |
| B | 0,0 | 1,2 |

**Table 2.2** Payoff matrix for the battle of companies game

*Example 2.3 (Company's Dilemma Problem).* This game is modeled on the lines of the popular prisoner's dilemma problem. Here again, we have two companies 1 and 2, each of which can produce two competing products A and B, but only one at a time. The companies are known for product A rather than for product B. Environmentalists have launched a negative campaign on product A branding it as non-eco friendly.

If both the companies produce product A, then, in spite of the negative campaign, their payoff is quite high since product A happens to be a niche product of both the companies. On the other hand, if both the companies produce product B, they still make some profit, but not as much as they would if they both produced product A.

On the other hand, if one company produces product A and the other company produces product B, then because of the negative campaign about product A, the company producing product A makes zero payoff while the other company captures all the market and makes a high payoff.

Table 2.3 depicts the payoff structure for this game. In our next example, we describe the classical *prisoner's dilemma problem.*

|   | 2 | |
|---|---|---|
| 1 | A | B |
| A | 6,6 | 0,8 |
| B | 8,0 | 3,3 |

**Table 2.3** Payoff matrix for the company's dilemma problem

*Example 2.4 (Prisoner's Dilemma Problem).* This is one of the most extensively studied problems in game theory, with many interesting interpretations cutting across disciplines. Two individuals are arrested for allegedly committing a crime and are lodged in separate prisons. The district attorney interrogates them separately. The attorney privately tells each prisoner that if he is the only one to confess, he will get a light sentence of 1 year in jail while the other would be sentenced to

10 years in jail. If both players confess, they would get 5 years each in jail. If neither confesses, then each would get 2 years in jail. The attorney also informs each prisoner what has been told to the other prisoner. Thus the payoff matrix is common knowledge. See Table 2.4.

| 1 | 2 NC | C |
|---|---|---|
| NC | -2, -2 | -10, -1 |
| C | -1, -10 | -5, -5 |

**Table 2.4** Payoff matrix for the prisoner's dilemma problem

How would the prisoners behave in such a situation? They would like to play a strategy that is best response to a (best) response strategy that the other player may adopt, the latter player also would like to play a best response to the other player's best response, and so on. First observe that C is each player's best strategy regardless of what the other player plays:

$$u_1(C,C) > u_1(NC,C); \quad u_1(C,NC) > u_1(NC,NC)$$

$$u_2(C,C) > u_2(C,NC); \quad u_2(NC,C) > u_2(NC,NC)$$

Thus (C,C) is a natural prediction for this game. However, the outcome (NC, NC) is the best outcome jointly for the players. Prisoner's Dilemma is a classic example of a game where rational, intelligent behavior does not lead to a socially optimal result. Also, each prisoner has a negative effect or externality on the other. When a prisoner moves away from (NC, NC) to reduce his jail term by 1 year, the jail term of the other player increases by 8 years.

*Example 2.5 (A Sealed Bid Auction).* There is a seller who wishes to allocate an indivisible item to one of $n$ prospective buyers in exchange for a payment. Here, $N = \{1,2,\ldots,n\}$ represents the set of buying agents. Let $v_1,v_2,\ldots,v_n$ be the valuations of the players for the object. The $n$ buying agents submit sealed bids and these bids need not be equal to the valuations. Assume that the sealed bid from player $i$ $(i = 1,\ldots,n)$ could be any real number greater than 0. Then the strategy sets of the players are: $S_i = (0,\infty)$ for $i = 1,\ldots,n$. Assume that the object is awarded to the agent with the lowest index among those who bid the highest. Let $b_1,\ldots,b_n$ be the bids from the $n$ players. Then the allocation function will be:

$$y_i(b_1,\ldots,b_n) = 1 \text{ if } b_i > b_j \text{ for } j = 1,2,\ldots,i-1 \text{ and}$$
$$b_i \geq b_j \text{ for } j = i+1,\ldots,n$$
$$= 0 \text{ else}$$

In the first price sealed bid auction, the winner pays an amount equal to his bid, and the losers do not pay anything. In the second price sealed bid auction, the winner pays an amount equal to the highest bid among the players who do not win, and as

usual the losers do not pay anything. The payoffs or utilities to the bidders in these
two auctions are of the form:

$$u_i(b_1,\ldots,b_n) = y_i(b_1,\ldots,b_n)(v_i - t_i(b_1,\ldots,b_n))$$

where $t_i(b_1,\ldots,b_n)$ is the amount to be paid by bidder $i$ in the auction. Suppose
$n = 4$, and suppose the values are $v_1 = 20$; $v_2 = 20$; $v_3 = 16$; $v_4 = 16$, and the
bids are $b_1 = 10$; $b_2 = 12$; $b_3 = 8$; $b_4 = 14$. Then for both first price and second
price auctions, we have the allocation $y_1(.) = 0$; $y_2(.) = 0$; $y_3(.) = 0$; $y_4(.) = 1$.
The payments for the first price auction are $t_1(.) = 0$; $t_2(.) = 0$; $t_3(.) = 0$; $t_4(.) =$
14 whereas the payments for the second price auction would be: $t_1(.) = 0$; $t_2(.) =$
$0$; $t_3(.) = 0$; $t_4(.) = 12$. The utilities can be easily computed from the values and the
payments.

An important question is: What will the strategies of the bidders be in these two
auctions. This question will be discussed at length in forthcoming sections.

*Example 2.6 (A Bandwidth Sharing Game).* This problem is based on an example
presented by Tardos and Vazirani [4]. There is a shared communication channel of
maximum capacity 1. There are $n$ users of this channel, and user $i$ wishes to send $x_i$
units of flow, where $x_i \in [0,1]$. We have

$$N = \{1,2,\ldots,n\}$$
$$S_1 = S_2 = \ldots = S_n = [0,1].$$

If $\sum_{i \in N} x_i \geq 1$, then the transmission cannot happen since the capacity is exceeded,
and the payoff to each player is zero. If $\sum_{i \in N} x_i < 1$, then assume that the following
is the payoff to user $i$:

$$u_i = x_i\left(1 - \sum_{j \in N} x_j\right)$$

The above expression models the fact that the payoff to a player is proportional to
the flow sent by the player but is negatively impacted by the total flow. The second
term captures the fact that the quality of transmission deteriorates with the total
bandwidth used. The above defines an $n$-player infinite game.

*Example 2.7 (A Duopoly Pricing Game).* This game model is due to Bertrand
(1883) [5]. Bertrand competition is a model of competition in a duopoly (that is an
economic environment with two competing companies), named after Joseph Louis
Franois Bertrand (1822-1900). There are two companies 1 and 2 that produce homogeneous products and that do not cooperate. The companies obviously wish to
maximize their profits. The quantity demanded for the product as a function of price
$p$ is given by a continuous and strictly decreasing function $x(p)$. The cost for producing each unit of product $= c > 0$. The companies simultaneously choose their
prices $p_1$ and $p_2$ and compete solely on price. We can see that the amount of sales
for each company is given by:

$$x_1(p_1,p_2) = x(p_1) \quad \text{if } p_1 < p_2$$
$$= \tfrac{1}{2}x(p_1) \text{ if } p_1 = p_2$$
$$= 0 \qquad \text{if } p_1 > p_2$$

$$x_2(p_1,p_2) = x(p_2) \quad \text{if } p_2 < p_1$$
$$= \tfrac{1}{2}x(p_2) \text{ if } p_2 = p_1$$
$$= 0 \qquad \text{if } p_2 > p_1.$$

It is assumed that the companies incur production costs only for an output level equal to their actual sales. Given prices $p_1$ and $p_2$, the utilities of the two companies would be:

$$u_1(p_1,p_2) = (p_1 - c)\,x_1(p_1,p_2)$$
$$u_2(p_1,p_2) = (p_2 - c)\,x_2(p_1,p_2).$$

Note that for this game, $N = \{1,2\}$ and $S_1 = S_2 = (0,\infty)$.

*Example 2.8 (A Procurement Exchange Game).* This example is adapted from an example presented by by Tardos and Vazirani [4]. Imagine a procurement exchange where buyers and sellers meet to match supply and demand for a particular product. Suppose that there are two sellers 1 and 2 and three buyers $A$, $B$, and $C$. Because of certain constraints such as logistics, assume that

- $A$ can only buy from seller 1.
- $C$ can only buy from seller 2.
- $B$ can buy from either seller 1 or seller 2.
- Each buyer has a maximum willingness to pay of 1 and wishes to buy one item.
- The sellers have enough items to sell.
- Each seller announces a price in the range $[0,1]$.

Let $s_1$ and $s_2$ be the prices announced. It is easy to see that buyer $A$ will buy an item from seller 1 at price $s_1$ and buyer $C$ will buy an item from seller 2 at price $s_2$. If $s_1 < s_2$, then buyer $B$ will buy an item from seller 1; otherwise buyer $B$ will buy from seller 2. Assume that buyer $B$ will buy from seller 1 if $s_1 = s_2$. The game can now be defined as follows:

$$N = \{1,2\}$$
$$S_1 = S_2 = [0,1]$$
$$u_1(s_1,s_2) = 2s_1 \text{ if } s_1 \leq s_2$$
$$= s_1 \text{ if } s_1 > s_2$$
$$u_2(s_1,s_2) = 2s_2 \text{ if } s_1 > s_2$$
$$= s_2 \text{ if } s_1 \leq s_2.$$

We now start analyzing strategic form games by looking at their equilibrium behavior. First we discuss the notion of dominant strategy equilibria. Next we introduce the notion of Nash equilibrium. The notation we use is summarized in Table 2.5.

| $N$ | A set of players, $\{1, 2, \ldots, n\}$ |
| $S_i$ | Set of actions or pure strategies of player $i$ |
| $S$ | Set of all action profiles $S_1 \times \ldots \times S_n$ |
| $s$ | A particular action profile, $s = (s_1, \ldots, s_n) \in S$ |
| $S_{-i}$ | Set of action profiles of all agents other than $i = S_1 \times \ldots S_{i-1} \times S_{i+1} \times \ldots \times S_n$ |
| $s_{-i}$ | A particular action profile of agents other than $i$, $s_{-i} = (s_1, \ldots, s_{i-1}, s_{i+1}, \ldots, s_n) \in S_{-i}$ |
| $(s_i, s_{-i})$ | Another representation for strategy profile $(s_1, \ldots, s_n)$ |
| $s_i^*$ | Equilibrium strategy of player $i$ |
| $u_i$ | Utility function of player $i$; $u_i : S \to \mathbb{R}$ |

**Table 2.5** Notation for a strategic form game

## 2.2 Dominant Strategy Equilibria

There are two notions of dominance that are aptly called strong dominance and weak dominance.

### 2.2.1 Strong Dominance

**Definition 2.2 (Strongly Dominated Strategy).** Given a game $\Gamma = \langle N, (S_i), (u_i) \rangle$, a strategy $s_i \in S_i$ is said to be strongly dominated if there exists another strategy $s_i' \in S_i$ such that

$$u_i(s_i', s_{-i}) > u_i(s_i, s_{-i}) \quad \forall s_{-i} \in S_{-i}.$$

We also say strategy $s_i'$ strongly dominates strategy $s_i$.

**Definition 2.3 (Strongly Dominant Strategy).** A strategy $s_i^* \in S_i$ is said to be a strongly dominant strategy for player $i$ if it strongly dominates every other strategy $s_i \in S_i$. That is, $\forall s_i \neq s_i^*$,

$$u_i(s_i^*, s_{-i}) > u_i(s_i, s_{-i}) \quad \forall s_{-i} \in S_{-i}.$$

**Definition 2.4 (Strongly Dominant Strategy Equilibrium).** A profile of strategies $(s_1^*, s_2^*, \ldots, s_n^*)$ is called a strongly dominant strategy equilibrium of the game $\Gamma = \langle N, (S_i), (u_i) \rangle$ if $\forall i = 1, 2, \ldots, n$, the strategy $s_i^*$ is a strongly dominant strategy for player $i$.

*Example 2.9 (Dominant Strategies in the Prisoner's Dilemma Problem).* Recall the prisoner's dilemma problem. Observe that the strategy NC is strongly dominated by C for player 1 since

$$u_1(C, NC) > u_1(NC, NC); \quad u_1(C, C) > u_1(NC, C).$$

Similarly, the strategy NC is strongly dominated by C for player 2 also, since

$$u_2(NC,C) > u_2(NC,NC); \quad u_2(C,C) > u_2(C,NC).$$

Thus $C$ is a strongly dominant strategy for both the players. Therefore $(C,C)$ is a strongly dominant strategy equilibrium for this game.

*Note 2.1.* If a player has a strongly dominant strategy then we should expect him to play it. On the other hand, if a player has a strongly dominated strategy, then we should expect him to not play it.

*Note 2.2.* A strongly dominant strategy equilibrium, if one exists, will be unique. The proof of this result is fairly straightforward.

### 2.2.2 Weak Dominance

**Definition 2.5 (Weakly Dominated Strategy).** Given a game $\Gamma = \langle N, (S_i), (u_i) \rangle$, a strategy $s_i \in S_i$ is said to be weakly dominated by a strategy $s_i' \in S_i$ for player $i$ if

$$u_i(s_i', s_{-i}) \geq u_i(s_i, s_{-i}) \ \forall s_{-i} \in S_{-i} \ \text{and} \ u_i(s_i', s_{-i}) > u_i(s_i, s_{-i}) \ \text{for some} \ s_{-i} \in S_{-i}.$$

The strategy $s_i'$ is said to weakly dominate strategy $s_i$.

**Definition 2.6 (Weakly Dominant Strategy).** A strategy $s_i^*$ is said to be a weakly dominant strategy for player $i$ if it weakly dominates every other strategy $s_i \in S_i$.

**Definition 2.7 (Weakly Dominant Strategy Equilibrium).** Given a game $\Gamma = \langle N, (S_i), (u_i) \rangle$, a strategy profile $(s_1^*, \ldots, s_n^*)$ is called a weakly dominant strategy equilibrium if for $i = 1, \ldots, n$, the strategy $s_i^*$ is a weakly dominant strategy for player $i$.

*Example 2.10 (A Modified Prisoner's Dilemma Problem).* Consider the payoff matrix of a slightly modified version of the prisoner's dilemma problem, shown in Table 2.6. Observe that the strategy $C$ is weakly dominant for player 1 since

|   1 |       | 2        |
|-----|-------|----------|
|     | NC    | C        |
| NC  | −2, −2 | −10, −2  |
| C   | −2, −10 | −5, −5  |

**Table 2.6** Payoff matrix of a modified prisoner's dilemma problem

$$u_1(C,NC) = u_1(NC,NC); \quad u_1(C,C) > u_1(NC,C).$$

Also, $C$ is weakly dominant for player 2 since

$$u_2(NC,C) = u_2(NC,NC); \quad u_2(C,C) > u_2(C,NC).$$

Therefore the strategy profile $(C,C)$ is a weakly dominant strategy equilibrium.

*Note 2.3.* It is to be noted that there could exist multiple weakly dominant strategies for a player, and therefore there could exist multiple weakly dominant strategy equilibria in a strategic form game.

*Example 2.11 (Second Price Sealed Bid Auction with Complete Information).* Consider the second price sealed bid auction for selling a single indivisible item discussed in Example 2.5. Let $b_1, b_2, \ldots, b_n$ be the bids (strategies), and we shall denote a bid profile (strategy profile) by $b = (b_1, b_2, \ldots, b_n)$. Assume that $v_i, b_i \in (0, \infty)$ for $i = 1, 2, \ldots, n$. Recall that the item is awarded to the bidder who has the lowest index among all the highest bidders. Recall the allocation function:

$$y_i(b_1, \ldots, b_n) \quad = 1 \text{ if } \quad b_i > b_j \text{ for } j = 1, 2, \ldots, i-1 \text{ and}$$
$$b_i \geq b_j \text{ for } j = i+1, \ldots, n$$
$$= 0 \text{ else.}$$

The payoff for each bidder is given by:

$$u_i(b_1, \ldots, b_n) = y_i(b_1, \ldots, b_n)(v_i - t_i(b_1, \ldots, b_n))$$

where $t_i(b_1, \ldots, b_n)$ is the amount paid by the winning bidder. Being second price auction, the winner pays only the next highest bid. We now show that the strategy profile $(b_1, \ldots, b_n) = (v_1, \ldots, v_n)$ is a weakly dominant strategy equilibrium for this game.

**Proof**: Consider bidder 1. His value is $v_1$ and bid is $b_1$. The other bidders have bids $b_2, \ldots, b_n$ and valuations $v_2, \ldots, v_n$. We consider the following cases.

> **Case 1**: $v_1 \geq \max(b_2, \ldots, b_n)$. There are two sub-cases here: $b_1 \geq \max(b_2, \ldots, b_n)$ and $b_1 < \max(b_2, \ldots, b_n)$.
> **Case 2**: $v_1 < \max(b_2, \ldots, b_n)$. There are two sub-cases here: $b_1 \geq \max(b_2, \ldots, b_n)$ and $b_1 < \max(b_2, \ldots, b_n)$.

We analyze these cases separately below.

**Case 1:** $v_1 \geq \max(b_2, \ldots, b_n)$.

We look at the following scenarios.

- Let $b_1 \geq \max(b_2, \ldots, b_n)$. This implies that bidder 1 is the winner, which implies that $u_1 = v_1 - \max(b_2, \ldots, b_n) \geq 0$.
- Let $b_1 < \max(b_2, \ldots, b_n)$. This means that bidder 1 is not the winner, which in turn means $u_1 = 0$.
- Let $b_1 = v_1$, then since $v_1 \geq \max(b_2, \ldots, b_n)$, we have $u_1 = v_1 - \max(b_2, \ldots, b_n)$.

Therefore, if $b_1 = v_1$, the utility $u_1$ is greater than or equal to the maximum utility obtainable. Thus, whatever the values of $b_2, \ldots, b_n$, it is a best response for player 1 to bid $v_1$. Thus $b_1 = v_1$ is a weakly dominant strategy for a bidder 1.

**Case 2:** $v_1 < \max(b_2, \ldots, b_n)$.

As before, we look at the following scenarios.

- Let $b_1 \geq \max(b_2, \ldots, b_n)$. This implies that bidder 1 is the winner and the payoff is given by:
$$u_1 = v_1 - \max(b_2, \ldots, b_n) \quad < 0.$$
- Let $b_1 < \max(b_2, \ldots, b_n)$. This means bidder 1 is not the winner. Therefore $u_1 = 0$.
- If $b_1 = v_1$, then bidder 1 is not the winner and therefore $u_1 = 0$.

From the above analysis, it is clear that $b_1 = v_1$ is a best response strategy for player 1 in Case 2 also. Combining our analysis of Case 1 and Case 2, we have that

$$u_1(v_1, b_2, \ldots, b_n) \geq u_1(\hat{b_1}, b_2, \ldots, b_n) \ \forall \, \hat{b_1} \in S_1 \ \ \forall \, b_2 \in S_2, \ \ldots, \ b_n \in S_n$$

Also, we can show (and this is left as an exercise) that, for any $b_1' \neq v_1$, we can always find $b_2 \in S_2, b_3 \in S_3, \ldots, b_n \in S_n$, such that

$$u_1(v_1, b_2, \ldots b_n) > u_1(b_1', b_2, \ldots, b_n).$$

Thus $b_1 = v_1$ is a weakly dominant strategy for a bidder 1. Using almost similar arguments, we can show that $b_i = v_i$ is a weakly dominant strategy for bidder $i$ where $i = 2, 3, \ldots, n$. Therefore $(v_1, \ldots, v_n)$ is a weakly dominant strategy equilibrium.

## 2.3 Pure Strategy Nash Equilibrium

Dominant strategy equilibria (strongly dominant, weakly dominant), if they exist, are very desirable but rarely do they exist because the conditions to be satisfied are too demanding. A dominant strategy equilibrium requires that each player's choice be a best response against all possible choices of all the other players. If we only insist that each player's choice is a best response against the best response strategies of the other players, we get the notion of Nash equilibrium. This solution concept derives its name from John Nash, one of the most celebrated game theorists of our times. In this section, we introduce and discuss the notion of pure strategy Nash equilibrium. In the following section, we discuss the notion of mixed strategy Nash equilibrium.

**John F. Nash, Jr.** is described by many as one of the most original mathematicians of the 20th Century. He was born in 1928 in Bluefield, West Virginia, USA. He completed his BS and MS in the same year (1948) at Carnegie Mellon University, majoring in Mathematics. He became a student of Professor Albert Tucker at Princeton University and completed his Ph.D. in Mathematics in 1950. His doctoral thesis (which had exactly 28 pages) proposed the brilliant notion of Nash Equilibrium, which helped expand the scope of game theory beyond two player zero sum games. His main result in his doctoral work settled the question of existence of a mixed strategy equilibrium in finite strategic form games. During his doctoral study, Nash also wrote a remarkable paper on the two player bargaining problem. He showed using a highly imaginative axiomatic approach that there exists a unique solution to the two person bargaining problem.

He worked as a professor of Mathematics at MIT in the Department of Mathematics where he did path-breaking work on algebraic geometry. He is also known for the Nash embedding theorem, which proves that any abstract Riemannian manifold can be isometrically realized as a sub-manifold of the Euclidean space. He is also known for his fundamental contributions to nonlinear parabolic partial differential equations.

The life and achievements of John Nash are fascinatingly captured in his biography *A Beautiful Mind* authored by Sylvia Nasar. This was later made into a popular movie with the same title. Professor Nash is currently at the Princeton University.

In 1994, John Nash was awarded the Nobel Prize in Economic Sciences, jointly with Professor John C. Harsanyi, University of California, Berkeley, CA, USA, and Professor Dr. Reinhard Selten, Rheinische Friedrich-Wilhelms-Universität, Bonn, Germany, for their pioneering analysis of equilibria in the theory of non-cooperative games.

**Definition 2.8 (Pure Strategy Nash Equilibrium).** Given a strategic form game $\Gamma = \langle N, (S_i), (u_i) \rangle$, the strategy profile $s^* = (s_1^*, s_2^*, \ldots, s_n^*)$ is said to be a pure strategy Nash equilibrium of $\Gamma$ if,

$$u_i(s_i^*, s_{-i}^*) \geq u_i(s_i, s_{-i}^*), \quad \forall s_i \in S_i, \quad \forall i = 1, 2, \ldots, n.$$

That is, each player's Nash equilibrium strategy is a best response to the Nash equilibrium strategies of the other players.

**Definition 2.9 (Best Response Correspondence for Player $i$).** Given a game $\Gamma = \langle N, (S_i), (u_i) \rangle$, the best response correspondence for player $i$ is the mapping $B_i : S_{-i} \to 2^{S_i}$ defined by

$$B_i(s_{-i}) = \{s_i \in S_i : u_i(s_i, s_{-i}) \geq u_i(s_i', s_{-i}) \ \forall s_i' \in S_i\}.$$

That is, given a profile $s_{-i}$ of strategies of the other players, $B_i(s_{-i})$ gives the set of all best response strategies of player $i$.

**Definition 2.10 (An Alternative Definition of Nash Equilibrium).** Given a strategic form game $\Gamma = \langle N, (S_i), (u_i) \rangle$, the strategy profile $(s_1^*, \ldots, s_n^*)$ is a Nash equilibrium iff,

$$s_i^* \in B_i(s_{-i}^*), \quad \forall i = 1, \ldots, n.$$

*Note 2.4.* Given a strategic form game $\Gamma = \langle N, (S_i), (u_i) \rangle$, a strongly dominant strategy equilibrium or a weakly dominant strategy equilibrium $(s_1^*, \ldots s_n^*)$ is also a Nash equilibrium. This can be shown easily. The intuitive explanation for this is as follows. In a dominant strategy equilibrium, the equilibrium strategy of each player is a best response irrespective of the strategies of the rest of the players. In a pure strategy Nash equilibrium, the equilibrium strategy of each player is a best response against the Nash equilibrium strategies of the rest of the players. Thus, the Nash equilibrium is a much weaker version of a dominant strategy equilibrium. It is also fairly obvious to note that a Nash equilibrium need not be a dominant strategy equilibrium.

### 2.3.1 Pure Strategy Nash Equilibria: Examples

*Example 2.12 (Battle of Companies).* Consider Example 2.2 from the previous section. There are two Nash equilibria here, namely (A,A) and (B,B). The profile (A,A) is Nash equilibrium because

$$u_1(A,A) > u_1(B,A); \quad u_2(A,A) > u_2(A,B).$$

The profile $(B,B)$ is a Nash equilibrium because

$$u_1(B,B) > u_1(A,B); \quad u_2(B,B) > u_2(B,A).$$

The best response sets are given by:

$$B_1(A) = \{A\}; \quad B_1(B) = \{B\}; \quad B_2(A) = \{A\}; \quad B_2(B) = \{B\}.$$

Since $A \in B_1(A)$ and $A \in B_2(A)$, $(A,A)$ is a Nash equilibrium. Similarly since $B \in B_1(B)$ and $B \in B_2(B)$, $(B,B)$ is a Nash equilibrium.

*Example 2.13 (Prisoner's Dilemma).* Consider the prisoner's dilemma problem introduced in Example 2.4. Note that $(C,C)$ is the unique pure strategy Nash equilibrium here. To see why, we look at the best response sets:

$$B_1(C) = \{C\}; \quad B_1(NC) = \{C\}; \quad B_2(C) = \{C\}; \quad B_2(NC) = \{C\}.$$

Since $s_i^* \in B_1(s_2^*)$ and $s_2^* \in B_2(s_1^*)$ for a Nash equilibrium, the only possible pure strategy Nash equilibrium here is $(C,C)$. In fact as already seen, this is a strongly dominant strategy equilibrium. Note that the strategy profile $(NC,NC)$ is not a Nash equilibrium, though, it is jointly the most desirable outcome for the two prisoners. Quite often, Nash equilibrium profiles are not necessarily the the best outcomes.

*Example 2.14 (Bandwidth Sharing Game).* Recall the bandwidth sharing game discussed in Example 2.6. We compute a Nash equilibrium for this game in the following way. Let $x_i$ be the amount of flow that player $i$ $(i = 1,2,\ldots,n)$ wishes to transmit on the channel and assume that

$$\sum_{i \in N} x_i < 1.$$

Consider player $i$ and define:

$$t = \sum_{j \neq i} x_j.$$

The payoff for the player $i$ is equal to

$$x_i(1 - t - x_i).$$

In order to maximize the above payoff, we have to choose

$$x_i^* = \arg \max_{x_i \in [0,1]} x_i(1 - t - x_i)$$

$$= \frac{1 - t}{2}$$

$$= \frac{1 - \sum_{j \neq i} x_j^*}{2}.$$

If this has to be satisfied for all $i \in N$, then we end up with $n$ simultaneous equations

$$x_i^* = \frac{1 - \sum_{j \neq i} x_j^*}{2} \quad i = 1, 2, \ldots, n.$$

A Nash equilibrium of this game is any solution to the above $n$ simultaneous equations. It can be shown that the above set of simultaneous equations has the unique solution:

$$x_i^* = \frac{1}{1 + n} \quad i = 1, 2, \ldots, n.$$

The profile $(x_1^*, \ldots, x_n^*)$ is thus a Nash equilibrium. The payoff for player $i$ in the above Nash equilibrium

$$= \left(\frac{1}{n+1}\right)\left(\frac{1}{n+1}\right).$$

Therefore the total payoff to all players combined

$$= \frac{n}{(n+1)^2}.$$

As shown below, the above is not a very happy situation. Consider the following non-equilibrium profile

$$\left(\frac{1}{2n}, \frac{1}{2n}, \ldots, \frac{1}{2n}\right).$$

This profile gives each player a payoff

$$= \frac{1}{2n}\left(1 - \frac{n}{2n}\right)$$

$$= \frac{1}{4n}.$$

Therefore the total payoff to all the players

$$= \frac{1}{4} > \frac{n}{(n+1)^2}.$$

Thus a non-equilibrium payoff $\left(\frac{1}{2n}, \frac{1}{2n}, \ldots, \frac{1}{2n}\right)$ provides more payoff than a Nash equilibrium payoff. This is referred to as a *tragedy of the commons*. In general, like in the prisoner's dilemma problem, the equilibrium payoffs may not be the best possible outcome for the players individually and also collectively. This lack of Pareto optimality is a property that Nash equilibrium payoffs often suffer from.

*Example 2.15 (Duopoly Pricing Game).* Recall the pricing game discussed in Example 2.7. There are two companies 1 and 2 that wish to maximize their profits by choosing their prices $p_1$ and $p_2$. The utilities of the two companies are:

$$u_1(p_1, p_2) = (p_1 - c) x_1(p_1, p_2)$$
$$u_2(p_1, p_2) = (p_2 - c) x_2(p_1, p_2).$$

Note that $u_1(c,c) = 0$ and $u_2(c,c) = 0$. Also, it can be easily noted that

$$u_1(c,c) \geq u_1(p_1,c) \ \forall p_1 \in S_1$$

$$u_2(c,c) \geq u_2(c,p_2) \ \forall p_2 \in S_2.$$

Therefore the strategy profile $(c,c)$ is a pure strategy Nash equilibrium. The implication of this result is that in the equilibrium, the companies set their prices equal to the marginal cost. The intuition behind this result is to imagine what would happen if both the companies set equal prices above marginal cost. Then the two companies would get half the market at a higher than marginal cost price. However, by lowering prices just slightly, a firm could gain the whole market, so both firms are tempted to lower prices as much as they can. It would be irrational to price below marginal cost, because the firm would make a loss. Therefore, both firms will lower prices until they reach the marginal cost limit.

*Example 2.16 (Game without a Pure Strategy Nash Equilibrium).* Recall the matching companies game (Example 2.1) and the payoff matrix for this game:

|   | 2 |  |
|---|---|---|
| 1 | A | B |
| A | $+1, -1$ | $-1, +1$ |
| B | $-1, +1$ | $+1, -1$ |

It is easy to see that this game does not have a pure strategy Nash equilibrium. This example shows that there is no guarantee that a pure strategy Nash equilibrium will exist. Later on in this section, we will state sufficient conditions under which a given strategic form game is guaranteed to have a pure strategy Nash equilibrium. In the next section, we will show that this game has a mixed strategy Nash equilibrium. We will now study another example that does not have a pure strategy Nash equilibrium.

*Example 2.17 (Procurement Exchange Game).* Recall Example 2.8. Let us investigate if this game has a pure strategy Nash equilibrium. First, we explore whether the strategy profile $(1, s_2)$ is a Nash equilibrium for any $s_2 \in [0, 1]$. Note that

$$
\begin{aligned}
u_1(1, s_2) &= 2 \text{ if } s_2 = 1 \\
&= 1 \text{ if } s_2 < 1 \\
u_2(1, s_2) &= 1 \text{ if } s_2 = 1 \\
&= 2s_2 \text{ if } s_2 < 1.
\end{aligned}
$$

It is easy to observe that $u_2(1, s_2)$ has a value $2s_2$ for $0 \le s_2 < 1$. Therefore $u_2(1, s_2)$ increases when $s_2$ increases from 0, until $s_2$ reaches 1 when it suddenly drops to 1. Thus it is clear that a profile of the form $(1, s_2)$ cannot be a Nash equilibrium for any $s_2 \in [0, 1]$. Similarly, no profile of the form $(s_1, 1)$ can be a Nash equilibrium for any $s_1 \in [0, 1]$.

Let us now explore if there exists any Nash equilibrium $(s_1^*, s_2^*)$, with $s_1^*, s_2^* \in [0, 1)$. There are two cases here.

- **Case 1**: If $s_1^* \le \frac{1}{2}$, then the best response for player 2 would be to bid $s_2 = 1$ since that would fetch him the maximum payoff. However bidding $s_2 = 1$ is not an option here since the range of values for $s_2$ is $[0, 1)$.
- **Case 2**: If $s_1^* > \frac{1}{2}$, there are two cases: (1) $s_1^* \le s_2^*$ (2) $s_1^* > s_2^*$. Suppose $s_1^* \le s_2^*$. Then

$$
\begin{aligned}
u_1(s_1^*, s_2^*) &= 2s_1^* \\
u_2(s_1^*, s_2^*) &= s_2^*.
\end{aligned}
$$

Choose $s_2$ such that $\frac{1}{2} < s_2 < s_1^*$. Then

$$
\begin{aligned}
u_2(s_1^*, s_2) &= 2s_2 \\
&> s_2^* \text{ since } 2s_2 > 1 \text{ and } s_2^* < 1 \\
&= u_2(s_1^*, s_2^*).
\end{aligned}
$$

Thus we are able to improve upon $(s_1^*, s_2^*)$ and hence $(s_1^*, s_2^*)$ is not a Nash equilibrium.

Now, suppose, $s_1^* > s_2^*$. Then

$$
\begin{aligned}
u_1(s_1^*, s_2^*) &= s_1^* \\
u_2(s_1^*, s_2^*) &= 2s_2^*.
\end{aligned}
$$

Now let us choose $s_1$ such that $1 > s_1 > s_1^*$. Then

$$
u_1(s_1, s_2^*) = s_1 > s_1^* = u_1(s_1^*, s_2^*).
$$

Thus we can always improve upon $(s_1^*, s_2^*)$. Therefore this game does not have a pure strategy Nash equilibrium. We wish to remark here that this game does not even have a mixed strategy Nash equilibrium.

## 2.3.2 Interpretations of Nash Equilibrium

Nash equilibrium is one of the most extensively discussed and debated topics in game theory. Many interpretations have been provided. Note that a Nash equilibrium is a profile of strategies, one for each of the $n$ players, that has the property that each player's choice is his best response given that the rest of the players play their Nash equilibrium strategies. By deviating from a Nash equilibrium strategy, a player will not be better off given that the other players stick to their Nash equilibrium strategies. The following discussion provides several interpretations put forward by game theorists.

A popular interpretation views a Nash equilibrium as a *prescription*. An adviser or a consultant to the $n$ players would essentially prescribe a Nash equilibrium strategy profile to the players. If the adviser recommends strategies that do not constitute a Nash equilibrium, then some players would find that it would be better for them to do differently than advised. If the adviser prescribes strategies that do constitute a Nash equilibrium, then the players are not unhappy because playing the equilibrium strategy is best under the assumption that the other players will play their equilibrium strategies.

Thus a logical, rational, adviser would recommend a Nash equilibrium profile to the players. The immediate caution, however, is that Nash equilibrium is an insurance against only unilateral deviations (that is, only one player at a time deviating from the equilibrium strategy). Two or more players deviating might result in players improving their payoffs compared to their equilibrium payoffs. For example, in the prisoner's dilemma problem, $(C,C)$ is a Nash equilibrium. If both the players decide to deviate, then the resulting profile is $(NC,NC)$, which is better for both the players. Note that $(NC,NC)$ is not a Nash equilibrium.

Another popular interpretation of Nash equilibrium is that of *prediction*. If the players are rational and intelligent, then a Nash equilibrium is a good prediction for the game. For example, a systematic elimination of strongly dominated strategies will lead to a reduced form that will include a Nash equilibrium. Often, iterated elimination of strongly dominated strategies leads to a unique prediction which would be invariably a Nash equilibrium.

An appealing interpretation of Nash equilibrium is that of *self-enforcing agreement*. A Nash equilibrium can be viewed as an implicit or explicit agreement between the players. Once this agreement is reached, it does not need any external means of enforcement because it is in the self-interest of each player to follow this agreement if the others do. In a noncooperative game, agreements cannot be enforced, hence, Nash equilibrium agreements are the only ones sustainable.

A natural, easily understood interpretation for Nash equilibrium has to do with *Evolution and Steady-State*. A Nash equilibrium is a potential stable point of a dynamic adjustment process in which players adjust their behavior to that of other players in the game, constantly searching for strategy choices that will give them the best results. This interpretation has been used to explain biological evolution. In this interpretation, Nash equilibrium is the outcome that results over time when a game is played repeatedly. A Nash equilibrium is like a stable social convention that people are happy to maintain forever.

Common knowledge was usually a standard assumption in determining conditions leading to a Nash equilibrium. More recently, it has been shown that the common knowledge assumption may not be required; instead, mutual knowledge is adequate. Suppose that each player is rational, knows his own payoff function, and knows the strategy choices of the others; then the strategy choices of the players will constitute a Nash equilibrium.

### 2.3.3 Existence of a Pure Strategy Nash Equilibrium

Consider a strategic form game $\Gamma = \langle N, (S_i), (u_i) \rangle$. We have seen examples (1) where only a unique pure strategy Nash equilibrium exists (2) where multiple pure strategy Nash equilibria exist, and (3) where a pure strategy Nash equilibrium does not exist. We now present a proposition that provides sufficient conditions under which a strategic form game is guaranteed to have at least one pure strategy Nash equilibrium. We do not delve into the technical aspects of this proposition; we refer the reader to consult [6] for more details.

**Proposition 2.1.** *Given a strategic form game $\Gamma = \langle N, (S_i), (u_i) \rangle$, a pure strategy Nash equilibrium exists if for all $i \in N$,*

- *$S_i$ is a non-empty, convex, compact subset of $\mathbb{R}^n$*
- *$u_i(s_1, \ldots, s_n)$ is continuous in $(s_1, \ldots, s_n)$ and quasi-concave in $s_i$*

In the next section where we consider mixed strategy Nash equilibria, we will state the celebrated Nash's Theorem, which guarantees the existence of a mixed strategy Nash equilibrium for a finite strategic form game.

### 2.3.4 Existence of Multiple Nash Equilibria

We have seen a few examples of strategic form games where multiple Nash equilibria exist. If a game has multiple Nash equilibria, then a fundamental question to ask is which of these would be implemented by the players? This question has been addressed by numerous game theorists, in particular, Thomas Schelling, who proposed the *focal point effect*. According to Schelling, anything that tends to focus the player's attention on one equilibrium may make them all expect it and hence fulfill it, like a self-fulfilling prophecy. Such a Nash equilibrium, which has some property that distinguishes it from all other equilibria is called a *focal equilibrium* or a *Schelling Point*.

As an example, consider the battle of companies game that we discussed in Example 2.2. Recall the payoff matrix of this game:

|   | 2   |     |
|---|-----|-----|
| 1 | A   | B   |
| A | 2,1 | 0,0 |
| B | 0,0 | 1,2 |

Here $(A,A)$ and $(B,B)$ are both Nash equilibria. If there is a special interest (or hype) created about product A, then $(A,A)$ may become the focal equilibrium. On the other hand, if there is a marketing blitz on product B, then $(B,B)$ may become the focal equilibrium.

 **Thomas Schelling** received, jointly with Robert Aumann, the 2005 Nobel Prize in Economic Sciences for contributions towards a clear understanding of conflict and cooperation through game theory analysis. Schelling's stellar contributions are best captured by several books that he has authored. The book *The Strategy of Conflict* that he wrote in 1960 is a classic work that has pioneered the study of bargaining and strategic behavior. It has been voted as one of the 100 most influential books since 1945. The notion of *focal point*, which is now called the *Schelling point* is introduced in this work to explain strategic behavior in the presence of multiple equilibria. Another book entitled *Arms and Influence* is also a popularly cited work. A highlight of Schelling's work has been to use simple game theoretic models in an imaginative way to obtain deep insights into global problems such as the cold war, nuclear arms race, war and peace, etc.

Schelling was born on April 14, 1921. He received his Doctorate in Economics from Harvard University in 1951. During 1950-53, Schelling was in the team of foreign policy advisers to the US President, and ever since, he has held many policy making positions in public service. He has played an influential role in the global warming debate also. He was at Harvard University from 1958 to 1990. Since 1990, he has been a Distinguished University Professor at the University of Maryland, in the Department of Economics and the School of Public policy.

## 2.4 Mixed Strategy Nash Equilibrium

### 2.4.1 Randomized Strategies or Mixed Strategies

Consider a strategic form game: $\Gamma = \langle N, (S_i), (u_i) \rangle$. The elements of $S_i$ are called the *pure strategies* of player $i$ $(i = 1, \ldots, n)$. If player $i$ randomly chooses one element of the set $S_i$, we have a mixed strategy or a randomized strategy. In the discussion that follows, we assume that $S_i$ is a finite for each $i = 1, 2, \ldots, n$.

**Definition 2.11 (Mixed Strategies).** Given a player $i$ with $S_i$ as the set of pure strategies, a mixed strategy $\sigma_i$ for player $i$ is a probability distribution over $S_i$. That is, $\sigma_i : S_i \to [0,1]$ assigns to each pure strategy $s_i \in S_i$, a probability $\sigma_i(s_i)$ such that

$$\sum_{s_i \in S_i} \sigma_i(s_i) = 1.$$

A pure strategy of a player, say $s_i \in S_i$, can be considered as a mixed strategy that assigns probability 1 to $s_i$ and probability 0 to all other strategies of player $i$. Such a mixed strategy is said to be a *degenerate mixed strategy* and is denoted by $e(s_i)$ or simply by $s_i$.

If $S_i = \{s_{i1}, s_{i2}, \ldots, s_{im}\}$, then clearly, the set of all mixed strategies of player $i$ is the set of all probability distributions on the set $S_i$. In other words, it is the simplex:

$$\Delta(S_i) = \left\{ (\sigma_{i1}, \ldots, \sigma_{im}) \in \mathbb{R}^m : \sigma_{ij} \geq 0 \text{ for } j = 1, \ldots, m \text{ and } \sum_{j=1}^{m} \sigma_{ij} = 1 \right\}.$$

The above simplex is called the *mixed extension* of $S_i$. Using the mixed extensions of strategy sets, we would like to define a mixed extension of the pure strategy game $\Gamma = \langle N, (S_i), (u_i) \rangle$. Let us denote the mixed extension of $\Gamma$ by

$$\Gamma_{ME} = \langle N, (\Delta(S_i)), (U_i) \rangle.$$

Note that, for $i = 1, 2, \ldots, n$,

$$U_i : \times_{i \in N} \Delta(S_i) \to \mathbb{R}.$$

Given $\sigma_i \in \Delta(S_i)$ for $i = 1, \ldots, n$, we compute $U_i(\sigma_1, \ldots, \sigma_n)$ as follows. First, we make the standard assumption that the randomizations of individual players are mutually independent. This implies that given a profile $(\sigma_1, \ldots, \sigma_n)$, the random variables $\sigma_1, \ldots, \sigma_n$ are mutually independent. Therefore the probability of a pure strategy profile $(s_1, \ldots, s_n)$ is given by

$$\sigma(s_1, \ldots, s_n) = \prod_{i \in N} \sigma_i(s_i).$$

The payoff functions $U_i$ are defined in a natural way as

$$U_i(\sigma_1, \ldots, \sigma_n) = \sum_{(s_1, \ldots, s_n) \in S} \sigma(s_1, \ldots, s_n) \, u_i(s_1, \ldots, s_n).$$

In the sequel, when there is no confusion, we will write $u_i$ instead of $U_i$. For example, instead of writing $U_i(\sigma_1, \ldots, \sigma_n)$, we will simply write $u_i(\sigma_1, \ldots, \sigma_n)$.

*Example 2.18 (Mixed Extension of the Battle of Companies Game).* Recall the game discussed in Example 2.2, having the following payoff matrix:

|   | 2 | |
|---|---|---|
| 1 | A | B |
| A | 2,1 | 0,0 |
| B | 0,0 | 1,2 |

Suppose $(\sigma_1, \sigma_2)$ is a mixed strategy profile. This means that $\sigma_1$ is a probability distribution on $S_1 = \{A, B\}$, and $\sigma_2$ is a probability distribution on $S_2 = \{A, B\}$. Let us represent

$$\sigma_1 = (\sigma_1(A), \sigma_1(B))$$
$$\sigma_2 = (\sigma_2(A), \sigma_2(B)).$$

We have

$$S = S_1 \times S_2 = \{(A,A), (A,B), (B,A), (B,B)\}.$$

We will now compute the payoff functions $u_1$ and $u_2$. Note that, for $i = 1, 2$,

$$u_i(\sigma_1, \sigma_2) = \sum_{(s_1, s_2) \in S} \sigma(s_1, s_2) \, u_i(s_1, s_2).$$

The function $u_1$ can be computed as

$$
\begin{aligned}
u_1(\sigma_1, \sigma_2) &= \sigma_1(A)\sigma_2(A)u_1(A,A) \\
&\quad + \sigma_1(A)\sigma_2(B)u_1(A,B) \\
&\quad + \sigma_1(B)\sigma_2(A)u_1(B,A) \\
&\quad + \sigma_1(B)\sigma_2(B)u_1(B,B) \\
&= 2\sigma_1(A)\sigma_2(A) + \sigma_1(B)\sigma_2(B) \\
&= 2\sigma_1(A)\sigma_2(A) + (1 - \sigma_1(A))(1 - \sigma_2(A)) \\
&= 1 + 3\sigma_1(A)\sigma_2(A) - \sigma_1(A) - \sigma_2(A).
\end{aligned}
$$

Similarly, we can show that

$$u_2(\sigma_1, \sigma_2) = 2 + 3\sigma_1(A)\sigma_2(A) - 2\sigma_1(A) - 2\sigma_2(A).$$

Suppose $\sigma_1 = \left(\frac{2}{3}, \frac{1}{3}\right)$ and $\sigma_2 = \left(\frac{1}{3}, \frac{2}{3}\right)$. Then it is easy to see that

$$u_1(\sigma_1, \sigma_2) = \frac{2}{3}; \quad u_2(\sigma_1, \sigma_2) = \frac{2}{3}.$$

### 2.4.2 Mixed Strategy Nash Equilibrium

We now define the notion of a mixed strategy Nash equilibrium.

**Definition 2.12 (Mixed Strategy Nash Equilibrium).** A mixed strategy profile $(\sigma_1^*, \ldots, \sigma_n^*)$ is called a Nash equilibrium if $\forall i \in N$,

$$u_i(\sigma_i^*, \sigma_{-i}^*) \geq u_i(\sigma_i, \sigma_{-i}^*) \quad \forall \sigma_i \in \Delta(S_i).$$

Define the best response functions $B_i(.)$ as follows.

$$B_i(\sigma_{-i}) = \{\sigma_i \in \Delta(S_i) : u_i(\sigma_i, \sigma_{-i}) \geq u_i(\sigma_i', \sigma_{-i}) \; \forall \sigma_i' \in \Delta(S_i)\}.$$

Then, clearly, a mixed strategy profile $(\sigma_1^*, \ldots, \sigma_n^*)$ is a Nash equilibrium iff

$$\sigma_i^* \in B_i(\sigma_{-i}^*) \quad \forall i = 1, 2, \ldots, n.$$

*Example 2.19 (Mixed Strategy Nash Equilibria for the Battle of Companies Game).*
Suppose $(\sigma_1, \sigma_2)$ is a mixed strategy profile. We have already shown that

$$u_1(\sigma_1, \sigma_2) = 1 + 3\sigma_1(A)\sigma_2(A) - \sigma_1(A) - \sigma_2(A)$$
$$u_2(\sigma_1, \sigma_2) = 2 + 3\sigma_1(A)\sigma_2(A) - 2\sigma_1(A) - 2\sigma_2(A).$$

Let $(\sigma_1^*, \sigma_2^*)$ be a mixed strategy equilibrium. Then

$$u_1(\sigma_1^*, \sigma_2^*) \geq u_1(\sigma_1, \sigma_2^*) \quad \forall \sigma_1 \in \Delta(S_1)$$

$$u_2(\sigma_1^*, \sigma_2^*) \geq u_2(\sigma_1^*, \sigma_2) \quad \forall \sigma_2 \in \Delta(S_2).$$

The above two equations are equivalent to:

$$3\sigma_1^*(A)\sigma_2^*(A) - \sigma_1^*(A) \geq 3\sigma_1(A)\sigma_2^*(A) - \sigma_1(A)$$

$$3\sigma_1^*(A)\sigma_2^*(A) - 2\sigma_2^*(A) \geq 3\sigma_1^*(A)\sigma_2(A) - 2\sigma_2(A).$$

The last two equations are equivalent to:

$$\sigma_1^*(A)\{3\sigma_2^*(A) - 1\} \geq \sigma_1(A)\{3\sigma_2^*(A) - 1\} \quad \forall \sigma_1 \in \Delta(S_1) \tag{2.1}$$

$$\sigma_2^*(A)\{3\sigma_1^*(A) - 2\} \geq \sigma_2(A)\{3\sigma_1^*(A) - 2\} \quad \forall \sigma_2 \in \Delta(S_2). \tag{2.2}$$

There are three possible cases.

- Case 1: $3\sigma_2^*(A) > 1$. This leads to the pure strategy Nash equilibrium $(A, A)$.
- Case 2: $3\sigma_2^*(A) < 1$. This leads to the pure strategy Nash equilibrium $(B, B)$.
- Case 3: $3\sigma_2^*(A) = 1$. This leads to the mixed strategy profile:

$$\sigma_1^*(A) = \frac{2}{3}; \ \sigma_1^*(B) = \frac{1}{3}; \ \sigma_2^*(A) = \frac{1}{3}; \ \sigma_2^*(B) = \frac{2}{3}.$$

This is a candidate mixed strategy Nash equilibrium. We will later on show that this is indeed a mixed strategy Nash equilibrium using a necessary and sufficient condition for a mixed strategy profile to be a Nash equilibrium, which we present next.

**Definition 2.13 (Support of a Mixed Strategy).** Let $\sigma_i$ be any mixed strategy of a player $i$. The support of $\sigma_i$, denoted by $\delta(\sigma_i)$, is the set of all pure strategies which have non-zero probabilities under $\sigma_i$. That is,

$$\delta(\sigma_i) = \{s_i \in S_i : \sigma_i(s_i) > 0\}.$$

**Theorem 2.1 (A Necessary and Sufficient Condition for Nash Equilibrium).** *The mixed strategy profile $(\sigma_1^*, \ldots, \sigma_n^*)$ is a mixed strategy Nash equilibrium iff for every player $i \in N$, we have*

*1. $u_i(s_i, \sigma_{-i}^*)$ is the same $\forall s_i \in \delta(\sigma_i^*)$.*

2. $u_i(s_i, \sigma_{-i}^*) \geq u_i(s_i', \sigma_{-i}^*) \ \forall s_i \in \delta(\sigma_i^*) \ \forall s_i' \notin \delta(\sigma_i^*)$

*(that is, the payoff of the player i for each pure strategy having non-zero probability is the same and is greater than or equal to the payoff for each pure strategy having zero probability).*

This theorem has much significance for computing Nash equilibria. Any standard textbook may be looked up (for example [1]) for a proof of this theorem. The theorem has the following implications.

1. In a mixed strategy Nash equilibrium, each player gets the same payoff (as in Nash equilibrium) by playing *any pure strategy* having positive probability in his Nash equilibrium strategy.
2. The above implies that the player can be indifferent about which of the pure strategies (with positive probability) he/she will play.
3. To verify whether or not a mixed strategy profile is a Nash equilibrium, it is enough to consider the effects of only pure strategy deviations.

*Example 2.20 (Mixed Strategy Nash Equilibrium for the Battle of Companies Game).* We now show, using the above theorem, that the strategy profile

$$\sigma_1^*(A) = \frac{2}{3}; \ \sigma_1^*(B) = \frac{1}{3}; \ \sigma_2^*(A) = \frac{1}{3}; \ \sigma_2^*(B) = \frac{2}{3}$$

is a mixed strategy Nash equilibrium. Note that $\delta(\sigma_1^*) = \{A, B\}$ and $\delta(\sigma_2^*) = \{A, B\}$. Note immediately that condition (2) of the above theorem is trivially satisfied for both the players. We will now investigate condition (1). Note that

$$u_1(A, \sigma_2^*) = u_1(B, \sigma_2^*) = 2 \times \frac{1}{3} + 0 \times \frac{2}{3} = \frac{2}{3}$$

$$u_2(\sigma_1^*, A) = u_2(\sigma_1^*, B) = 1 \times \frac{2}{3} + 0 \times \frac{1}{3} = = \frac{2}{3}.$$

Therefore, condition (2) is also satisfied for both the players. Thus the above profile is a mixed strategy Nash equilibrium.

*Example 2.21 (Mixed Strategy Nash Equilibrium for Matching Companies).* For this game (Example 2.1), we have seen that there does not exist a pure strategy Nash equilibrium. The mixed strategy profile $(\sigma_1^*, \sigma_2^*)$ defined by

$$\sigma_1^*(A) = \frac{1}{2}; \ \sigma_1^*(B) = \frac{1}{2}; \ \sigma_2^*(A) = \frac{1}{2}; \ \sigma_2^*(B) = \frac{1}{2}$$

can be easily shown to be a mixed strategy Nash equilibrium.

### 2.4.3 Existence of Mixed Strategy Nash Equilibrium

This is an important question that confronted game theorists for a long time. The first important result on this question was resolved by Von Neumann in 1928, who showed that a two person zero sum game is guaranteed to have a randomized saddle point (which is the same as a mixed strategy Nash equilibrium for two person zerosum games). In 1950, Nash introduced his notion of equilibrium for multiperson games and established a significant result that is stated in the following proposition.

**Theorem 2.2 (Nash's Theorem).** *Let* $\Gamma = \langle N, (S_i), (u_i) \rangle$ *be a finite strategic form game (that is, N is finite and $S_i$ is finite for each $i \in N$). Then $\Gamma$ has at least one mixed strategy Nash equilibrium.*

The proof of Nash's theorem is based on Kakutani's fixed point theorem. The books [6, 1] may be consulted for a proof of this theorem. Nash's theorem, however, does not guarantee the existence of a Nash equilibrium for an infinite game. For example, the pricing game in a procurement exchange (Example 2.8), which is an infinite game, does not have a mixed strategy Nash equilibrium.

### 2.4.4 Computation of Nash Equilibria

The problem of computing Nash equilibria is one of the fundamental computational problems in game theory. In fact, this problem has been listed as one of the important current challenges in the area of algorithms and complexity. The problem has generated intense interest since the 1950s. A famous algorithm for computing a Nash equilibrium in bimatrix games (two player non-zero sum games) is due to Lemke and Howson (1964). Another famous algorithm for bimatrix games is due to Mangasarian (1964). Rosenmuller (1971) extended the Lemke-Howson algorithm to $n$-person finite games. Wilson (1971) and Scarf (1973) developed efficient algorithms for $n$-person games. McKelvey and McLennan [7] have provided an excellent survey of various algorithms for computing Nash equilibria.

In the case of two person zero-sum games, computing a mixed strategy Nash equilibrium can be accomplished by solving a linear program. This is a consequence of the famous minimax theorem and the linear programming duality based formulation provided by Von Neumann. The running time, therefore, is worst case polynomial time for two person zerosum games.

In the case of two person non-zero sum games, the complexity is unknown in the general case. There are no known complexity results even for symmetric two player games or even for pure strategy Bayesian Nash equilibria. According to Papadimitriou [8], proof of NP-completeness seems impossible. Papadimitriou has defined a complexity class called TFNP and shows that the following problems belong to TFNP: (a) computing NE, (b) factoring, (c) fixed point problems, (d) local optimization. Computing equilibria with certain special properties (such as maximal payoff, maximal support) have been shown to be NP-hard.

Complexity results for $n$-person games are mostly open and this is currently an active area of research. The articles by Papadimitriou [8] and Stengel [9] present a survey of the current art in this area.

## 2.5 Bayesian Games

We have so far studied strategic form games with complete information. We will now study games with incomplete information, which are crucial to the theory of mechanism design. A game with *incomplete information* is one in which, at the first point in time when the players can begin to plan their moves in the game, some players have *private information* about the game that other players do not know. In contrast, in *complete information* games, there is no such private information, and all information is publicly known to everybody. Clearly, incomplete information games are more realistic, more practical.

The initial private information that a player has, just before making a move in the game, is called the *type* of the player. For example, in an auction involving a single indivisible item, each player would have a valuation for the item, and typically the player himself would know this valuation deterministically while the other players may only have a guess about how much this player values the item.

John Harsanyi (Joint Nobel Prize winner in Economic Sciences in 1994 with John Nash and Reinhard Selten) proposed in 1968, *Bayesian form* games to represent games with incomplete information.

In 1994, **John Charles Harsanyi** was awarded the Nobel Prize in Economic Sciences, jointly with Professor John Nash and Professor Reinhard Selten, for their pioneering analysis of equilibria in the theory of non-cooperative games. Harsanyi is best known for his work on games with incomplete information and in particular Bayesian games, which he published as a series of three celebrated papers titled *Games with incomplete information played by Bayesian players* in the Management Science journal in 1967 and 1968. His work on analysis of Bayesian games is of foundational value to mechanism design since mechanisms crucially use the framework of games with incomplete information. Harsanyi is also acclaimed for his intriguing work on *utilitarian ethics*, where he applied game theory and economic reasoning in political and moral philosophy. Harsanyi's collaboration with Reinhard Selten on the topic of equilibrium analysis resulted in a celebrated book entitled *A General Theory of Equilibrium Selection in Games* (MIT Press, 1988).

John Harsanyi was born in Budapest, Hungary, on May 29, 1920. He got two doctoral degrees — the first one in philosophy from the University of Budapest in 1947 and the second one in economics from Stanford University in 1959. His adviser at Stanford University was Professor Kenneth Arrow, who got the Economics Nobel Prize in 1972. Harsanyi worked at the University of California, Berkeley, from 1964 to 1990 when he retired. He died on August 9, 2000, in Berkeley, California.

 **Reinhard Selten** was jointly awarded the Nobel Prize in Economic Sciences in 1994, with Professor John Nash and Professor John Harsanyi. Harsanyi's analysis of Bayesian games was helped by the suggestions of Selten, and in fact Harsanyi refers to the type agent representation of Bayesian games as the *Selten game*. Selten is best known for his fundamental work on extensive form games and their transformation to strategic form through a representation called the agent normal form. Selten is also widely known for his deep work on bounded rationality. Furthermore, he is regarded as a pioneer of experimental economics. Harsanyi and Selten, in their remarkable book *A General Theory of Equilibrium Selection in Games* (MIT Press, 1988), develop a general framework to identify a unique equilibrium as the solution of a given finite strategic form game. Their solution can be thought of as a limit of an evolutionary process.

Selten was born in Breslau (currently in Poland but formerly in Germany) on October 5, 1930. He earned a doctorate in Mathematics from Frankfurt University, working with Professor Ewald Burger and Wolfgang Franz. He is currently a Professor Emeritus at the University of Bonn, Germany.

**Definition 2.14 (Bayesian Game).** A Bayesian game $\Gamma$ is defined as a tuple

$$\Gamma = \langle N, (\Theta_i), (S_i), (p_i), (u_i) \rangle$$

where

- $N = \{1, 2, \ldots, n\}$ is a set of players.
- $\Theta_i$ is the set of types of player $i$ where $i = 1, 2, \ldots, n$.
- $S_i$ is the set of actions or pure strategies of player $i$ where $i = 1, 2, \ldots, n$.
- The probability function $p_i$ is a function from $\Theta_i$ into $\Delta(\Theta_{-i})$, the set of probability distributions over $\Theta_{-i}$. That is, for any possible type $\theta_i \in \Theta_i$, $p_i$ specifies a probability distribution $p_i(.|\theta_i)$ over the set $\Theta_{-i}$ representing what player $i$ would believe about the types of the other players if his own type were $\theta_i$;
- The payoff function $u_i : \Theta \times S \to \mathbb{R}$ is such that, for any profile of actions and any profile of types $(\theta, s) \in \Theta \times S$, $u_i(\theta, s)$ specifies the payoff that player $i$ would get, in some *von Neumann – Morgenstern utility scale*, if the players' actual types were all as in $\theta$ and the players all chose their actions according to $s$.

The notation for Bayesian games is described in Table 2.7.

*Note 2.5.* When we study a Bayesian game, we assume that

1. Each player $i$ knows the entire structure of the game as defined above.
2. Each player knows his own type $\theta_i \in \Theta_i$.
3. The above facts are common knowledge among all the players in $N$.
4. The exact type of a player is not known deterministically to the other players who however have a probabilistic guess of what this type is. The belief functions $p_i$ describe these conditional probabilities. Note that the belief functions $p_i$ are also common knowledge among the players.

*Note 2.6.* The phrases *actions* and *strategies* are used differently in the Bayesian game context. A strategy for a player $i$ in Bayesian games is defined as a mapping

| $N$ | A set of players, $\{1,2,\ldots,n\}$ |
|---|---|
| $\Theta_i$ | Set of types of player $i$ |
| $S_i$ | Set of actions or pure strategies of player $i$ |
| $\Theta$ | Set of all type profiles $= \Theta_1 \times \Theta_2 \times \ldots \times \Theta_n$ |
| $\theta$ | $\theta = (\theta_1,\ldots,\theta_n) \in \Theta$; a type profile |
| $\Theta_{-i}$ | Set of type profiles of agents except $i = \Theta_1 \times \ldots \Theta_{i-1} \times \Theta_{i+1} \times \ldots \times \Theta_n$ |
| $\theta_{-i}$ | $\theta_{-i} \in \Theta_{-i}$; a profile of types of agents except $i$ |
| $S$ | Set of all action profiles $= S_1 \times S_2 \times \ldots \times S_n$ |
| $p_i$ | A probability (belief) function of player $i$ |
|  | A function from $\Theta_i$ into $\Delta(\Theta_{-i})$ |
| $u_i$ | Utility function of player $i$; $u_i : \Theta \times S \to \mathbb{R}$ |

**Table 2.7** Notation for a Bayesian game

from $\Theta_i$ to $S_i$. A strategy $s_i$ of a player $i$, therefore, specifies a pure action for each type of player $i$; $s_i(\theta_i)$ for a given $\theta_i \in \Theta_i$ would specify the pure action that player $i$ would play if his type were $\theta_i$. The notation $s_i(.)$ is used to refer to the pure action of player $i$ corresponding to an arbitrary type from his type set .

**Definition 2.15 (Consistency of Beliefs).** We say beliefs $(p_i)_{i \in N}$ in a Bayesian game are *consistent* if there is some common prior distribution over the set of type profiles $\Theta$ such that each player's beliefs given his type are just the conditional probability distributions that can be computed from the prior distribution by the Bayes' formula.

*Note 2.7.* If the game is finite, beliefs are consistent if there exists some probability distribution $P \in \Delta(\Theta)$ such that

$$p_i(\theta_{-i}|\theta_i) = \frac{P(\theta_i, \theta_{-i})}{\sum_{t_{-i} \in \Theta_{-i}} P(\theta_i, t_{-i})}$$

$$\forall \theta_i \in \Theta_i; \ \forall \theta_{-i} \in \Theta_{-i}; \ \forall i \in N.$$

Consistency simplifies the definition of the model. The common prior on $\Theta$ determines all the probability functions.

### 2.5.1 Examples of Bayesian Games

*Example 2.22 (A Two Player Bargaining Game).* This example is taken from the book by Myerson [1]. There are two players, player 1 and player 2. Player 1 is the seller of some object, and player 2 is a potential buyer. Each player knows what the object is worth to himself but thinks that its value to the other player may be any integer from 1 to 100 with probability $\frac{1}{100}$. Assume that each player will simultaneously announce a bid between 0 and 100 for trading the object. If the buyer's bid is greater than or equal to the seller's bid they will trade the object at a price equal to the average of their bids; otherwise no trade occurs. For this game:

$$N = \{1,2\}$$
$$\Theta_1 = \Theta_2 = \{1,2,\ldots,100\}$$
$$S_1 = S_2 = \{0,1,2,\ldots,100\}$$
$$p_i(\theta_{-i}|\theta_i) = \frac{1}{100} \quad \forall i \in N \quad \forall(\theta_i,\theta_{-i}) \in \Theta$$
$$u_1(\theta_1,\theta_2,s_1,s_2) = \frac{s_1+s_2}{2} - \theta_1 \quad \text{if } s_2 \geq s_1$$
$$= 0 \quad \text{if } s_2 < s_1$$
$$u_2(\theta_1,\theta_2,s_1,s_2) = \theta_2 - \frac{s_1+s_2}{2} \quad \text{if } s_2 \geq s_1$$
$$= 0 \quad \text{if } s_2 < s_1.$$

Note that the type of the seller indicates the willingness to sell (minimum price at which the seller is prepared to sell the item), and the type of the buyer indicates the willingness to pay (maximum price the buyer is prepared to pay for the item).

Also, note that the beliefs are consistent with the prior:

$$P(\theta_1,\theta_2) = \frac{1}{10000} \quad \forall \theta_1 \in \Theta_1 \quad \forall \theta_2 \in \Theta_2$$

where

$$\Theta_1 \times \Theta_2 = \{1,\ldots,100\} \times \{1,\ldots,100\}.$$

*Example 2.23 (Sealed Bid Auction).* Consider a seller who wishes to sell an indivisible item through an auction. Let there be two prospective buyers who bid for this item. The buyers have their individual valuations for this item. These valuations could be considered as the types of the buyers. Here the game consists of the two bidders, namely the buyers, so $N = \{1,2\}$. The two bidders submit bids, say $s_1$ and $s_2$ for the item. Let us say that the one who bids higher is awarded the item with a tie resolved in favor of bidder 1. The winner determination function therefore is:

$$f_1(s_1,s_2) = 1 \quad \text{if } s_1 \geq s_2$$
$$= 0 \quad \text{if } s_1 < s_2$$

$$f_2(s_1,s_2) = 1 \quad \text{if } s_1 < s_2$$
$$= 0 \quad \text{if } s_1 \geq s_2.$$

Assume that the valuation set for each buyer is the real interval $[0,1]$ and also that the strategy set for each buyer is again $[0,1]$. This means $\Theta_1 = \Theta_2 = [0,1]$ and $S_1 = S_2 = [0,1]$. If we assume that each player believes that the other player's valuation is chosen according to an independent uniform distribution, then note that

$$p_i([x,y]|\theta_i) = y - x \quad \forall \ 0 \leq x \leq y \leq 1; \quad i = 1,2.$$

In a first price auction, the winner will pay what is bid by her, and therefore the utility function of the players is given by

$$u_i(\theta_1, \theta_2, s_1, s_2) = f_i(s_1, s_2)(\theta_i - s_i); \quad i = 1, 2.$$

This completes the definition of the Bayesian game underlying a first price auction involving two bidders. One can similarly develop the Bayesian game for the second price sealed bid auction.

### 2.5.2 Type Agent Representation and the Selten Game

This is a representation of Bayesian games that enables a Bayesian game to be transformed to a strategic form game (with complete information). Given a Bayesian game

$$\Gamma = \langle N, (\Theta_i), (S_i), (p_i), (u_i) \rangle$$

the Selten game is an equivalent strategic form game

$$\Gamma^s = \langle N^s, (S_j^s), (U_j) \rangle.$$

The idea used in formulating a Selten game is to have *type agents*. Each player in the original Bayesian game is now replaced with a number of type agents; in fact, a player is replaced by exactly as many type agents as the number of types in the type set of that player. We can safely assume that the type sets of the players are mutually disjoint. The set of players in the Selten game is given by:

$$N^s = \bigcup_{i \in N} \Theta_i.$$

Note that each type agent of a particular player can play precisely the same actions as the player himself. This means that for every $\theta_i \in \Theta_i$,

$$S_{\theta_i}^s = S_i.$$

From now on, we will use $S^s$ and $S$ interchangeably whenever there is no confusion.

The payoff function $U_{\theta_i}$ for each $\theta_i \in \Theta_i$ is the conditionally expected utility to player $i$ in the Bayesian game given that $\theta_i$ is his actual type. It is a mapping with the following domain and co-domain:

$$U_{\theta_i} : \underset{i \in N}{\times} \underset{\theta_i \in \Theta_i}{\times} S_i \rightarrow \mathbb{R}.$$

We will explain the way $U_{\theta_i}$ is derived using an example. This example is developed, based on the illustration in the book by Myerson [1].

*Example 2.24 (Selten Game for a Bayesian Pricing Game).* Consider two firms, company 1 and company 2. Company 1 produces a product $x_1$ whereas company 2 produces either product $x_2$ or product $y_2$. The product $x_2$ is somewhat similar to product $x_1$ while the product $y_2$ is a different line of product. The product to be produced by company 2 is a closely guarded secret, so it can be taken as private information of company 2. We thus have $N = \{1,2\}$, $\Theta_1 = \{x_1\}$, and $\Theta_2 = \{x_2,y_2\}$. Each firm has to choose a price for the product it produces, and this is the strategic decision to be taken by the company. Company 1 has the choice of choosing a low price $a_1$ or a high price $b_1$ whereas company 2 has the choice of choosing a low price $a_2$ or a high price $b_2$. We therefore have $S_1 = \{a_1,b_1\}$ and $S_2 = \{a_2,b_2\}$. The type of company 1 is common knowledge since $\Theta_1$ is a singleton. Therefore, the belief probabilities of company 2 about company 1 are given by $p_2(x_1|x_2) = 1$ and $p_2(x_1|y_2) = 1$. Let us assume the belief probabilities of company 1 about company 2 to be $p_1(x_2|x_1) = 0.6$ and $p_1(y_2|x_1) = 0.4$. To complete the definition of the Bayesian game, we now have to specify the utility functions. Let the utility functions for the two possible type profiles $\theta_1 = x_1$, $\theta_2 = x_2$ and $\theta_1 = x_1$, $\theta_2 = y_2$ be given as in Tables 2.8 and 2.9.

|   |   | 2 |
|---|---|---|
| 1 | $a_2$ | $b_2$ |
| $a_1$ | 1,2 | 0, 1 |
| $b_1$ | 0,4 | 1, 3 |

**Table 2.8** $u_1$ and $u_2$ for $\theta_1 = x_1$; $\theta_2 = x_2$

|   |   | 2 |
|---|---|---|
| 1 | $a_2$ | $b_2$ |
| $a_1$ | 1,3 | 0, 4 |
| $b_1$ | 0,1 | 1, 2 |

**Table 2.9** $u_1$ and $u_2$ for $\theta_1 = x_1$; $\theta_2 = y_2$

This completes the description of the Bayesian game. We now derive the equivalent Selten game:

$$\langle N^s, (S_{\theta_i})_{\substack{\theta_i \in \Theta_i \\ i \in N}}, (U_{\theta_i})_{\substack{\theta_i \in \Theta_i \\ i \in N}} \rangle.$$

We have

$$N^s = \Theta_1 \cup \Theta_2 = \{x_1, x_2, y_2\}$$
$$S_{x_1} = S_1 = \{a_1, b_1\}$$
$$S_{x_2} = S_{y_2} = S_2 = \{a_2, b_2\}.$$

Note that

$$U_{\theta_i} : S_1 \times S_2 \times S_2 \rightarrow \mathbb{R} \ \forall \theta_i \in \Theta_i, \forall i \in N$$

$$S_1 \times S_2 \times S_2 = \{(a_1,a_2,a_2),(a_1,a_2,b_2),(a_1,b_2,a_2),(a_1,b_2,b_2),(b_1,a_2,a_2),$$
$$(b_1,a_2,b_2),(b_1,b_2,a_2),(b_1,b_2,b_2)\}.$$

The above set gives the set of all strategy profiles of all the type agents. A typical strategy profile can be represented as $(s_{x_1}, s_{x_2}, s_{y_2})$. This could also be represented as $(s_1(.),s_2(.))$ where the strategy $s_1$ is a mapping from $\Theta_1$ to $S_1$, and the strategy $s_2$ is a mapping from $\Theta_2$ to $S_2$. In general, for an $n$ player Bayesian game, a pure strategy profile is of the form

$$((s_{\theta_1})_{\theta_1 \in \Theta_1}, (s_{\theta_2})_{\theta_2 \in \Theta_2}, \ldots, (s_{\theta_n})_{\theta_n \in \Theta_n}).$$

Another way to write this would be $(s_1(.),s_2(.),\ldots,s_n(.))$, where $s_i$ is a mapping from $\Theta_i$ to $S_i$ for $i = 1,2,\ldots,n$.

The payoffs for type agents (in the Selten game) are obtained as conditional expectations over the type profiles of the rest of the agents. For example, let us compute the payoff $U_{x_1}(a_1,a_2,a_2)$, which is the expected payoff obtained by type agent $x_1$ (belonging to player 1 ) when this type agent plays action $a_1$ and the type agents $x_2$ and $y_2$ of player 2 play the actions $a_2$ and $a_2$ respectively. In this case, the type of player 1 is known, but the type of player could be $x_2$ or $y_2$ with probabilities given by the belief function $p_1(.|x_1)$. The following conditional expectation gives the required payoff.

$$\begin{aligned}
U_{x_1}(a_1,a_2,a_2) &= p_1(x_2|x_1)u_1(x_1,x_2,a_1,a_2) \\
&\quad + p_1(y_2|x_1)u_1(x_1,y_2,a_1,a_2) \\
&= (0.6)(1) + (0.4)(1) \\
&= 0.6 + 0.4 \\
&= 1.
\end{aligned}$$

Similarly, the payoff $U_{x_1}(a_1,a_2,b_2)$ can be computed as follows.

$$\begin{aligned}
U_{x_1}(a_1,a_2,b_2) &= p_1(x_2|x_1)u_1(x_1,x_2,a_1,a_2) \\
&\quad + p_1(y_2|x_1)u_1(x_1,y_2,a_1,b_2) \\
&= (0.6)(1) + (0.4)(0) \\
&= 0.6.
\end{aligned}$$

It can be similarly shown that

$$U_{x_1}(b_1, a_2, a_2) = 0$$
$$U_{x_1}(b_1, a_2, b_2) = 0.4$$
$$U_{x_2}(a_1, a_2, b_2) = 2$$
$$U_{x_2}(a_1, b_2, b_2) = 1$$
$$U_{y_2}(a_1, a_2, b_2) = 4$$
$$U_{y_2}(a_1, a_2, a_2) = 3.$$

(2.3)

From the above, we see that

$$U_{x_1}(a_1, a_2, b_2) > U_{x_1}(b_1, a_2, b_2)$$
$$U_{x_2}(a_1, a_2, b_2) > U_{x_2}(a_1, b_2, b_2)$$
$$U_{y_2}(a_1, a_2, b_2) > U_{y_2}(a_1, a_2, a_2).$$

(2.4)

From this, we can conclude that the action profile $(a_1, a_2, b_2)$ is a Nash equilibrium of the type agent representation.

### 2.5.2.1 Payoff Computation in Selten Game

From now on, when there is no confusion, we will use $u$ instead of $U$. In general, given: **(1)** a Bayesian game $\Gamma = \langle N, (\Theta_i), (S_i), (p_i), (u_i) \rangle$, **(2)** its equivalent Selten game $\Gamma^s = \langle N^s, (S_{\theta_i}), (U_{\theta_i}) \rangle$, and **(3)** an action profile in the type agent representation of the form

$$((s_{\theta_1})_{\theta_1 \in \Theta_1}, (s_{\theta_2})_{\theta_2 \in \Theta_2}, \ldots, (s_{\theta_n})_{\theta_n \in \Theta_n}),$$

the payoffs $u_{\theta_i}$ for $\theta_i \in \Theta_i$ ($i \in N$) are computed as follows.

$$u_{\theta_i}(s_{\theta_i}, s_{-\theta_i}) = \sum_{t_{-i} \in \Theta_{-i}} p_i(t_{-i} | \theta_i) u_i(\theta_i, t_{-i}, s_{\theta_i}, s_{t_{-i}})$$

where $s_{t_{-i}}$ is the strategy profile corresponding to the type agents in $t_{-i}$. A concise way of writing the above would be:

$$u_{\theta_i}(s_{\theta_i}, s_{-\theta_i}) = E_{\theta_{-i}}[u_i(\theta_i, \theta_{-i}, s_{\theta_i}, s_{\theta_{-i}})].$$

The notation $u_{\theta_i}$ refers to the utility of player $i$ conditioned on the type being equal to $\theta_i$. We will be using this notation frequently in this section. With this setup, we now look into the notion of an equilibrium in Bayesian games.

## 2.5.3 Equilibria in Bayesian Games

**Definition 2.16 (Pure Strategy Bayesian Nash Equilibrium).** A pure strategy Bayesian Nash equilibrium in a Bayesian game

$$\Gamma = \langle N, (\Theta_i), (S_i), (p_i), (u_i) \rangle$$

can be defined in a natural way as a pure strategy Nash equilibrium of the equivalent Selten game. That is, a profile of type agent strategies

$$s^* = ((s^*_{\theta_1})_{\theta_1 \in \Theta_1}, (s^*_{\theta_2})_{\theta_2 \in \Theta_2}, \ldots, (s^*_{\theta_n})_{\theta_n \in \Theta_n})$$

is said to be a pure strategy Bayesian Nash equilibrium of $\Gamma$ if $\forall i \in N$, $\forall \theta_i \in \Theta_i$,

$$u_{\theta_i}(s^*_{\theta_i}, s^*_{-\theta_i}) \geq u_{\theta_i}(s_i, s^*_{-\theta_i}) \ \forall s_i \in S_i.$$

Alternatively, a strategy profile $(s^*_1(.), s^*_2(.), \ldots, s^*_n(.))$ is said to be a Bayesian Nash equilibrium if

$$u_{\theta_i}(s^*_i(\theta_i), s^*_{-i}(\theta_{-i})) \geq u_{\theta_i}(s_i, s^*_{-i}(\theta_{-i})) \ \forall s_i \in S_i \ \forall \theta_i \in \Theta_i \ \forall \theta_{-i} \in \Theta_{-i} \ \forall i \in N.$$

*Example 2.25 (Pure Strategy Bayesian Nash Equilibrium).* Consider the example being discussed. We make the following observations.

- When $\theta_2 = x_2$, the strategy $b_2$ is strongly dominated by $a_2$. Thus player 2 chooses $a_2$ when $\theta_2 = x_2$.
- When $\theta_2 = y_2$, the strategy $a_2$ is strongly dominated by $b_2$ and therefore player 2 chooses $b_2$ when $\theta_2 = y_2$.
- When the action profiles are $(a_1, a_2)$ or $(b_1, b_2)$, player 1 has payoff 1 regardless of the type of player 2. In all other profiles, payoff of player 1 is zero.
- Since $p_1(x_2|x_1) = 0.6$ and $p_1(y_2|x_1) = 0.4$, player 1 thinks that the type $x_2$ of player 2 is more likely than type $y_2$.

The above arguments show that the unique pure strategy Bayesian Nash equilibrium in the above example is given by:

$$(s^*_{x_1} = a_1, s^*_{x_2} = a_2, s^*_{y_2} = b_2)$$

thus validating what we have already shown. Note that the equilibrium strategy for company 1 is always to price the product low whereas for company 2, the equilibrium strategy is to price it low if it produces $x_2$ and to price it high if it produces $y_2$.

The above example also illustrates the danger of analyzing each matrix separately. If it is common knowledge that player 2's type is $x_2$, then the unique Nash equilibrium is $(a_1, a_2)$. If it is common knowledge that player 2 has type $y_2$, then we get $(b_1, b_2)$ as the unique Nash equilibrium. However, in a Bayesian game, the type of player 2 is not common knowledge, and hence the above prediction based on analyzing the matrices separately would be wrong.

*Example 2.26 (Bayesian Nash Equilibrium of First Price Sealed Bid Auction).* Consider an auctioneer or a seller and two potential buyers as in Example 2.23. Here each buyer submits a sealed bid, $s_i \geq 0$ $(i = 1, 2)$. The sealed bids are looked at, and the buyer with the higher bid is declared the winner. If there is a tie, buyer 1 is declared the winner. The winning buyer pays to the seller an amount equal to his bid. The losing bidder does not pay anything.

Let us make the following assumptions:

1. $\theta_1, \theta_2$ are independently drawn from the uniform distribution on $[0, 1]$.
2. The sealed bid of buyer $i$ takes the form $s_i(\theta_i) = \alpha_i \theta_i$, where $\alpha_i \in [0, 1]$. This assumption implies that player $i$ bids a fraction $\alpha_i$ of his value; this is a reasonable assumption that implies a linear relationship between the bid and the value.

Buyer 1's problem is now to bid in a way to maximize his expected payoff:

$$\max_{s_1 \geq 0} (\theta_1 - s_1) P\{s_2(\theta_2) \leq s_1\}.$$

Since the bid of player 2 is $s_2(\theta_2) = \alpha_2 \theta_2$ and $\theta_2 \in [0, 1]$, the maximum bid of buyer 2 is $\alpha_2$. Buyer 1 knows this and therefore $s_1 \in [0, \alpha_2]$. Also,

$$\begin{aligned} P\{s_2(\theta_2) \leq s_1\} &= P\{\alpha_2 \theta_2 \leq s_1\} \\ &= P\{\theta_2 \leq \frac{s_1}{\alpha_2}\} \\ &= \frac{s_1}{\alpha_2} \text{ (since } \theta_2 \text{ is uniform over } [0, 1]). \end{aligned}$$

Thus buyer 1's problem is:

$$\max_{s_1 \in [0, \alpha_2]} (\theta_1 - s_1) \frac{s_1}{\alpha_2}.$$

The solution to this problem is

$$s_1(\theta_1) = \begin{cases} \frac{1}{2}\theta_1 & \text{if } \frac{1}{2}\theta_1 \leq \alpha_2 \\ \alpha_2 & \text{if } \frac{1}{2}\theta_1 > \alpha_2. \end{cases}$$

We can show on similar lines that

$$s_2(\theta_2) = \begin{cases} \frac{1}{2}\theta_2 & \text{if } \frac{1}{2}\theta_2 \leq \alpha_1 \\ \alpha_1 & \text{if } \frac{1}{2}\theta_2 > \alpha_1. \end{cases}$$

Let $\alpha_1 = \alpha_2 = \frac{1}{2}$. Then we get

$$s_1(\theta_1) = \frac{\theta_1}{2} \qquad \forall \theta_1 \in \Theta_1 = [0, 1]$$

$$s_2(\theta_2) = \frac{\theta_2}{2} \qquad \forall \theta_2 \in \Theta_2 = [0, 1].$$

Note that if $s_2(\theta_2) = \frac{\theta_2}{2}$, the best response of buyer 1 is $s_1(\theta_1) = \frac{\theta_1}{2}$ and vice-versa. Hence the profile $\left(\frac{\theta_1}{2}, \frac{\theta_2}{2}\right)$ is a Bayesian Nash equilibrium.

### 2.5.4 Dominant Strategy Equilibria

The dominant strategy equilibria of Bayesian games can again be defined using the Selten game representation.

**Definition 2.17 (Strongly Dominant Strategy Equilibrium).** Given a Bayesian game

$$\Gamma = \langle N, (\Theta_i), (S_i), (p_i), (u_i) \rangle$$

a profile of type agent strategies $(s_1^*(.), s_2^*(.), \ldots, s_n^*(.))$ is said to be a strongly dominant strategy equilibrium if

$$u_{\theta_i}(s_i^*(\theta_i), s_{-i}(\theta_{-i})) > u_{\theta_i}(s_i, s_{-i}(\theta_{-i}))$$

$$\forall s_i \in S_i \setminus \{s_i^*(\theta_i)\}, \ \forall s_{-i}(\theta_{-i}) \in S_{-i}, \ \forall \theta_i \in \Theta_i, \ \forall \theta_{-i} \in \Theta_{-i}, \ \forall i \in N.$$

**Definition 2.18 (Weakly Dominant Strategy Equilibrium).** A profile of type agent strategies $(s_1^*(.), s_2^*(.), \ldots, s_n^*(.))$ is said to be a weakly dominant strategy equilibrium if

$$u_{\theta_i}(s_i^*(\theta_i), s_{-i}(\theta_{-i})) \geq u_{\theta_i}(s_i, s_{-i}(\theta_{-i}))$$

$$\forall s_i \in S_i, \ \forall s_{-i}(\theta_{-i}) \in S_{-i}, \ \forall \theta_i \in \Theta_i, \ \forall \theta_{-i} \in \Theta_{-i}, \ \forall i \in N$$

and strict inequality satisfied for at least one $s_i \in S_i$.

*Note 2.8.* The notion of dominant strategy equilibrium is independent of the belief functions, and this is what makes it a very powerful notion and a very strong property. The notion of a weakly dominant strategy equilibrium is used extensively in mechanism design theory to define *dominant strategy implementation* of mechanisms.

*Example 2.27 (Weakly Dominant Strategy Equilibrium of Second Price Auction).* We have shown above that the first price sealed bid auction has a Bayesian Nash equilibrium. Now we consider the second price sealed bid auction with two bidders and show that it has a weakly dominant strategy equilibrium. Let us say buyer 2 announces his bid as $\hat{\theta}_2$. There are two cases.

1. $\theta_1 \geq \hat{\theta}_2$.
2. $\theta_1 < \hat{\theta}_2$.

**Case 1:** $\theta_1 \geq \hat{\theta}_2$

Let $\hat{\theta}_1$ be the announcement of buyer 1. Here there are two cases.

- If $\hat{\theta}_1 \geq \hat{\theta}_2$, then the payoff for buyer 1 is $\theta_1 - \hat{\theta}_2 \geq 0$.
- If $\hat{\theta}_1 < \hat{\theta}_2$, then the payoff for buyer 1 is 0.
- Thus in this case, the maximum payoff possible is $\theta_1 - \hat{\theta}_2 \geq 0$.

If $\hat{\theta}_1 = \theta_1$ (that is, buyer 1 announces his true valuation), then payoff for buyer 1 is $\theta_1 - \hat{\theta}_2$, which happens to be the maximum possible payoff as shown above. Thus announcing $\theta_1$ is a best response to buyer 1 whatever the announcement of buyer 2.

**Case 2:** $\theta_1 < \hat{\theta}_2$

Here again there are two cases: $\hat{\theta}_1 \geq \hat{\theta}_2$ and $\hat{\theta}_1 < \hat{\theta}_2$.

- If $\hat{\theta}_1 > \hat{\theta}_2$, then the payoff for buyer 1 is $\theta_1 - \hat{\theta}_2$, which is negative.
- If $\hat{\theta}_1 < \hat{\theta}_2$, then buyer 1 does not win and payoff for him is zero.
- Thus in this case, the maximum payoff possible is 0.

If $\hat{\theta}_1 = \theta_1$, payoff for buyer 1 is 0. By announcing $\hat{\theta}_1 = \theta_1$, his true valuation, buyer 1 gets zero payoff, which in this case is a best response.

We can now make the following observations about this example.

- Bidding his true valuation is optimal for buyer 1 regardless of what buyer 2 announces.
- Similarly bidding his true valuation is optimal for buyer 2 whatever the announcement of buyer 1.
- This means truth revelation is a weakly dominant strategy for each player, and $(\theta_1, \theta_2)$ is a weakly dominant strategy equilibrium.

## 2.6 The Mechanism Design Environment

Mechanism design is concerned with how to implement *system-wide solutions* to problems that involve *multiple self-interested agents*, each with *private information* about their preferences. A mechanism could be viewed as an institution or a framework of protocols that would prescribe particular ways of interaction among the agents so as to ensure a socially desirable outcome from this interaction. Without the mechanism, the interaction among the agents may lead to an outcome that is far from socially optimal. One can view mechanism design as an approach to solving a well-formulated but *incompletely specified optimization problem* where *some of the inputs to the problem are held by the individual agents*. So in order to solve the problem, the *social planner* needs to elicit these private values from the individual agents.

The following provides a general setting for formulating, analyzing, and solving mechanism design problems.

- There are $n$ agents, $1, 2, \ldots, n$, with $N = \{1, 2, \ldots, n\}$. The agents are rational and intelligent.

- $X$ is a set of *alternatives* or *outcomes*. The agents are required to make a collective choice from the set $X$.
- Prior to making the collective choice, each agent privately observes his preferences over the alternatives in $X$. This is modeled by supposing that agent $i$ privately observes a parameter or signal $\theta_i$ that determines his preferences. The value of $\theta_i$ is known to agent $i$ and is not known to the other agents. $\theta_i$ is called a private value or type of agent $i$.
- We denote by $\Theta_i$ the set of private values of agent $i$, $i = 1, 2, \ldots, n$. The set of all type profiles is given by $\Theta = \Theta_1 \times \ldots \times \Theta_n$. A typical type profile is represented as $\theta = (\theta_1, \ldots, \theta_n)$.
- It is assumed that there is a common prior distribution $\Phi \in \Delta(\Theta)$. To maintain consistency of beliefs, individual belief functions $p_i$ that describe the beliefs that player $i$ has about the type profiles of the rest of the players can all be derived from the common prior.
- Individual agents have preferences over outcomes that are represented by a utility function $u_i : X \times \Theta_i \to \mathbb{R}$. Given $x \in X$ and $\theta_i \in \Theta_i$, the value $u_i(x, \theta_i)$ denotes the payoff that agent $i$ having type $\theta_i \in \Theta_i$ receives from a decision $x \in X$. In the more general case, $u_i$ depends not only on the outcome and the type of player $i$, but could depend on the types of the other players also, so $u_i : X \times \Theta \to \mathbb{R}$. We restrict our attention to the former case in this monograph since most real-world situations fall into the former category.
- The set of outcomes $X$, the set of players $N$, the type sets $\Theta_i$ $(i = 1, \cdots, n)$, the common prior distribution $\Phi \in \Delta(\Theta)$, and the payoff functions $u_i$ $(i = 1, \cdots, n)$ are assumed to be *common knowledge* among all the players. The specific value $\theta_i$ observed by agent $i$ is private information of agent $i$.

**Social Choice Functions**

Since the agents' preferences depend on the realization of their types $\theta = (\theta_1, \cdots, \theta_n)$, it is natural to make the collective decision to depend on $\theta$. This leads to the definition of a social choice function.

**Definition 2.19 (Social Choice Function).** Given a set of agents $N = \{1, 2, \ldots, n\}$, their type sets $\Theta_1, \Theta_2, \ldots, \Theta_n$, and a set of outcomes $X$, a social choice function is a mapping

$$f : \Theta_1 \times \cdots \times \Theta_n \to X$$

that assigns to each possible type profile $(\theta_1, \theta_2, \ldots, \theta_n)$ a collective choice from the set of alternatives.

*Example 2.28 (Shortest Path Problem with Incomplete Information).* Consider a connected directed graph with a source vertex and destination vertex identified. Let the graph have $n$ edges, each owned by a rational and intelligent agent. Let the set of agents be denoted by $N = \{1, 2, \ldots, n\}$. Assume that the cost of the edge is private information of the agent owning the edge and let $\theta_i$ be this private information for agent $i$ $(i = 1, 2, \ldots, n)$. Let us say that a social planner is interested in finding a shortest path from the source vertex to the destination vertex. The social planner knows everything about the graph except the costs of the edges. So, the social planner first needs to extract this information from each agent and then find a shortest path from the source vertex to the destination vertex. Thus there are two problems facing the social planner, which are described below.

### Preference Elicitation Problem

Consider a social choice function $f : \Theta_1 \times \ldots \times \Theta_n \to X$. The types $\theta_1, \cdots, \theta_n$ of the individual agents are private information of the agents. Hence for the social choice $f(\theta_1, \cdots, \theta_n)$ to be chosen when the individual types are $\theta_1, \cdots, \theta_n$, each agent must disclose its true type to the social planner. However, given a social choice function $f$, a given agent may not find it in its best interest to reveal this information truthfully. This is called the *preference elicitation* problem or the *information revelation* problem. In the shortest path problem with incomplete information, the preference elicitation problem is to elicit the true values of the costs of the edges from the respective edge owners.

### Preference Aggregation Problem

Once all the agents report their types, the profile of reported types has to be transformed to an outcome, based on the social choice function. Let $\theta_i$ be the true type and $\hat{\theta}_i$ the reported type of agent $i$ $(i = 1, \ldots, n)$. The process of computing $f(\hat{\theta}_1, \ldots, \hat{\theta}_n)$ is called the *preference aggregation* problem. In the shortest path problem with incomplete information, the preference aggregation problem is to compute a shortest path from the source vertex to the destination vertex, given the structure of the graph and the (reported) costs of the edges. The preference aggregation problem is usually an optimization problem. Figure 2.1 provides a pictorial representation of all the elements making up the mechanism design environment.

### Direct and Indirect Mechanisms

One can view mechanism design as the process of solving an incompletely specified optimization problem where the specification is first elicited and then the underlying optimization problem is solved. Specification elicitation is basically the preference elicitation or type elicitation problem. To elicit the type information from the agents in a truthful way, there are broadly two kinds of mechanisms, which are aptly called *indirect mechanisms* and *direct mechanisms*. We define these below. In these definitions, we assume that the set of agents $N$, the set of outcomes $X$, the sets of types $\Theta_1, \ldots, \Theta_n$, a common prior $\Phi \in \Delta(\Theta)$, and the utility functions $u_i : X \times \Theta_i \to \mathbb{R}$ are given and are common knowledge.

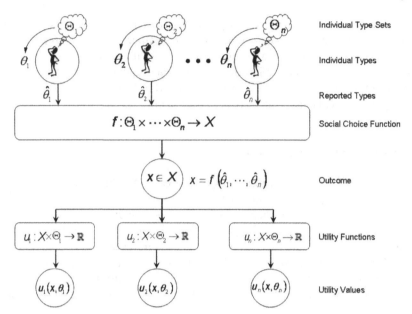

**Fig. 2.1** Mechanism design environment

**Definition 2.20 (Direct Mechanism).** Given a social choice function $f : \Theta_1 \times \Theta_2 \times \ldots \times \Theta_n \to X$, a direct (revelation) mechanism consists of the tuple $(\Theta_1, \Theta_2, \ldots, \Theta_n, f(.))$.

The idea of a direct mechanism is to *directly* seek the type information from the agents by asking them to reveal their true types.

**Definition 2.21 (Indirect Mechanism).** An indirect (revelation) mechanism consists of a tuple $(S_1, S_2, \ldots, S_n, g(.))$ where $S_i$ is a set of possible actions for agent $i$ $(i = 1, 2, \ldots, n)$ and $g : S_1 \times S_2 \times \ldots \times S_n \to X$ is a function that maps each action profile to an outcome.

The idea of an indirect mechanism is to provide a choice of actions to each agent and specify an outcome for each action profile. This induces a game among the players and the strategies played by the agents in an equilibrium of this game will indirectly reflect their original types.

   In the next four sections of this chapter, we will understand the process of mechanism design in the following way. First, we provide an array of examples to understand social choice functions and to appreciate the need for mechanisms. Next, we understand the process of implementing social choice functions through mechanisms. Following this, we will introduce the important notion of incentive compatibility and present a fundamental result in mechanism design, the *revelation theorem*. Then we will look into different properties that we would like a social choice function to satisfy.

## 2.7 Examples of Social Choice Functions

*Example 2.29 (Technology Driven Supplier Selection).* Suppose there is a buyer who wishes to procure a certain volume of an item that is produced by two suppliers, call them 1 and 2. We have $N = \{1,2\}$. Supplier 1 is known to use technology $a_1$ to produce these items, while supplier 2 uses one of two possible technologies, a high end technology $a_2$ and a low end technology $b_2$. The technology $a_2$ is known to be superior to $a_1$ also. The technology elements could be taken as the types of the suppliers, so we have $\Theta_1 = \{a_1\}$; $\Theta_2 = \{a_2, b_2\}$. See Figure 2.2.

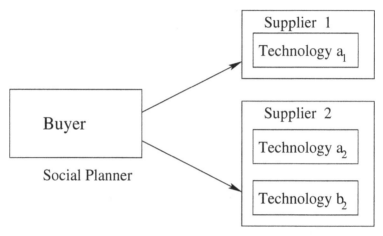

**Fig. 2.2** A sourcing scenario with one buyer and two suppliers

Let us define three outcomes (alternatives) $x, y, z$ for this situation. The alternative $x$ means that the entire volume required is sourced from supplier 1 while the alternative $z$ means that the entire volume required is sourced from supplier 2. The alternative $y$ indicates that 50% of the requirement is sourced from supplier 1 and the rest is sourced from supplier 2.

Since the buyer already has a long-standing relationship with supplier 1, this supplier is the preferred one. However, because of the superiority of technology $a_2$ over $a_1$ and $b_2$, the buyer would like to certainly source some quantity from supplier 2 if it is known that supplier 2 is guaranteed to use technology $a_2$. To reflect these facts, we assume the payoff functions to be given by:

$$u_1(x, a_1) = 100; \quad u_1(y, a_1) = 50; \quad u_1(z, a_1) = 0$$

$$u_2(x, a_2) = 0; \quad u_2(y, a_2) = 50; \quad u_2(z, a_2) = 100$$

$$u_2(x, b_2) = 0; \quad u_2(y, b_2) = 50; \quad u_2(z, b_2) = 25.$$

Note that $\Theta = \{(a_1,a_2),(a_1,b_2)\}$. Consider the social choice function $f(a_1,a_2) = y$ and $f(a_1,b_2) = x$. This means that when it is guaranteed that supplier 2 will use technology $a_2$, the buyer would like to procure from both the suppliers, whereas if it is known that supplier 2 uses technology $b_2$, the buyer would rather source the entire requirement from supplier 1.

Likewise, there are eight other social choice functions that one can define here. These are:

$$f(a_1,a_2) = x; \quad f(a_1,b_2) = x$$
$$f(a_1,a_2) = x; \quad f(a_1,b_2) = y$$
$$f(a_1,a_2) = x; \quad f(a_1,b_2) = z$$
$$f(a_1,a_2) = y; \quad f(a_1,b_2) = y$$
$$f(a_1,a_2) = y; \quad f(a_1,b_2) = z$$
$$f(a_1,a_2) = z; \quad f(a_1,b_2) = x$$
$$f(a_1,a_2) = z; \quad f(a_1,b_2) = y$$
$$f(a_1,a_2) = z; \quad f(a_1,b_2) = z.$$

It would be interesting to look at the implications of these social choice functions. We will return to this example later on, in many different contexts.

*Example 2.30 (Procurement of a Single Indivisible Resource).* Procurement is a ubiquitous activity in any organization. Every organization procures a variety of direct and indirect materials. For example, a factory procures raw material or sub-assemblies from a pool of suppliers. In a computational grid, a grid user procures computational or storage resources from the grid. A network user procures network resources. A dynamic supply chain planner procures supply chain service providers. Every organization procures indirect materials such as office supplies and services.

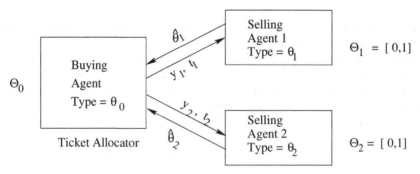

**Fig. 2.3** Procurement with two suppliers

The generic procurement situation involves a buyer, a pool of suppliers, and items to be procured. We consider a simple abstraction of the problem by considering a buying agent (call the agent 0) and two selling agents (call them 1 and 2), so we have $N = \{0, 1, 2\}$. See Figure 2.3. An indivisible item or resource is to be procured from one of the sellers in return for a monetary consideration. An outcome here can be represented by $x = (y_0, y_1, y_2, t_0, t_1, t_2)$. For $i = 0$, we have

$$y_0 = 0 \quad \text{if the buyer buys the good}$$
$$\phantom{y_0} = 1 \quad \text{otherwise}$$
$$t_0 = \text{monetary transfer received by the buyer.}$$

For $i = 1, 2$, we have

$$y_i = 1 \quad \text{if agent } i \text{ supplies the good to the buyer}$$
$$\phantom{y_i} = 0 \quad \text{if agent } i \text{ does not supply the good}$$
$$t_i = \text{monetary transfer received by the agent } i.$$

The set $X$ of all feasible outcomes is given by

$$X = \{(y_0, y_1, y_2, t_0, t_1, t_2) : y_i \in \{0, 1\}, \sum_{i=0}^{2} y_i = 1, t_i \in \mathbb{R}, \sum_{i=0}^{2} t_i \leq 0\}.$$

The constraint $\sum_i t_i \leq 0$ implies that the total money received by all the agents is less than or equal to zero. That is, total money paid by all the agents is greater than or equal to zero (that is, buyer pays at least as much as the sellers receive. The excess between the payment and receipts is the surplus). For $x = (y_0, y_1, y_2, t_0, t_1, t_2)$, define the utilities to be of the form:

$$u_i(x, \theta_i) = u_i((y_0, y_1, y_2, t_0, t_1, t_2), \theta_i) = -y_i \theta_i + t_i ; \quad i = 1, 2$$

where $\theta_i \in \mathbb{R}$ can be viewed as seller $i$'s valuation of the good. Such utility functions are said to be of *quasilinear* form (because it is linear in some of the variables and possibly non-linear in the other variables). We will be studying such utility forms quite extensively in this chapter.

We make the following assumptions regarding valuations.

- The buyer has a *known* value $\underline{\theta_0}$ for the good. This valuation does not depend on the choice of the seller from whom the item is purchased.
- Let $\Theta_i$ be the real interval $[\underline{\theta_i}, \overline{\theta_i}]$. The types $\theta_1$ and $\theta_2$ of the sellers are drawn independently from the interval $[\underline{\theta_i}, \overline{\theta_i}]$ and this fact is *common knowledge* among all the players. The type of a seller is to be viewed as the *willingness to sell* (minimum price below which the seller is not interested in selling the item).

Consider the following social choice function.

- The buyer buys the good from the seller with the lowest willingness to sell. If both the sellers have the same type, the buyer will buy the object from seller 1.

- The buyer pays the selected selling agent his willingness to sell.

The above social choice function $f(\theta) = (y_0(\theta), y_1(\theta), y_2(\theta), t_0(\theta), t_1(\theta), t_2(\theta))$ can be precisely written as

$$
\begin{aligned}
y_0(\theta) &= 0 \quad \forall \theta \\
y_1(\theta) &= 1 \quad \text{if } \theta_1 \leq \theta_2 \\
&= 0 \quad \text{if } \theta_1 > \theta_2 \\
y_2(\theta) &= 1 \quad \text{if } \theta_1 > \theta_2 \\
&= 0 \quad \text{if } \theta_1 \leq \theta_2
\end{aligned}
$$

$$
\begin{aligned}
t_1(\theta) &= \theta_1\, y_1(\theta) \\
t_2(\theta) &= \theta_2\, y_2(\theta) \\
t_0(\theta) &= -(t_1(\theta) + t_2(\theta)).
\end{aligned}
$$

We will refer to the above SCF as SCF-PROC1 in the sequel.

Suppose we consider another social choice function, which has the same allocation rule as the one we have just studied but has a different payment rule: The buyer now pays the winning seller a payment equal to the second lowest willingness to sell (as usual, the losing seller does not receive any payment). The new social choice function, which we will call SCF-PROC2, will be the following.

$$
\begin{aligned}
y_0(\theta) &= 0 \quad\quad \forall\, \theta \\
y_1(\theta) &= 1 \quad\quad \text{if } \theta_1 \leq \theta_2 \\
&= 0 \quad\quad \text{otherwise} \\
y_2(\theta) &= 1 \quad\quad \text{if } \theta_1 > \theta_2 \\
&= 0 \quad\quad \text{otherwise} \\
t_1(\theta) &= \theta_2\, y_1(\theta) \\
t_2(\theta) &= \theta_1\, y_2(\theta) \\
t_0(\theta) &= -(t_1(\theta) + t_2(\theta)).
\end{aligned}
$$

Let us define one more SCF, which we call SCF-PROC3, in the following way. SCF-PROC3 has the allocation rule as SCF-PROC1 and SCF-PROC2, but the payments are defined as:

$$
\begin{aligned}
t_1(\theta) &= \frac{(1 + \theta_1)}{2}\, y_1(\theta) \\
t_2(\theta) &= \frac{(1 + \theta_2)}{2}\, y_2(\theta) \\
t_0(\theta) &= -(t_1(\theta) + t_2(\theta)).
\end{aligned}
$$

The reason for defining the payment rule in the above way will become clear in the next section, where we will discuss the implementability of SCF-PROC1, SCF-PROC2, and SCF-PROC3.

*Example 2.31 (Funding a Public Project).* There is a set of agents $N = \{1, 2, \ldots, n\}$ who have a stake in a common infrastructure, for example, a bridge, community building, Internet infrastructure, etc. For example, the agents could be firms forming a business cluster and interested in creating a shared infrastructure. The cost of the project is to be shared by the agents themselves since there is no source of external funding. Let $k = 1$ indicate that the project is taken up, with $k = 0$ indicating that the project is dropped. Let $t_i \in \mathbb{R}$ denote the payment received by agent $i$ (which means $-t_i$ is the payment made by agent $i$) for each $i \in N$. Let the cost of the project be $C$. Since the agents have to fund the project themselves,

$$C \leq -\sum_{i \in N} t_i \quad \text{if } k = 1$$

If $k = 0$, we have

$$0 \leq -\sum_{i \in N} t_i$$

Combining the above two possibilities, we get the condition

$$\sum_{i \in N} t_i \leq -kC \; ; \; k \in \{0, 1\}.$$

Thus, a natural set of outcomes for this problem is:

$$X = \{(k, t_1, \ldots, t_n) : k \in \{0, 1\}, \, t_i \in \mathbb{R} \, \forall i \in N, \, \sum_{i \in N} t_i \leq -kC.\}$$

We assume the utility of agent $i$, when its type is $\theta_i$ corresponding to an outcome $(k, t_1, t_2, \ldots, t_n)$ to be given by

$$u_i((k, t_1, \ldots, t_n), \theta_i) = k\theta_i + t_i.$$

The type $\theta_i$ of agent $i$ has the natural interpretation of being the willingness to pay of agent $i$ (maximum amount that agent $i$ is prepared to pay) towards the project. A social choice function in this context is $f(\theta) = (k(\theta), t_1(\theta), \ldots, t_n(\theta))$ given by

$$k(\theta) = \begin{cases} 1 \text{ if } \sum_{i \in N} \theta_i \geq C \\ 0 \text{ otherwise} \end{cases}$$

$$t_i(\theta) = -\left(\frac{k(\theta)C}{n}\right).$$

The way $k(\theta)$ is defined ensures that the project is taken up only if the combined willingness to pay of all the agents is at least the cost of the project. The definition of $t_i(\theta)$ above follows the egalitarian principle, namely that the agents share the cost of the project equally among themselves.

*Example 2.32 (Bilateral Trade).* Consider two agents 1 and 2 where agent 1 is the seller of an indivisible private good and agent 2 is a prospective buyer of the good. See Figure 2.4. An outcome here is of the form $x = (y_1, y_2, t_1, t_2)$ where $y_i = 1$ if

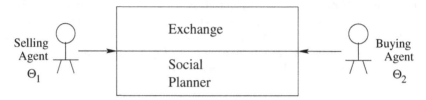

**Fig. 2.4** A bilateral trade environment

agent $i$ gets the good and $t_i$ denotes the payment received by agent $i$ $(i = 1, 2)$. A natural set of outcomes here is

$$X = \{(y_1, y_2, t_1, t_2) : y_1 + y_2 = 1;\ y_1, y_2 \in \{0, 1\},\ t_1 + t_2 \le 0\}.$$

The condition $t_1 + t_2 \le 0$ indicates that the the amount paid by the buyer should be at least equal to the amount received by the seller (the surplus could perhaps be retained by a market maker or mediator). The utility of the agent $i$ $(i = 1, 2)$ would be of the form

$$u_i((y_1, y_2, t_1, t_2), \theta_i) = y_i \theta_i + t_i.$$

The type $\theta_1$ of agent 1 (seller) can be interpreted as the willingness to sell of the agent (minimum price at which agent 1 is willing to sell). The type $\theta_2$ of agent 2 (buyer) has the natural interpretation of willingness to pay (maximum price the buyer is willing to pay). A social choice function here would be $f(\theta) = (y_1(\theta), y_2(\theta), t_1(\theta), t_2(\theta))$ defined as

$$y_1(\theta_1, \theta_2) = 1 \quad \theta_1 > \theta_2$$
$$= 0 \quad \theta_1 \le \theta_2$$
$$y_2(\theta_1, \theta_2) = 1 \quad \theta_1 \le \theta_2$$
$$= 0 \quad \theta_1 > \theta_2$$
$$t_1(\theta_1, \theta_2) = y_2(\theta_1, \theta_2) \frac{\theta_1 + \theta_2}{2}$$
$$t_2(\theta_1, \theta_2) = -y_2(\theta_1, \theta_2) \frac{\theta_1 + \theta_2}{2}.$$

*Example 2.33 (Network Formation Problem).* Consider a supply chain network scenario where a supply chain planner (SCP) is interested in forming an optimal network for delivering products/services. Multiple service providers or supply chain partners are needed for executing the end-to-end process. The directed graph shown

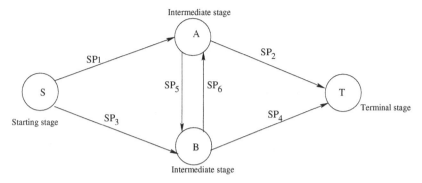

**Fig. 2.5** A graph representation of network formation

in Figure 2.5 describes different possible ways in which the supply chain can be formed. The node $S$ denotes the starting stage, and the node $T$ denotes the terminal stage in the supply chain. $SP_1, SP_2, \ldots, SP_6$ denote the service providers. They own the respective edges of the directed graph. The nodes $A$ and $B$ are two intermediate stages in the supply chain process. Each path in the network from $S$ to $T$ corresponds to a particular way in which the supply chain network can be formed. For example, $(SP_1, SP_5, SP_4)$ is a path from $S$ to $T$ and this is one possible solution to the supply chain formation problem. Each service provider has a service cost that is private information of the provider. The problem facing the supply chain planner here is to determine the network of service providers who will enable forming the supply chain at the least cost. Since the private values of the service providers are not known, the supply chain planner is faced with a mechanism design problem.

A situation such as described above occurs in many other settings. The first example is a logistics network scenario where $S$ denotes the source, $T$ denotes the destination, and the nodes $A$ and $B$ denote two separate logistics hubs. The edges denote the connectivity between logistics hubs, the source, and the destination. The service providers here are logistics providers who own the transportation service on the edges. Another example is a procurement network scenario where the service providers correspond to suppliers. One more example is that of a telecom network where the service providers could correspond to Internet/bandwidth service providers, and the nodes correspond to cities.

For this example, we have $N = \{SCP, SP_1, \ldots, SP_6\}$. For brevity, we will call this set $\{0, 1, 2, 3, 4, 5, 6\}$, where 0 is the supply chain planner and $1, 2, \ldots, 6$ are the service providers. Let us assume the type sets to be:

$$\Theta_0 = \{\theta_0\}; \quad \Theta_i = \left[\underline{\theta_i}, \bar{\theta_i}\right] \subset \mathbb{R} \ \forall i \in \{1, 2, \ldots, 6\}.$$

An outcome is of the form $(y_0, y_1, \ldots, y_6, t_0, t_1, \ldots, t_6)$ where

$\quad\quad y_0 = 1$ if the supply chain planner buys a path from $S$ to $T$
$\quad\quad\ \ = 0$ otherwise.

For $i = 1,\ldots,6$,

$$y_i = 1 \text{ if } SP_i \text{ is part of the supply chain formed}$$
$$= 0 \text{ otherwise}$$
$$t_i = \text{ payment received by service provider } SP_i.$$

Note that $-t_0$, the payment required to be made by the supply chain planner is just the sum of $t_1,\ldots,t_6$. The set of outcomes in this example is given by

$$X = \{(y_0,y_1,\ldots,y_6,t_0,t_1,\ldots,t_6) : y_i \in \{0,1\} \; \forall i \in N;$$
$$\{SP_i \in N \setminus \{0\} : y_i = 1\} \text{ forms a path from S to T};$$
$$t_i \in \mathbb{R} \; \forall i \in N; t_0 = -\textstyle\sum_{i=1}^{6} t_i\}.$$

A social choice function in this problem is given by

$$f(\theta) = (y_0(\theta),y_1(\theta),\ldots,y_6(\theta),t_0(\theta),t_1(\theta),\ldots,t_6(\theta))$$

$$y_0(\theta) = 1 \; \forall \theta \in \Theta.$$

For $i = 1,2,\ldots,6, \; \forall \theta \in \Theta$,

$$
\begin{aligned}
y_i(\theta) \quad &= 1 \text{ if } SP_i \text{ is part of a shortest cost path}\\
&\quad\text{ from } S \text{ to } T \text{ selected by the supply chain planner}\\
&= 0 \text{ otherwise}\\
t_i(\theta) \quad &= \text{ payment received by } SP_i; \; i = 1,2,\ldots,6\\
t_0(\theta) \quad &= -\textstyle\sum_{i=1}^{6} t_i(\theta)\\
u_i(f(\theta),\theta_i) &= u_i(y_0(\theta),\ldots,y_6(\theta),t_0(\theta),\ldots,t_6(\theta);\theta_i)\\
&= t_i(\theta) - y_i(\theta)\theta_i.
\end{aligned}
$$

$\theta_i$ is to be viewed as the willingness to sell of the agent $i$ where $i = 1,2,\ldots,6$.

## 2.8 Implementation of Social Choice Functions

In the preceding section, we have seen a series of examples of social choice functions. In this section, we motivate, through illustrative examples, the concept of implementation of social choice functions. We then formally define the notion of implementation through direct mechanisms and indirect mechanisms.

### 2.8.1 Implementation Through Direct Mechanisms

We first provide examples to motivate implementation by direct mechanisms.

*Example 2.34 (Technology driven Supplier Selection).* Recall Example 2.29 where $N = \{1,2\}$; $\Theta_1 = \{a_1\}$; , $\Theta_2 = \{a_2,b_2\}$ and $\Theta = \{(a_1,a_2),(a_1,b_2)\}$. Suppose the social planner (in this case, the buyer) wishes to implement the social choice function $f$ with $f(a_1,a_2) = y$ and $f(a_1,b_2) = x$. Announcing this as the social choice function, let us say the social planner asks the agents to reveal their types. Agent 1 has nothing to reveal since his type is common knowledge (as his type set is a singleton). We will now check whether agent 2 would be willing to truthfully reveal its type.

- If $\theta_2 = a_2$, then, because $f(a_1, a_2) = y$ and $f(a_1, b_2) = x$ and $u_2(y, a_2) > u_2(x, a_2)$, agent 2 is happy to reveal $a_2$ as its type.
- However if $\theta_2 = b_2$, then because $u_2(y, b_2) > u_2(x, b_2)$ and $f(a_1, b_2) = x$, agent 2 would wish to lie and claim that its type is $a_2$ and not $b_2$.

Thus though the social planner would like to implement an SCF $f(\cdot)$, the social planner would be unable to implement the above SCF since one of the agents (in this case agent 2) does not find it in his best interest to reveal his true type.

On the other hand, let us say the social planner wishes to implement the social choice function $f$ given by $f(a_1, a_2) = z$ and $f(a_1, b_2) = y$. One can show in this case that the SCF can be implemented. Table 2.10 lists all the nine SCFs and their implementability.

| SCF | | Implementable |
|---|---|---|
| $f(a_1,a_2)$ | $f(a_1,b_2)$ | |
| $x$ | $x$ | ✓ |
| $x$ | $y$ | ✗ |
| $x$ | $z$ | ✗ |
| $y$ | $x$ | ✗ |
| $y$ | $y$ | ✓ |
| $y$ | $z$ | ✗ |
| $z$ | $x$ | ✗ |
| $z$ | $y$ | ✓ |
| $z$ | $z$ | ✓ |

**Table 2.10** Social choice functions and their implementability

*Example 2.35 (Implementability of SCF-PROC1).* Recall the social choice function SCF-PROC1 that we introduced in the context of procurement of a single indivisible resource (Example 2.30). Recall the definition of SCF-PROC1:

$$y_0(\theta) = 0 \quad \forall \theta$$
$$y_1(\theta) = 1 \quad \text{if } \theta_1 \le \theta_2$$
$$\quad\quad = 0 \quad \text{if } \theta_1 > \theta_2$$
$$y_2(\theta) = 1 \quad \text{if } \theta_1 > \theta_2$$
$$\quad\quad = 0 \quad \text{if } \theta_1 \le \theta_2$$

$$t_1(\theta) = \theta_1 \, y_1(\theta)$$
$$t_2(\theta) = \theta_2 \, y_2(\theta)$$
$$t_0(\theta) = -(t_1(\theta) + t_2(\theta)).$$

We note that the social choice function is very attractive to the buyer since the buyer will capture all of the consumption benefits that are generated by the good. We assume that $\theta_1$ and $\theta_2$ are drawn independently from a uniform distribution over $[0, 1]$. Now we ask the question: Can we implement this social choice function? The answer for this question is *no*. The following analysis shows why.

Let us say seller 2 announces his true value $\theta_2$. Suppose the valuation of seller 1 is $\theta_1$, and he announces $\hat{\theta}_1$. If $\theta_2 \geq \hat{\theta}_1$, then seller 1 is the winner, and his utility will be $\hat{\theta}_1 - \theta_1$. If $\theta_2 < \hat{\theta}_1$, then seller 2 is the winner, and seller 1's utility is zero. Since seller 1 wishes to maximize his expected utility he solves the problem

$$\max_{\hat{\theta}_1}(\hat{\theta}_1 - \theta_1)\, P\{\theta_2 \geq \hat{\theta}_1\}.$$

Since $\theta_2$ is uniformly distributed on $[0,1]$,

$$P\{\theta_2 \geq \hat{\theta}_1\} = 1 - P\{\theta_2 < \hat{\theta}_1\} = 1 - \hat{\theta}_1.$$

Thus seller 1 tries to solve the problem:

$$\max_{\hat{\theta}_1}(\hat{\theta}_1 - \theta_1)(1 - \hat{\theta}_1).$$

This problem has the solution

$$\hat{\theta}_1 = \frac{1 + \theta_1}{2}.$$

Thus if seller 2 announces his true valuation, then the best response for seller 1 is to announce $\frac{1+\theta_1}{2}$.

Similarly if seller 1 announces his true valuation $\theta_1$, then the best response of seller 2 is to announce $\frac{1+\theta_2}{2}$. Thus there is no incentive for the sellers to announce their true valuations. So, a social planner who wishes to realize the above social choice function finds the rational players will not report their true private values. Thus the social choice function cannot be implemented through a direct mechanism.

*Example 2.36 (Implementability of SCF-PROC2).* Recall the social choice function SCF-PROC2, again in the context of procurement of a single indivisible resource (Example 2.30):

$$y_0(\theta) = 0 \quad \forall \theta$$
$$\begin{aligned} y_1(\theta) &= 1 \quad \text{if } \theta_1 \leq \theta_2 \\ &= 0 \quad \text{if } \theta_1 > \theta_2 \\ y_2(\theta) &= 1 \quad \text{if } \theta_1 > \theta_2 \\ &= 0 \quad \text{if } \theta_1 \leq \theta_2 \end{aligned}$$

$$\begin{aligned} t_1(\theta) &= \theta_2\, y_1(\theta) \\ t_2(\theta) &= \theta_1\, y_2(\theta) \\ t_0(\theta) &= -(t_1(\theta) + t_2(\theta)). \end{aligned}$$

We now show that the function SCF-PROC2 can be implemented. Let us say seller 2 announces his valuation as $\hat{\theta}_2$. There are two cases.

1. $\theta_1 \leq \hat{\theta}_2$
2. $\theta_1 > \hat{\theta}_2$.

**Case 1:** $\theta_1 \leq \hat{\theta}_2$

Let $\hat{\theta}_1$ be the announcement of seller 1. Here there are two cases.

- If $\hat{\theta}_1 \leq \hat{\theta}_2$, then the payoff for seller 1 is $\hat{\theta}_2 - \theta_1 \geq 0$.
- If $\hat{\theta}_1 > \hat{\theta}_2$, then the payoff for seller 1 is 0.
- Thus in this case, the maximum payoff possible is $\hat{\theta}_2 - \theta_1 \geq 0$.

If $\hat{\theta}_1 = \theta_1$ (that is, seller 1 announces his true valuation), then payoff for seller 1 is $\hat{\theta}_2 - \theta_1$, which happens to be the maximum possible payoff as shown above. Thus announcing $\theta_1$ is a best response to seller 1 whatever the announcement of seller 2.

**Case 2:** $\theta_1 > \hat{\theta}_2$

Here again there are two cases: $\hat{\theta}_1 \leq \hat{\theta}_2$ and $\hat{\theta}_1 > \hat{\theta}_2$.

- If $\hat{\theta}_1 \leq \hat{\theta}_2$, then the payoff for seller 1 is $\hat{\theta}_2 - \theta_1$, which is negative.
- If $\hat{\theta}_1 > \hat{\theta}_2$, then seller 1 does not win, and payoff for him is zero.
- Thus in this case, the maximum payoff possible is 0.

If $\hat{\theta}_1 = \theta_1$, payoff for seller 1 is 0. By announcing $\hat{\theta}_1 = \theta_1$, his true valuation, seller 1 gets zero payoff, which in this case is a best response.

We can now make the following observations about this example.

- Revealing his true valuation is optimal for seller 1 regardless of what seller 2 announces.
- Similarly, announcing his true valuation is optimal for seller 2 whatever the announcement of seller 1.
- More formally, truth revelation is a weakly dominant strategy for each player.
- Thus this social choice function can be implemented even though the valuations are private values. We simply ask each seller to report his type and then we choose $f(\theta)$.

## 2.8.2 Implementation Through Indirect Mechanisms

The examples above have shown us a possible way in which to try to implement a social choice function. The protocol we followed for implementing the social choice functions was:

- Ask each agent to reveal his or her types $\theta_i$;
- Given the announcements $(\hat{\theta}_1, \ldots, \hat{\theta}_n)$, choose the outcome $x = f(\hat{\theta}_1, \ldots, \hat{\theta}_n) \in X$.

Such a method of trying to implement an SCF is referred to as a *direct revelation mechanism*. Another approach to implementing a social choice function is the *indirect way*. Here the mechanism makes the agents interact through an institutional framework in which there are rules governing the actions the agents would be allowed to play and in which there is a way of transforming these actions into a social outcome. The actions the agents choose will depend on their private values and become the strategies of the players. Auctions provide an example of *indirect mechanisms*. We provide an example right away.

*Example 2.37 (First Price Procurement Auction).* Consider an auctioneer or a buyer and two potential sellers as before. Here each seller submits a sealed bid, $b_i \geq 0$ $(i = 1, 2)$. The sealed bids are examined and the seller with the lower bid is declared the winner. If there is a tie, seller 1 is declared the winner. The winning seller receives an amount equal to his bid from the buyer. The losing seller does not receive anything.

Note that there is a subtle difference between the situations in Example 2.35 and Example 2.37. In Example 2.35 (direct mechanism), each seller is asked to announce his type, whereas in Example 2.37 (indirect mechanism), each seller is asked to submit a bid. The bid submitted may (and will) of course depend on the type. Based on the type, the seller has a strategy for bidding. So it becomes a game.

Let us make the following assumptions:

1. $\theta_1, \theta_2$ are independently drawn from the uniform distribution on $[0, 1]$.
2. The sealed bid of seller $i$ takes the form $b_i(\theta_i) = \alpha_i \theta_i + \beta_i$, where $\alpha_i \in [0, 1], \beta_i \in [0, 1 - \alpha_i]$. He has to make sure that $b_i \in [0, 1]$. The term $\beta_i$ is like a fixed cost whereas $\alpha_i \theta_i$ indicates a fraction of the true cost.

Seller 1's problem is now to bid in a way to maximize his payoff:

$$\max_{1 \geq b_1 \geq 0} (b_1 - \theta_1) P\{b_2(\theta_2) \geq b_1\}$$

$$
\begin{aligned}
P\{b_2(\theta_2) \geq b_1\} &= 1 - P\{b_2(\theta_2) < b_1\} \\
&= 1 - P\{\alpha_2 \theta_2 + \beta_2 < b_1\} \\
&= 1 - \frac{b_1 - \beta_2}{\alpha_2} \text{ if } b_1 \geq \beta_2 \qquad (2.5) \\
&\quad \text{since } \theta_2 \text{ is uniform over } [0, 1] \\
&= 1 \text{ if } b_1 < \beta_2. \qquad (2.6)
\end{aligned}
$$

Thus seller 1's problem is:

$$\max_{b_1 \geq \beta_2} (b_1 - \theta_1)\left(1 - \frac{b_1 - \beta_2}{\alpha_2}\right).$$

The solution to this problem is

$$b_1(\theta_1) = \frac{\alpha_2 + \beta_2}{2} + \frac{\theta_1}{2}. \qquad (2.7)$$

We can show on similar lines that

$$b_2(\theta_2) = \frac{\alpha_1 + \beta_1}{2} + \frac{\theta_2}{2}. \qquad (2.8)$$

As the bid of seller $i$ takes the form $b_i(\theta_i) = \alpha_i \theta_i + \beta_i$, where $\alpha_i \in [0, 1], \beta_i \in [0, 1 - \alpha_i]$, from the equations (2.7) and (2.8), we obtain $\alpha_1 = \alpha_2 = \frac{1}{2}$. As the goal of each seller is to maximize the profit and $\beta_i \in [0, 1 - \alpha_i], \beta_1 = \beta_2 = \frac{1}{2}$. Then we get

$$b_1(\theta_1) = \frac{1 + \theta_1}{2} \qquad \forall \ \theta_1 \in \Theta_1 = [0, 1]$$

$$b_2(\theta_2) = \frac{1 + \theta_2}{2} \qquad \forall \ \theta_2 \in \Theta_2 = [0, 1].$$

Note that if $b_2(\theta_2) = \frac{1+\theta_2}{2}$, the best response of seller 1 is $b_1(\theta_1) = \frac{1+\theta_1}{2}$ and vice-versa. Hence the profile $\left( \frac{1+\theta_1}{2}, \frac{1+\theta_2}{2} \right)$ is a Bayesian Nash equilibrium of an underlying Bayesian game. In other words, there is a Bayesian Nash equilibrium of an underlying game (induced by the indirect mechanism called the first price procurement auction) that (indirectly) yields the outcome

$$f(\theta) = (y_0(\theta), y_1(\theta), y_2(\theta), t_0(\theta), t_1(\theta), t_2(\theta))$$

such that

$$y_0(\theta) = 0 \qquad \forall \ \theta \in \Theta$$
$$y_1(\theta) = 1 \qquad \text{if } \theta_1 \leq \theta_2$$
$$\qquad = 0 \qquad \text{else}$$
$$y_2(\theta) = 1 \qquad \text{if } \theta_1 > \theta_2$$
$$\qquad = 0 \qquad \text{else}$$
$$t_1(\theta) = \frac{1 + \theta_1}{2} y_1(\theta)$$
$$t_2(\theta) = \frac{1 + \theta_2}{2} y_2(\theta)$$
$$t_0(\theta) = -(t_1(\theta) + t_2(\theta)).$$

The above SCF is precisely SCF-PROC3 that we had introduced in Example 2.30.

*Example 2.38 (Second Price Procurement Auction).* Here, each seller is asked to submit a sealed bid $b_i \geq 0$. The bids are examined, and the seller with the lower bid is declared the winner. In case there is a tie, seller 1 is declared the winner. The winning seller receives as payment from the buyer an amount equal to the second lowest bid. The losing bidder does not receive anything. In this case, we can show that $b_i(\theta_i) = \theta_i$ for $i = 1, 2$ constitutes a weakly dominant strategy for each player. The arguments are identical to those in Example 2.36.

Thus the game induced by the indirect mechanism second price procurement auction has a weakly dominant strategy in which the social choice function SCF-PROC2 is implemented.

We can summarize the findings of the current section so far in the following way.

- The function SCF-PROC1 cannot be implemented.
- The function SCF-PROC2 can be implemented in dominant strategies by a direct mechanism. Also, the indirect mechanism, namely second price procurement auction, implements SCF-PROC2 in dominant strategies.
- The function SCF-PROC3 is implemented in Bayesian Nash equilibrium by an indirect mechanism, the first price procurement auction.

### 2.8.3 Bayesian Game Induced by a Mechanism

Recall that a mechanism is an institution or a framework with a set of rules that prescribe the actions available to players and specify how these action profiles are transformed into outcomes. A mechanism specifies an action set for each player. The outcome function gives the rule for obtaining outcomes from action profiles. Given:

1. a set of agents $\{1, 2, \ldots, n\}$,
2. type sets $\Theta_1, \ldots, \Theta_n$,
3. a common prior $\phi \in \Delta(\Theta)$,
4. a set of outcomes $X$,
5. utility functions $u_1, \ldots, u_n$, with $u_i : X \times \Theta_i \rightarrow \mathbb{R}$,

a mechanism $M = (S_1, \ldots, S_n, g(.))$ induces a Bayesian game

$$(N, (\Theta_i), (S_i), (p_i), (U_i))$$

among the players where

$$U_i(\theta_1, \ldots, \theta_n, s_1, \ldots, s_n) = u_i(g(s_1, \ldots, s_n), \theta_i).$$

**Strategies in the Induced Bayesian Game**

A strategy $s_i$ for an agent $i$ in the induced Bayesian game is a function $s_i : \Theta_i \rightarrow S_i$. Thus, given a private value $\theta_i \in \Theta_i$, $s_i(\theta_i)$ will give the action of player $i$. The strategy $s_i(.)$ will specify actions corresponding to private values. In the auction scenario, the bid $b_i$ of player $i$ is a function of his valuation $\theta_i$. For example, $b_i(\theta_i) = \alpha_i \theta_i + \beta_i$ is a particular strategy for player $i$.

Figure 2.6 captures the idea behind an indirect mechanism and the Bayesian game that is induced by an indirect mechanism.

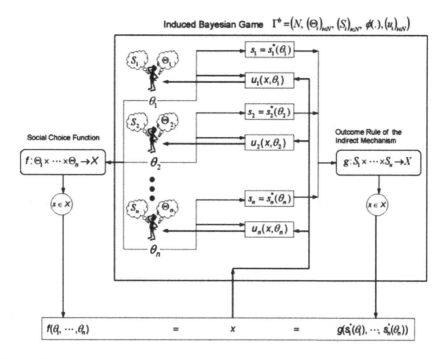

**Fig. 2.6** The idea behind implementation by an indirect mechanism

*Example 2.39 (Bayesian Game Induced by First Price Procurement Auction).* First, note that $N = \{0,1,2\}$. The type sets are $\Theta_0, \Theta_1, \Theta_2$, and the common prior is $\phi \in \Delta(\Theta)$. The set of outcomes is

$$X = \{(y_0, y_1, y_2, t_0, t_1, t_2) : y_i \in \{0,1\}, y_0 + y_1 + y_2 = 1, t_i \in \mathbb{R}, t_0 + t_1 + t_2 \leq 0\}$$

$$u_i((y_0, y_1, y_2, t_0, t_1, t_2), \theta_i) = -\theta_i y_i + t_i \quad i = 1, 2$$

$$u_0((y_0, y_1, y_2, t_0, t_1, t_2), \theta_0) = \theta_0 y_0 + t_0$$

$$S_1 = \mathbb{R}_+ \; ; \; S_2 = \mathbb{R}_+$$

$$g(b_0, b_1, b_2) = (y_0(b_0, b_1, b_2), y_1(b_0, b_1, b_2), y_2(b_0, b_1, b_2),$$
$$t_0(b_0, b_1, b_2), t_1(b_0, b_1, b_2), t_2(b_0, b_1, b_2))$$

such that

$$y_0(b_0, b_1, b_2) = 0 \qquad \forall\, b_0, b_1, b_2$$
$$y_1(b_0, b_1, b_2) = 1 \qquad \text{if } b_1 \leq b_2$$
$$\qquad\qquad\quad = 0 \qquad \text{if } b_1 > b_2$$
$$y_2(b_0, b_1, b_2) = 1 \qquad \text{if } b_1 > b_2$$
$$\qquad\qquad\quad = 0 \qquad \text{if } b_1 \leq b_2$$
$$t_1(b_0, b_1, b_2) = b_1 y_1(b_0, b_1, b_2)$$
$$t_2(b_0, b_1, b_2) = b_2 y_2(b_0, b_1, b_2)$$
$$t_0(b_0, b_1, b_2) = -(t_1(b_0, b_1, b_2) + t_2(b_0, b_1, b_2)).$$

The game induced by the second price procurement auction will be similar except for appropriate changes in $t_1$ and $t_2$.

## 2.8.4 Implementation of a Social Choice Function by a Mechanism

We now formalize the notion of implementation of a social choice function by a mechanism.

**Definition 2.22 (Implementation of an SCF).** We say that a mechanism $\mathcal{M} = ((S_i)_{i \in N}, g(\cdot))$ implements the social choice function $f(\cdot)$ if there is a pure strategy equilibrium $s^*(\cdot) = (s_1^*(\cdot), \ldots, s_n^*(\cdot))$ of the Bayesian game $\Gamma^b$ induced by $\mathcal{M}$ such that $g(s_1^*(\theta_1), \ldots, s_n^*(\theta_n)) = f(\theta_1, \ldots, \theta_n) \ \forall (\theta_1, \ldots, \theta_n) \in \Theta$.

Figure 2.6 explains the idea behind a mechanism implementing a social choice function. Depending on the nature of the underlying equilibrium, two ways of implementing an SCF $f(\cdot)$ are standard in the literature.

**Definition 2.23 (Implementation in Dominant Strategies).** We say that a mechanism $\mathcal{M} = ((S_i)_{i \in N}, g(\cdot))$ implements the social choice function $f(\cdot)$ in dominant strategy equilibrium if there is a weakly dominant strategy equilibrium $s^*(\cdot) = (s_1^*(\cdot), \ldots, s_n^*(\cdot))$ of the game $\Gamma^b$ induced by $\mathcal{M}$ such that

$$g(s_1^*(\theta_1), \ldots, s_n^*(\theta_n)) = f(\theta_1, \ldots, \theta_n) \ \forall (\theta_1, \ldots, \theta_n) \in \Theta.$$

*Note 2.9.* Since a strongly dominant strategy equilibrium is automatically a weakly dominant strategy equilibrium, the above definition applies to the strongly dominant case also. In the latter case, we could say the implementation is in strongly dominant strategy equilibrium. It is worth recalling that there could exist multiple weakly dominant strategy equilibria whereas a strongly dominant strategy equilibrium, if it exists, is unique.

**Definition 2.24 (Implementation in Bayesian Nash Equilibrium).** We say that a mechanism $\mathcal{M} = ((S_i)_{i \in N}, g(\cdot))$ implements the social choice function $f(\cdot)$ in Bayesian Nash equilibrium if there is a pure strategy Bayesian Nash equilibrium $s^*(\cdot) = (s_1^*(\cdot), \ldots, s_n^*(\cdot))$ of the game $\Gamma^b$ induced by $\mathcal{M}$ such that

$$g\left(s_1^*(\theta_1), \ldots, s_n^*(\theta_n)\right) = f\left(\theta_1, \ldots, \theta_n\right) \; \forall \left(\theta_1, \ldots, \theta_n\right) \in \Theta.$$

*Note 2.10.* In the definition, what is implicitly implied is a pure strategy Bayesian Nash equilibrium. Such an equilibrium may or may not exist, but we implicitly assume that such an equilibrium exists.

*Note 2.11.* The game $\Gamma^b$ induced by the mechanism $\mathcal{M}$ may have more than one equilibrium, but the above definition requires only that *one of them* induces outcomes in accordance with the SCF $f(\cdot)$. Implicitly, then, the above definition assumes that, if multiple equilibria exist, the agents will play the equilibrium that the mechanism designer (social planner) wants. This is an extremely important problem that is addressed by a theory called *implementation theory*. A brief idea about implementation theory will be provided in Section 2.21.

*Note 2.12.* Another implicit assumption of the above definition is that the game induced by the mechanism is a simultaneous move game, that is all the agents, after learning their types, choose their actions simultaneously.

## 2.9 Incentive Compatibility and the Revelation Theorem

The notion of incentive compatibility is perhaps the most fundamental concept in mechanism design, and the revelation theorem is perhaps the most fundamental result in mechanism design. We have already seen that mechanism design involves the preference revelation (or elicitation) problem and the preference aggregation problem. The preference revelation problem involves eliciting truthful information from the agents about their types. In order to elicit truthful information, there is a need to somehow make truth revelation a best response for the agents, consistent with rationality and intelligence assumptions. Offering incentives is a way of doing this; incentive compatibility essentially refers to offering the right amount of incentive to induce truth revelation by the agents. There are broadly two types of incentive compatibility: (1) Truth revelation is a best response for each agent irrespective of what is reported by the other agents; (2) Truth revelation is a best response for each agent whenever the other agents also reveal their true types. The first one is called dominant strategy incentive compatibility (DSIC), and the second one is called Bayesian Nash incentive compatibility (BIC). Since truth revelation is always with respect to types, only direct revelation mechanisms are relevant when formalizing the notion of incentive compatibility. The notion of incentive compatibility was first introduced by Leonid Hurwicz [10].

**Leonid Hurwicz**, Eric Maskin, and Roger Myerson were jointly awarded the Nobel prize in economic sciences in 2007 for having laid the foundations of mechanism design theory. Hurwicz, born in 1917, has become the oldest winner of the Nobel prize. It was Hurwicz who first introduced the notion of mechanisms with his work in 1960 [11]. He defined a mechanism as a communication system in which participants send messages to each other and perhaps to a *message center* and a prespecified rule assigns an outcome (such as allocation of goods and payments to be made) for every collection of received messages. Hurwicz [10] introduced the key notion of incentive compatibility in 1972. This notion allowed mechanism design to incorporate the incentives of rational players and opened up the area of mechanism design. The notion of incentive compatibility plays a central role in the revelation theorem, which is a fundamental result in mechanism design theory. Hurwicz is also credited with many important possibility and impossibility results in mechanism design. For example, he showed that, in a standard exchange economy, no incentive compatible mechanism that satisfies individual rationality can produce Pareto optimal outcomes. Hurwicz's work in game theory and mechanism design demonstrated, beyond doubt, the value of using analytical methods in modeling economic institutions.

Hurwicz was, until his demise on June 24, 2008, Regents Professor Emeritus in the Department of Economics at the University of Minnesota. He taught there in the areas of welfare economics, public economics, mechanisms and institutions, and mathematical economics.

**Eric Maskin** is a joint winner with Leonid Hurwicz and Roger Myerson, of the Nobel prize in Economic Sciences in 2007. One of his most creative contributions was his work on implementation theory, which addresses the following problem: Given a social goal, can we characterize when we can design a mechanism whose equilibrium outcomes coincide with the outcomes that are desirable according to that goal? Maskin [34] gave a general solution to this problem. He brilliantly showed that if social goals are to be implementable, then they must satisfy a certain kind of *monotonicity* which is now famously called *Maskin Monotonicity*. He also showed that monotonicity guarantees implementation under certain mild conditions (at least three players and no veto power). He has also made major contributions to dynamic games. One of his early contributions was to formalize the Revelation Theorem to the setting of Bayesian incentive compatible mechanisms.

Maskin was born on December 12, 1950, in New York city. He earned an A.B. in Mathematics and a Ph. D. in Applied Mathematics from Harvard University in 1976. He taught at the Massachusetts Institute of Technology during 1977–1984 and at Harvard University during 1985–2000. Since 2000, he is the Albert O. Hirschman Professor of Social Science at the Institute for Advanced Study in Princeton, NJ, USA.

 **Roger Bruce Myerson** jointly received the Nobel Prize in Economic Sciences in 2007, with Leonid Hurwicz and Eric Maskin, for having laid the foundations of mechanism design theory. Myerson has straddled several subareas in game theory and mechanism design, and his contributions have left a deep impact in the area. He was instrumental in conceptualizing and proving the revelation theorem in mechanism design for Bayesian implementations in its most generality. His work on optimal auctions in 1981 is a landmark result and has led to a phenomenal amount of further work in the area of optimal auctions. He has also made major contributions in bargaining with incomplete information and cooperative games with incomplete information. His textbook *Game Theory: Analysis of Conflict* is a scholarly and comprehensive reference text that embodies all important results in game theory in a rigorous, yet insightful way. Myerson has also worked on economic analysis of political institutions and written several influential papers in this area including recently on democratization and the Iraq war.

Myerson was born on March 29, 1951. He received his A.B., S.M., and Ph.D., all in Applied Mathematics from Harvard University. He completed his Ph.D. in 1976, working with the legendary Kenneth Arrow. He was a Professor of Economics at the Kellogg School of Management in Northwestern University during 1976-2001. Since 2001, he has been the Glen A. Lloyd Distinguished Service Professor of Economics at the University of Chicago.

## 2.9.1 Incentive Compatibility (IC)

**Definition 2.25 (Incentive Compatibility).** A social choice function $f : \Theta_1 \times \ldots \times \Theta_n \to X$ is said to be incentive compatible (or truthfully implementable) if the Bayesian game induced by the direct revelation mechanism $\mathscr{D} = ((\Theta_i)_{i \in N}, f(\cdot))$ has a pure strategy equilibrium $s^*(\cdot) = (s_1^*(\cdot), \ldots, s_n^*(\cdot))$ in which $s_i^*(\theta_i) = \theta_i, \forall \theta_i \in \Theta_i, \forall i \in N$.

That is, truth revelation by each agent constitutes an equilibrium of the game induced by $\mathscr{D}$. It is easy to infer that if an SCF $f(\cdot)$ is incentive compatible then the direct revelation mechanism $\mathscr{D} = ((\Theta_i)_{i \in N}, f(\cdot))$ can implement it. That is, directly asking the agents to report their types and using this information in $f(\cdot)$ to get the social outcome will solve both the problems, namely, preference elicitation and preference aggregation.

Based on the type of equilibrium concept used, two types of incentive compatibility are defined.

**Definition 2.26 (Dominant Strategy Incentive Compatibility (DSIC)).** A social choice function $f : \Theta_1 \times \ldots \times \Theta_n \to X$ is said to be dominant strategy incentive compatible (or truthfully implementable in dominant strategies) if the direct revelation mechanism $\mathscr{D} = ((\Theta_i)_{i \in N}, f(\cdot))$ has a *weakly dominant strategy equilibrium* $s^*(\cdot) = (s_1^*(\cdot), \ldots, s_n^*(\cdot))$ in which $s_i^*(\theta_i) = \theta_i, \forall \theta_i \in \Theta_i, \forall i \in N$.

That is, truth revelation by each agent constitutes a dominant strategy equilibrium of the game induced by $\mathscr{D}$. Strategy-proof, cheat-proof, straightforward are the alternative phrases used for this property.

*Example 2.40 (Dominant Strategy Incentive Compatibility of Second Price Procurement Auction).* It is easy to see that the social choice function implemented by the second price auction is dominant strategy incentive compatible.

Using the definition of a dominant strategy equilibrium in Bayesian games (Section 2.5), the following necessary and sufficient condition for an SCF $f(\cdot)$ to be dominant strategy incentive compatible can be easily derived:

$$u_i\left(f(\theta_i, \theta_{-i}), \theta_i\right) \geq u_i(f(\hat{\theta}_i, \theta_{-i}), \theta_i), \forall i \in N, \forall \theta_i \in \Theta_i, \forall \theta_{-i} \in \Theta_{-i}, \forall \hat{\theta}_i \in \Theta_i. \quad (2.9)$$

The above condition says that if the SCF $f(.)$ is DSIC, then, irrespective of what the other agents report, it is always a best response for agent $i$ to report his true type $\theta_i$.

**Definition 2.27 (Bayesian Incentive Compatibility (BIC)).** A social choice function $f : \Theta_1 \times \ldots \times \Theta_n \to X$ is said to be Bayesian incentive compatible (or truthfully implementable in Bayesian Nash equilibrium) if the direct revelation mechanism $\mathscr{D} = ((\Theta_i)_{i \in N}, f(\cdot))$ has a *Bayesian Nash equilibrium* $s^*(\cdot) = (s_1^*(\cdot), \ldots, s_n^*(\cdot))$ in which $s_i^*(\theta_i) = \theta_i, \forall \theta_i \in \Theta_i, \forall i \in N$.

That is, truth revelation by each agent constitutes a Bayesian Nash equilibrium of the game induced by $\mathscr{D}$.

*Example 2.41 (Bayesian Incentive Compatibility of First Price Procurement Auction).* We have seen that the first price procurement auction for a single indivisible item implements the following social choice function:

$$f(\theta) = (y_0(\theta), y_1(\theta), y_2(\theta), t_0(\theta), t_1(\theta), t_2(\theta))$$

with

$$\begin{aligned}
y_0(\theta) &= 0 \quad \forall \theta \in \Theta \\
y_1(\theta) &= 1 \quad \text{if } \theta_1 \leq \theta_2 \\
&= 0 \quad \text{otherwise} \\
y_2(\theta) &= 1 \quad \text{if } \theta_1 > \theta_2 \\
&= 0 \quad \text{otherwise} \\
t_1(\theta) &= \frac{1 + \theta_1}{2} y_1(\theta) \\
t_2(\theta) &= \frac{1 + \theta_2}{2} y_2(\theta) \\
t_0(\theta) &= -(t_1(\theta) + t_2(\theta)).
\end{aligned}$$

If seller 1 has type $\theta_1$, then his optimal bid $\hat{\theta}_1$ is obtained by solving

$$\max_{\hat{\theta}_1} \left(\frac{1 + \hat{\theta}_1}{2} - \theta_1\right) P\{\theta_2 \geq \hat{\theta}_1\}.$$

This is the same as

$$\max_{\hat{\theta}_1} \left(\frac{1 + \hat{\theta}_1}{2} - \theta_1\right) (1 - \hat{\theta}_1).$$

This yields $\hat{\theta}_1 = \theta_1$. Thus it is optimal for seller 1 to reveal his true private value if seller 2 reveals his true value. The same situation applies to seller 2. This implies that the social choice function is Bayesian Nash incentive compatible (since the equilibrium involved is a Bayesian Nash equilibrium).

Using the definition of a Bayesian Nash equilibrium in Bayesian games (Section 2.5), the following necessary and sufficient condition for an SCF $f(\cdot)$ to be Bayesian incentive compatible can be easily derived:

$$E_{\theta_{-i}}[u_i(f(\theta_i, \theta_{-i}), \theta_i)|\theta_i] \geq E_{\theta_{-i}}[u_i(f(\hat{\theta}_i, \theta_{-i}), \theta_i)|\theta_i], \forall i \in N, \forall \theta_i \in \Theta_i, \forall \hat{\theta}_i \in \Theta_i \quad (2.10)$$

where the expectation is taken over the type profiles of agents other than agent $i$.

*Note 2.13.* If a social choice function $f(\cdot)$ is dominant strategy incentive compatible then it is also Bayesian incentive compatible. The proof of this follows trivially from the fact that a weakly dominant strategy equilibrium is necessarily a Bayesian Nash equilibrium.

## 2.9.2 The Revelation Principle for Dominant Strategy Equilibrium

The revelation principle basically illustrates the relationship between an indirect mechanism $\mathcal{M}$ and a direct revelation mechanism $\mathcal{D}$ with respect to a given SCF $f(\cdot)$. This result enables us to restrict our inquiry about truthful implementation of an SCF to the class of direct revelation mechanisms only.

**Theorem 2.3.** *Suppose that there exists a mechanism $\mathcal{M} = (S_1, \ldots, S_n, g(\cdot))$ that implements the social choice function $f(\cdot)$ in dominant strategy equilibrium. Then $f(\cdot)$ is dominant strategy incentive compatible.*

**Proof:** If $\mathcal{M} = (S_1, \ldots, S_n, g(\cdot))$ implements $f(\cdot)$ in dominant strategies, then there exists a profile of strategies $s^*(\cdot) = (s_1^*(\cdot), \ldots, s_n^*(\cdot))$ such that

$$g(s_1^*(\theta_1), \ldots, s_n^*(\theta_n)) = f(\theta_1, \ldots, \theta_n) \ \forall (\theta_1, \ldots, \theta_n) \in \Theta \quad (2.11)$$

and

$$u_i(g(s_i^*(\theta_i), s_{-i}(\theta_{-i})), \theta_i) \geq u_i(g(s_i'(\theta_i), s_{-i}(\theta_{-i})), \theta_i)$$
$$\forall i \in N, \forall \theta_i \in \Theta_i, \forall \theta_{-i} \in \Theta_{-i}, \forall s_i'(\cdot) \in S_i, \forall s_{-i}(\cdot) \in S_{-i}. \quad (2.12)$$

Condition (2.12) implies, in particular, that

$$u_i(g(s_i^*(\theta_i), s_{-i}^*(\theta_{-i})), \theta_i) \geq u_i(g(s_i^*(\hat{\theta}_i), s_{-i}^*(\theta_{-i})), \theta_i)$$
$$\forall i \in N, \forall \theta_i \in \Theta_i, \forall \hat{\theta}_i \in \Theta_i, \forall \theta_{-i} \in \Theta_{-i}. \quad (2.13)$$

Conditions (2.11) and (2.13) together imply that

$$u_i(f(\theta_i, \theta_{-i}), \theta_i) \geq u_i(f(\hat{\theta}_i, \theta_{-i}), \theta_i), \forall i \in N, \forall \theta_i \in \Theta_i, \forall \theta_{-i} \in \Theta_{-i}, \forall \hat{\theta}_i \in \Theta_i.$$

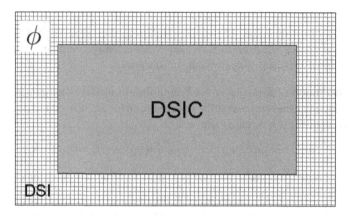

DSI: Dominant Strategy Implementable
DSIC: Dominant Strategy Incentive Compatible
DSI \ DSIC = $\phi$

**Fig. 2.7** Revelation principle for dominant strategy equilibrium

But this is precisely condition (2.9), the condition for $f(\cdot)$ to be truthfully implementable in dominant strategies.

*Q.E.D.*

The idea behind the revelation principle can be understood with the help of Figure 2.7. In this picture, **DSI** represents the set of all social choice functions that are implementable in dominant strategies and **DSIC** is the set of all social choice functions that are dominant strategy incentive compatible. The picture depicts the obvious fact that **DSIC** is a subset of **DSI** and illustrates the revelation theorem by showing that the set difference between these two sets is the empty set, thus implying that **DSIC** is precisely the same as **DSI**.

### 2.9.3 The Revelation Principle for Bayesian Nash Equilibrium

**Theorem 2.4.** *Suppose that there exists a mechanism* $\mathcal{M} = (S_1,\ldots,S_n,g(\cdot))$ *that implements the social choice function* $f(\cdot)$ *in Bayesian Nash equilibrium. Then* $f(\cdot)$ *is truthfully implementable in Bayesian Nash equilibrium (Bayesian incentive compatible).*

**Proof:** If $\mathcal{M} = (S_1,\ldots,S_n,g(\cdot))$ implements $f(\cdot)$ in Bayesian Nash equilibrium, then there exists a profile of strategies $s^*(\cdot) = (s_1^*(\cdot),\ldots,s_n^*(\cdot))$ such that

$$g\left(s_1^*(\theta_1),\ldots,s_n^*(\theta_n)\right) = f\left(\theta_1,\ldots,\theta_n\right) \ \forall \left(\theta_1,\ldots,\theta_n\right) \in \Theta \qquad (2.14)$$

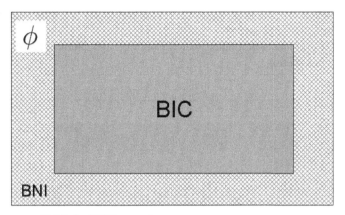

BNI \ BIC $= \phi$
BNI: Bayesian Nash Implementable
BIC: Bayesian Incentive Compatible

**Fig. 2.8** Revelation principle for Bayesian Nash equilibrium

and

$$E_{\theta_{-i}}\left[u_i(g(s_i^*(\theta_i), s_{-i}^*(\theta_{-i})), \theta_i)|\theta_i\right] \geq E_{\theta_{-i}}\left[u_i(g(s_i'(\theta_i), s_{-i}^*(\theta_{-i})), \theta_i)|\theta_i\right]$$

$$\forall i \in N, \forall \theta_i \in \Theta_i, \forall s_i'(\cdot) \in S_i. \qquad (2.15)$$

Condition (2.15) implies, in particular, that

$$E_{\theta_{-i}}\left[u_i(g(s_i^*(\theta_i), s_{-i}^*(\theta_{-i})), \theta_i)|\theta_i\right] \geq E_{\theta_{-i}}\left[u_i(g(s_i^*(\hat{\theta}_i), s_{-i}^*(\theta_{-i})), \theta_i)|\theta_i\right]$$

$$\forall i \in N, \forall \theta_i \in \Theta_i, \forall \hat{\theta}_i \in \Theta_i. \qquad (2.16)$$

Conditions (2.14) and (2.16) together imply that

$$E_{\theta_{-i}}\left[u_i(f(\theta_i, \theta_{-i}), \theta_i)|\theta_i\right] \geq E_{\theta_{-i}}\left[u_i(f(\hat{\theta}_i, \theta_{-i}), \theta_i)|\theta_i\right], \forall i \in N, \forall \theta_i \in \Theta_i, \forall \hat{\theta}_i \in \Theta_i.$$

But this is precisely condition (2.10), the condition for $f(\cdot)$ to be truthfully implementable in Bayesian Nash equilibrium.

*Q.E.D.*

In a way similar to the revelation principle for dominant strategy equilibrium, the revelation principle for Bayesian Nash equilibrium can be explained with the help of Figure 2.8. In this picture, **BNI** represents the set of all social choice functions which are implementable in Bayesian Nash equilibrium and **BIC** is the set of all social choice functions which are Bayesian incentive compatible. The picture depicts the fact that **BIC** is a subset of **BNI** and illustrates the revelation theorem by showing

that the set difference between these two sets is the empty set, thus implying that
**BIC** is precisely the same as **BNI**.

Figure 2.9 provides a combined view of both the revelation theorems that we
have seen in this section.

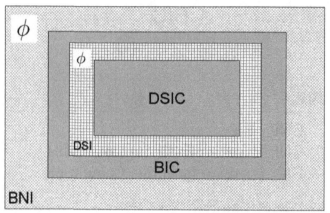

BNI \ BIC = $\phi$                                    DSI: Dominant Strategy Implementable
BNI: Bayesian Nash Implementable     DSIC: Dominant Strategy Incentive Compatible
BIC: Bayesian Incentive Compatible    DSI \ DSIC = $\phi$

**Fig. 2.9** Combined view of revelation theorems for dominant strategy equilibrium and Bayesian
Nash equilibrium

## 2.10 Properties of Social Choice Functions

We have seen that a mechanism provides a solution to both the preference elicitation
problem and preference aggregation problem, if the mechanism can implement the
desired social choice function $f(\cdot)$. It is obvious that some SCFs are implementable
and some are not. Before we look into the question of characterizing the space of
implementable social choice functions, it is important to know which social choice
function ideally a social planner would wish to implement. In this section, we high-
light a few properties of an SCF that ideally a social planner would wish the SCF to
have.

### 2.10.1 Ex-Post Efficiency

**Definition 2.28 (Ex-Post Efficiency).** The SCF $f : \Theta \rightarrow X$ is said to be ex-post efficient (or Paretian) if for every profile of agents' types, $\theta \in \Theta$, the outcome $f(\theta)$ is a Pareto optimal outcome. The outcome $f(\theta_1, \ldots, \theta_n)$ is Pareto optimal if there does not exist any $x \in X$ such that:

$$u_i(x, \theta_i) \geq u_i(f(\theta), \theta_i) \; \forall \; i \in N \text{ and } u_i(x, \theta_i) > u_i(f(\theta), \theta_i) \text{ for some } i \in N.$$

*Example 2.42 (Supplier Selection Problem).* Consider the supplier selection problem (Example 2.29). Let the social choice function $f$ be given by

$$f(a_1, a_2) = x$$
$$f(a_1, b_2) = x.$$

The outcome $f(a_1, a_2) = x$ is Pareto optimal since the other outcomes $y$ and $z$ are such that

$$u_1(y, a_1) < u_1(x, a_1)$$
$$u_1(z, a_1) < u_1(x, a_1).$$

The outcome $f(a_1, b_2) = x$ is Pareto optimal since the other outcomes $y$ and $z$ are such that

$$u_1(y, a_1) < u_1(x, a_1)$$
$$u_1(z, a_1) < u_1(x, a_1).$$

Thus SCF 1 is ex-post efficient.

*Example 2.43 (Procurement of a Single Indivisible Item).* We have looked at three social choice functions, SCF-PROC1, SCF-PROC2, SCF-PROC3, in the previous section. One can show that all these SCFs are ex-post efficient.

### 2.10.2 Dictatorship in SCFs

We define this through a dictatorial social choice function.

**Definition 2.29 (Dictatorship).** A social choice function $f : \Theta \rightarrow X$ is said to be dictatorial if there exists an agent $d$ (called dictator) who satisfies the following property:

$$\forall \; \theta \in \Theta, \; f(\theta) \text{ is such that } u_d(f(\theta), \theta_d) \geq u_d(x, \theta_d) \; \forall x \in X.$$

A social choice function that is not dictatorial is said to be nondictatorial.

In a dictatorial SCF, every outcome that is picked by the SCF is such that it is a most favored outcome for the dictator.

*Example 2.44 (Supplier Selection Problem).* Let the social choice function $f$ be given by

$$f(a_1, a_2) = x; \quad f(a_1, b_2) = x.$$

It is easy to see that agent 1 is a dictator and hence this is a dictatorial SCF. On the other hand, consider the following SCF:

$$f(a_1, a_2) = x; \quad f(a_1, b_2) = y.$$

One can verify that this is not a dictatorial SCF.

## 2.10.3 Individual Rationality

Individual rationality is also often referred to as voluntary participation property. Individual rationality of a social choice function essentially means that each agent gains a nonnegative utility by participating in a mechanism that implements the social choice function. There are three stages at which individual rationality constraints (also called participation constraints) may be relevant in a mechanism design situation.

### 2.10.3.1 Ex-Post Individual Rationality

These constraints become relevant when any agent $i$ is given a choice to withdraw from the mechanism at the ex-post stage, that is, after all the agents have announced their types and an outcome in $X$ has been chosen. Let $\overline{u_i}(\theta_i)$ be the utility that agent $i$ receives by withdrawing from the mechanism when his type is $\theta_i$. Then, to ensure agent $i$'s participation, we must satisfy the following *ex-post participation (or individual rationality) constraints*

$$u_i(f(\theta_i, \theta_{-i}), \theta_i) \geq \overline{u_i}(\theta_i) \ \forall \ (\theta_i, \theta_{-i}) \in \Theta.$$

### 2.10.3.2 Interim Individual Rationality

Let the agent $i$ be allowed to withdraw from the mechanism only at an interim stage that arises after the agents have learned their type but before they have chosen their actions in the mechanism. In such a situation, the agent $i$ will participate in the mechanism only if his interim expected utility $U_i(\theta_i | f) = E_{\theta_{-i}}[u_i(f(\theta_i, \theta_{-i}), \theta_i) | \theta_i]$ from social choice function $f(\cdot)$, when his type is $\theta_i$, is greater than $\overline{u_i}(\theta_i)$. Thus, *interim participation (or individual rationality) constraints* for agent $i$ require that

$$U_i(\theta_i|f) = E_{\theta_{-i}}[u_i(f(\theta_i, \theta_{-i}), \theta_i)|\theta_i] \geq \overline{u}_i(\theta_i) \ \forall \ \theta_i \in \Theta_i.$$

### 2.10.3.3 Ex-Ante Individual Rationality

Let agent $i$ be allowed to refuse to participate in a mechanism only at ex-ante stage, that is, before the agents learn their type. In such a situation, the agent $i$ will participate in the mechanism only if his ex-ante expected utility $U_i(f) = E_\theta[u_i(f(\theta_i, \theta_{-i}), \theta_i)]$ from social choice function $f(\cdot)$ is at least $E_{\theta_i}[\overline{u}_i(\theta_i)]$. Thus, *ex-ante participation (or individual rationality) constraints* for agent $i$ require that

$$U_i(f) = E_\theta[u_i(f(\theta_i, \theta_{-i}), \theta_i)] \geq E_{\theta_i}[\overline{u}_i(\theta_i)].$$

The following proposition establishes a relationship among the three different participation constraints discussed above. The proof is left as an exercise.

**Proposition 2.2.** *For any social choice function $f(\cdot)$, we have*

$$f(\cdot) \text{ is ex-post IR} \Rightarrow f(\cdot) \text{ is interim IR} \Rightarrow f(\cdot) \text{ is ex-ante IR.}$$

## 2.10.4 Efficiency

We have seen the notion of ex-post efficiency already. Depending on the epoch at which we look into the game, we have three notions of efficiency, on the lines of individual rationality. These notions were introduced by Holmstrom and Myerson [12]. Let $F$ be any collection of social choice functions that are of interest.

**Definition 2.30 (Ex-Ante Efficiency).** For any given set of social choice functions $F$, and any member $f(\cdot) \in F$, we say that $f(\cdot)$ is ex-ante efficient in $F$ if there is no other $\hat{f}(\cdot) \in F$ having the following two properties:

$$E_\theta[u_i(\hat{f}(\theta), \theta_i)] \geq E_\theta[u_i(f(\theta), \theta_i)] \ \forall \ i = 1, \ldots, n,$$
$$E_\theta[u_i(\hat{f}(\theta), \theta_i)] > E_\theta[u_i(f(\theta), \theta_i)] \text{ for some } i.$$

**Definition 2.31 (Interim Efficiency).** For any given set of social choice functions $F$, and any member $f(\cdot) \in F$, we say that $f(\cdot)$ is interim efficient in $F$ if there is no other $\hat{f}(\cdot) \in F$ having the following two properties:

$$E_{\theta_{-i}}[u_i(\hat{f}(\theta), \theta_i)|\theta_i] \geq E_{\theta_{-i}}[u_i(f(\theta), \theta_i)|\theta_i] \ \forall \ i = 1, \ldots, n, \ \forall \ \theta_i \in \Theta_i,$$
$$E_{\theta_{-i}}[u_i(\hat{f}(\theta), \theta_i)|\theta_i] > E_{\theta_{-i}}[u_i(f(\theta), \theta_i)|\theta_i] \text{ for some } i \text{ and some } \theta_i \in \Theta_i.$$

**Definition 2.32 (Ex-Post Efficiency).** For any given set of social choice functions $F$, and any member $f(\cdot) \in F$, we say that $f(\cdot)$ is ex-post efficient in $F$ if there is no other $\hat{f}(\cdot) \in F$ having the following two properties:

$$u_i(\hat{f}(\theta), \theta_i) \geq u_i(f(\theta), \theta_i) \ \forall \, i = 1, \ldots, n, \ \forall \, \theta \in \Theta,$$

$$u_i(\hat{f}(\theta), \theta_i) > u_i(f(\theta), \theta_i) \text{ for some } i \text{ and some } \theta \in \Theta.$$

Using the above definition of ex-post efficiency, we can say that a social choice function $f(\cdot)$ is ex-post efficient in the sense of Definition 2.28 if and only if it is ex-post efficient in the sense of Definition 2.32 when we take $F = \{f : f \text{ is a mapping from } \Theta \text{ to } X\}$.

The following proposition establishes a relationship among these three different notions of efficiency.

**Proposition 2.3.** *Given any set of feasible social choice functions F and $f(\cdot) \in F$, we have*

$$f(\cdot) \text{ is ex-ante efficient} \Rightarrow f(\cdot) \text{ is interim efficient} \Rightarrow f(\cdot) \text{ is ex-post efficient.}$$

For a proof of the above proposition, refer to Proposition 23.F.1 of [6]. Also, compare the above proposition with the Proposition 2.2.

## 2.11 The Gibbard–Satterthwaite Impossibility Theorem

We have seen in the last section that dominant strategy incentive compatibility is an extremely desirable property of social choice functions. However the DSIC property, being a strong one, precludes certain other desirable properties to be satisfied. In this section, we discuss the Gibbard–Satterthwaite impossibility theorem (G–S theorem, for short), which shows that the DSIC property will force an SCF to be dictatorial if the utility environment is an unrestricted one. In fact, in the process, even ex-post efficiency will have to be sacrificed. One can say that the G–S theorem has shaped the course of research in mechanism design during the 1970s and beyond, and is therefore a landmark result in mechanism design theory. The G–S theorem is credited independently to Gibbard in 1973 [13] and Satterthwaite in 1975 [14]. The G–S theorem is a brilliant reinterpretation of the famous Arrow's impossibility theorem (which we discuss in the next section). We start our discussion of the G–S theorem with a motivating example.

**Allan Gibbard** is currently Richard B. Brandt Distinguished University Professor of Philosophy at the University of Michigan. His classic paper *Manipulation of Voting Schemes: A General Result* published in Econometrica (Volume 41, Number 4) in 1973 presents the famous Gibbard Satterthwaite theorem, which was also independently proposed by Mark Satterthwaite. Professor Gibbard's current research interests are in ethical theory. He is the author of two widely popular books: *Thinking How to Live* (2003 - Harvard University Press) and *Wise Choices, Apt Feelings* (1990 - Harvard University Press and Oxford University Press).

**Mark Satterthwaite** is currently A.C. Buehler Professor in Hospital and Health Services Management and Professor of Strategic Management and Managerial Economics at the Kellogg School of Management, Northwestern university. He is a microeconomic theorist with keen interest in how health care markets work. His paper *Strategy-proofness and Arrow's Conditions: Existence and Correspondence Theorems for Voting Procedures and Social Welfare Functions* published in the Journal of Economic Theory (Volume 10, April 1975) presented a brilliant reinterpretation of Arrow's impossibility theorem, which is now famously known as the Gibbard–Satterthwaite Theorem. He has authored a large number of scholarly papers in the areas of dynamic matching in markets, organizational dynamics, and mechanism design.

*Example 2.45 (Supplier Selection Problem).* We have seen this example earlier (Example 2.29). We have $N = \{1,2\}$, $X = \{x,y,z\}$, $\Theta_1 = \{a_1\}$, and $\Theta_2 = \{a_2,b_2\}$. Consider the following utility functions (note that these are different from the ones considered in Example 2.29):

$$u_1(x,a_1) = 100; \ u_1(y,a_1) = 50; \ u_1(z,a_1) = 0$$
$$u_2(x,a_2) = 0; \ u_2(y,a_2) = 50; \ u_2(z,a_2) = 100$$
$$u_2(x,b_2) = 30; \ u_2(y,b_2) = 60; \ u_2(z,b_2) = 20.$$

We observe for this example that the DSIC and BIC notions are identical since the type of player 1 is common knowledge and hence player 1 always reports the true type (since the type set is a singleton). Consider the social choice function $f$ given by $f(a_1,a_2) = x$; $f(a_1,b_2) = x$. It can be seen that this SCF is ex-post efficient.

To investigate DSIC, suppose the type of player 2 is $a_2$. If player 2 reports his true type, then the outcome is $x$. If he misreports his type as $b_2$, then also the outcome is $x$. Hence there is no incentive for player 2 to misreport. A similar situation presents itself when the type of player 2 is $b_2$. Thus $f$ is DSIC.

In both the type profiles, the outcome happens to be the most favorable one for player 1, that is, $x$. Therefore, player 1 is a dictator and $f$ is dictatorial. Thus the above function is ex-post efficient and DSIC but dictatorial.

Now, let us consider a different SCF $h$ defined by $h(a_1,a_2) = y; h(a_1,b_2) = x$. Following similar arguments as above, $h$ can be shown to be ex-post efficient and nondictatorial but not DSIC. Table 2.11 lists all the nine possible social choice functions in this scenario and the combination of properties each function satisfies.

Note that the situation is quite desirable with the following SCFs.

$$f_5(a_1,a_2) = y; \ f_5(a_1,b_2) = y$$
$$f_7(a_1,a_2) = z; \ f_7(a_1,b_2) = x.$$

The reason is these functions are ex-post efficient, DSIC, and also nondictatorial. Unfortunately however, such desirable situations do not occur in general. In the present case, the desirable situations do occur because of certain reasons that will

| $i$ | $f_i(a_1,a_2)$ | $f_i(a_1,b_2)$ | EPE | DSIC | NON-DICT |
|-----|------|------|------|------|----------|
| 1 | $x$ | $x$ | ✓ | ✓ | × |
| 2 | $x$ | $y$ | ✓ | × | ✓ |
| 3 | $x$ | $z$ | × | × | ✓ |
| 4 | $y$ | $x$ | ✓ | × | ✓ |
| 5 | $y$ | $y$ | ✓ | ✓ | ✓ |
| 6 | $y$ | $z$ | × | × | ✓ |
| 7 | $z$ | $x$ | ✓ | ✓ | ✓ |
| 8 | $z$ | $y$ | ✓ | ✓ | × |
| 9 | $z$ | $z$ | × | ✓ | ✓ |

**Table 2.11** Social choice functions and properties satisfied by them

become clear soon. In a general setting, ex-post efficiency, DSIC, and nondictatorial properties can never be satisfied simultaneously. In fact, even DSIC and nondictatorial properties cannot coexist. This is the implication of the powerful Gibbard–Satterthwaite theorem.

### 2.11.1 The G–S Theorem

We will build up some notation before presenting the theorem. We have already seen that the preference of an agent $i$, over the outcome set $X$, when its type is $\theta_i$ can be described by means of a *utility function* $u_i(\cdot, \theta_i) : X \to \mathbb{R}$, which assigns a real number to each element in $X$. A utility function $u_i(\cdot, \theta_i)$ always induces a *unique* preference relation $\succsim$ on $X$ which can be described in the following manner

$$x \succsim y \Leftrightarrow u_i(x, \theta_i) \geq u_i(y, \theta_i).$$

The above preference relation is often called a rational preference relation and it is formally defined as follows.

**Definition 2.33 (Rational Preference Relation).** We say that a relation $\succsim$ on the set $X$ is called a rational preference relation if it possesses the following three properties:

1. Reflexivity: $\forall\ x \in X$, we have $x \succsim x$.
2. Completeness: $\forall\ x, y \in X$, we have that $x \succsim y$ or $y \succsim x$ (or both).
3. Transitivity: $\forall\ x, y, z \in X$, if $x \succsim y$ and $y \succsim z$, then $x \succsim z$.

The following proposition establishes the relationship between these two ways of expressing the preferences of an agent $i$ over the set $X$.

**Proposition 2.4.**

1. *If a preference relation $\succsim$ on $X$ is induced by some utility function $u_i(\cdot, \theta_i)$, then it will be a rational preference relation.*

2. *For every preference relation $\succsim$ on X, there may not exist a utility function that induces it. However, when the set X is finite, given any preference relation, there will exist a utility function that induces it.*
3. *For a given preference relation $\succsim$ on X, there might be several utility functions that induce it. Indeed, if the utility function $u_i(\cdot, \theta_i)$ induces $\succsim$, then $u_i'(x, \theta_i) = f(u_i(x, \theta_i))$ is another utility function that will also induce $\succsim$, where $f : \mathbb{R} \to \mathbb{R}$ is a strictly increasing function.*

## Strict Total Preference Relations

We now define a special class of rational preference relations that satisfy the anti-symmetry property also.

**Definition 2.34 (Strict-total Preference Relation).** We say that a rational preference relation $\succsim$ is strict-total if it possesses the antisymmetry property, in addition to reflexivity, completeness, and transitivity. By antisymmetry, we mean that, for any $x, y \in X$ such that $x \neq y$, we have either $x \succsim y$ or $y \succsim x$, but not both.

The strict-total preference relation is also known as a *linear order relation* because it satisfies the properties of the usual *greater than or equal to* relationship on the real line. Let us denote the set of all rational preference relations and strict-total preference relations on the set $X$ by $\mathscr{R}$ and $\mathscr{P}$, respectively. It is easy to see that $\mathscr{P} \subset \mathscr{R}$.

## Ordinal Preference Relations

In a mechanism design problem, for agent $i$, the preference over the set $X$ is described in the form of a utility function $u_i : X \times \Theta_i \to \mathbb{R}$. That is, for every possible type $\theta_i \in \Theta_i$ of agent $i$, we can define a utility function $u_i(\cdot, \theta_i)$ over the set $X$. Let this utility function induce a rational preference relation $\succsim_i (\theta_i)$ over $X$. The set $\mathscr{R}_i = \{\succsim : \succsim = \succsim_i (\theta_i) \text{ for some } \theta_i \in \Theta_i\}$ is known as the set of ordinal preference relations for agent $i$. It is easy to see that $\mathscr{R}_i \subset \mathscr{R} \;\; \forall i \in N$.

With all the above notions in place, we are now in a position to state the G–S theorem.

**Theorem 2.5 (Gibbard–Satterthwaite Impossibility Theorem).** *Consider a social choice function $f : \Theta \to X$. Suppose that*

1. *The outcome set X is finite and contains at least three elements,*
2. *$\mathscr{R}_i = \mathscr{P} \;\; \forall i \in N$,*
3. *$f(\cdot)$ is an onto mapping, that is, the image of SCF $f(\cdot)$ is the set X.*

*Then the social choice function $f(\cdot)$ is dominant strategy incentive compatible iff it is dictatorial.*

For a proof of this theorem, the reader is referred to Proposition 23.C.3 of the book by Mas-Colell, Whinston, and Green [6]. We only provide a brief outline of the proof. To prove the necessity, we assume that the social choice function $f(\cdot)$ is dictatorial and it is shown that $f(\cdot)$ is DSIC. This can be shown in a fairly straightforward way. The proof of the sufficiency part of the theorem starts with the assumption that $f(\cdot)$ is DSIC and proceeds in three steps:

1. It is shown using the second condition of the theorem ($\mathscr{R}_i = \mathscr{P} \ \forall \ i \in N$) that $f(\cdot)$ is monotonic.
2. Next using conditions (2) and (3) of the theorem, it is shown that monotonicity implies ex-post efficiency.
3. Finally, it is shown that a SCF $f(\cdot)$ that is monotonic and ex-post efficient is necessarily dictatorial.

Figure 2.10 shows a pictorial representation of the G–S theorem. The figure depicts two classes $F_1$ and $F_2$ of social choice functions. The class $F_1$ is the set of all SCFs that satisfy conditions (1) and (2) of the theorem while the class $F_2$ is the set of all SCFs that satisfy conditions (1) and (3) of the theorem. The class $GS$ is the set of all SCFs in the intersection of $F_1$ and $F_2$ which are DSIC. The functions in the class GS have to be necessarily dictatorial.

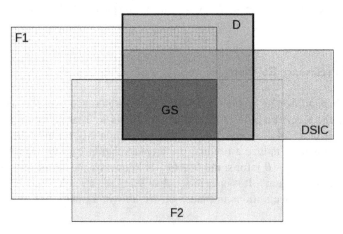

$F_1$ : Set of all SCFs for which $|X| \geqslant 3$
and $\mathscr{R}_i = \mathscr{P} \ \forall \ i \in N$

$F_2$ : Set of all onto SCF

DSIC: Dominant strategy incentive
compatible SCFs

$D = $ Dictatorial SCFs

$GS = F_1 \cap F_2 \cap DSIC$

**Fig. 2.10** An illustration of the Gibbard–Satterthwaite Theorem

## 2.11.2 Implications of the G–S Theorem

One way to get around the impossible situation described by the G–S Theorem is to hope that at least one of the conditions (1), (2), and (3) of the theorem does not hold. We discuss each one of these below.

- Condition (1) asserts that $|X| \geq 3$. This condition is violated only if $|X| = 1$ or $|X| = 2$. The case $|X| = 1$ corresponds to a trivial situation and is not of interest. The case $|X| = 2$ is more interesting but is of only limited interest. A public project problem where only a go or no-go decision is involved and no payments by agents are involved corresponds to this situation.
- Condition (2) asserts that $\mathscr{R}_i = \mathscr{P} \ \forall \ i \in N$. This means that the preferences of each agent cover the entire space of strict total preference relations on $X$. That is, each agent has an extremely rich set of preferences. If we are able to somehow restrict the preferences, we can hope to violate this condition. One can immediately note that this condition was violated in the motivating example (Example 2.45, the supplier selection problem). The celebrated class of VCG mechanisms has been derived by restricting the preferences to the quasilinear domain. This will be discussed in good detail in a later section.
- Condition (3) asserts that $f$ is an onto function. Note that this condition also was violated in Example 2.45. This provides one more route for getting around the G–S Theorem.

Another way of escaping from the jaws of the G–S Theorem is to settle for a weaker form of incentive compatibility than DSIC. We have already discussed Bayesian incentive compatibility (BIC) which only guarantees that reporting true types is a best response for each agent whenever all other agents also report their true types. Following this route leads us to Bayesian incentive compatible mechanisms. These are discussed in good detail in a future section.

The G–S Theorem is an influential result that defined the course of mechanism design research in the 1970s and 1980s. As already stated, the theorem happens to be an ingenious reinterpretation, in the context of mechanism design, of the celebrated Arrow's impossibility theorem, which is discussed next.

## 2.12  Arrow's Impossibility Theorem

This famous impossibility theorem is due to Kenneth Arrow (1951), Nobel laureate in Economic Sciences in 1972. This result has shaped the discipline of social choice theory in many significant ways.

**Kenneth Joseph Arrow** received the Nobel Prize in Economic Sciences in 1972, jointly with John R. Hicks, for their pioneering contributions to general economic equilibrium theory and welfare theory. Arrow is regarded as one of the most influential economists of all time. With his path-breaking contributions in social choice theory, general equilibrium analysis, endogenous growth theory, and economics of information behind him, Kenneth Arrow is truly a legend of economics. Three of his doctoral students, John Harsanyi, Michael Spencer, and Roger Myerson are also Economics Nobel laureates. The famous Arrow impossibility theorem was one of the outstanding results included in his classic book in 1951 *Social Choice and Individual Values*, which itself was inspired by his doctoral work. This theorem is perhaps the most important result in welfare economics and also has far-reaching ramifications for mechanism design theory (in fact, the Gibbard–Satterthwaite theorem is an ingenious reinterpretation of the Arrow Impossibility Theorem).

Kenneth Arrow was born in the New York City on August 23, 1921. He earned his doctorate from Columbia University in 1951, working with Professor Harold Hotelling. He is a recipient of the von Neumann Theory Prize in 1986, and he was awarded in 2004 the National Medal of Science, the highest scientific honor in the United States. His joint work with Gerard Debreu on general equilibrium theory is also a major landmark that was prominently noted in the Nobel Prize awarded to Gerard Debreu in 1983. Arrow is currently the Joan Kenney Professor of Economics and Professor of Operations Research, Emeritus, at Stanford University.

Before discussing this result, we first set up some relevant notation. Consider a set of agents $N = \{1, 2, \ldots, n\}$ and a set of outcomes $X$. Let $\succsim_i$ be a rational preference relation of agent $i$ ($i \in N$). Subscript $i$ in $\succsim_i$ indicates that the relation corresponds to agent $i$. For example, $\succsim_i$ could be induced by $u_i(., \theta_i)$ where $\theta_i$ is a certain type of agent $i$. Each agent is thus naturally associated with a set $\mathscr{R}_i$ of rational preference relations derived from the utility functions $u_i(., \theta_i)$ where $\theta_i \in \Theta_i$.

Given a rational preference relation $\succsim_i$, let us denote by $\succ_i$ the relation defined by

$$(x, y) \in \succ_i \text{ iff } (x, y) \in \succsim_i \text{ and } (y, x) \notin \succsim_i.$$

The relation $\succ_i$ is said to be the *strict total preference relation* derived from $\succsim_i$. Note that $\succ_i = \succsim_i$ if $\succsim_i$ itself is a strict total preference relation. Given an outcome set $X$, a strict total preference relation can be simply represented as an ordered tuple of elements of $X$. Given $\succsim_i$, let us denote by $\sim_i$ the relation defined by

$$(x, y) \in \sim_i \text{ iff } (x, y) \in \succsim_i \text{ and } (y, x) \in \succsim_i.$$

The relation $\sim_i$ is said to be the *indifference relation* derived from $\succsim_i$.

As usual $\mathscr{R}$ and $\mathscr{P}$ denote, respectively, the set of all rational preference relations and strict total preference relations on the set $X$. Let $\mathscr{A}$ be any nonempty subset of $\mathscr{R}^n$. We define a social welfare functional as a mapping from $\mathscr{A}$ to $\mathscr{R}$.

**Definition 2.35.** [Social Welfare Functional] Given a set of agents $N = \{1, 2, \ldots, n\}$, an outcome set $X$, and a set of profiles $\mathscr{A}$ of rational preference relations of the agents, $\mathscr{A} \subset \mathscr{R}^n$, a social welfare functional is a mapping $W : \mathscr{A} \longrightarrow \mathscr{R}$.

Note that a social welfare functional $W$ assigns a rational preference relation $W(\succsim_1, \ldots, \succsim_n)$ to a given profile of rational preference relations $(\succsim_1, \ldots, \succsim_n) \in \mathscr{A}$.

*Example 2.46 (Social Welfare Functional).* Consider the example of the supplier selection problem discussed in Example 2.45, where $N = \{1, 2\}, X = \{x, y, z\}, \Theta_1 = \{a_1\}$, and $\Theta_2 = \{a_2, b_2\}$. Recall the utility functions:

$$u_1(x, a_1) = 100; \ u_1(y, a_1) = 50; \ u_1(z, a_1) = 0$$
$$u_2(x, a_2) = 0; \ u_2(y, a_2) = 50; \ u_2(z, a_2) = 100$$
$$u_2(x, b_2) = 30; \ u_2(y, b_2) = 60; \ u_2(z, b_2) = 20.$$

The utility function $u_1$ leads to the following strict preference relation:

$$\succsim_{a_1} = (x, y, z).$$

The utility function $u_2$ leads to the strict total preference relations:

$$\succsim_{a_2} = (z, y, x); \ \succsim_{b_2} = (y, x, z).$$

Let the set $\mathscr{A}$ be defined as

$$\mathscr{A} = \{(\succsim_{a_1}, \succsim_{a_2}), (\succsim_{a_1}, \succsim_{b_2})\}.$$

An example of a social welfare functional here would be the mapping $W_1$ given by

$$W_1(\succsim_{a_1}, \succsim_{a_2}) = (x, y, z); \ W_1(\succsim_{a_1}, \succsim_{b_2}) = (y, x, z).$$

Another example would be the mapping $W_2$ given by

$$W_2(\succsim_{a_1}, \succsim_{a_2}) = (x, y, z); \ W_2(\succsim_{a_1}, \succsim_{b_2}) = (z, y, x).$$

Note the difference between a social choice function and a social welfare functional. In the case of a social choice function, the preferences are summarized in terms of types and each type profile is mapped to a social outcome. On the other hand, a social welfare functional maps a profile of individual preferences to a social preference relation. Recall that the type of an agent determines a preference relation on the set $X$ through the utility function.

We now define three properties of a social welfare functional: *unanimity* (also called *Paretian property*); *pairwise independence* (also called *independence of irrelevant alternatives* (IIA)), and *dictatorship*.

**Definition 2.36 (Unanimity).** A social welfare functional $W : \mathscr{A} \longrightarrow \mathscr{R}$ is said to be unanimous if $\forall \ (\succsim_1, \ldots, \succsim_n) \in \mathscr{A}$ and $\forall x, y \in X$,

$$(x, y) \in \succsim_i \forall i \in N \Longrightarrow (x, y) \in W_p(\succsim_1 \ldots, \succsim_n)$$

where $W_p(\succsim_1 \ldots, \succsim_n)$ is the strict preference relation derived from $W(\succsim_1 \ldots, \succsim_n)$.

The above definition means that, for all pairs $x, y \in X$, whenever $x$ is preferred to $y$ for every agent, then $x$ is also socially preferred to $y$.

*Example 2.47 (Unanimity).* For the problem being discussed, let

$$W_1(\succsim_{a_1}, \succsim_{a_2}) = W_1((x,y,z),(z,y,x)) = (x,y,z)$$

$$W_1(\succsim_{a_1}, \succsim_{b_2}) = W_1((x,y,z),(y,x,z)) = (y,x,z).$$

This is unanimous because

- $(y,z) \in \succsim_{a_1}, (y,z) \in \succsim_{b_2}$, and $(y,z) \in W_1(\succsim_{a_1}, \succsim_{b_2})$; and
- $(x,z) \in \succsim_{a_1}, (x,z) \in \succsim_{b_2}$, and $(x,z) \in W_1(\succsim_{a_1}, \succsim_{b_2})$.

On the other hand, let

$$W_2((x,y,z),(z,y,x)) = (x,y,z); \quad W_2((x,y,z),(y,x,z)) = (z,y,x)$$

Here $(y,z) \in \succsim_{a_1}$ and $(y,z) \in \succsim_{b_2}$ but $(y,z) \notin W_2(\succsim_{a_1}, \succsim_{b_2})$. So $W_2$ is not unanimous.

**Definition 2.37 (Pairwise Independence).** The social welfare functional $W : \mathscr{A} \longrightarrow \mathscr{R}$ is said to satisfy pairwise independence if $\forall x, y \in X$, the social preference between $x$ and $y$ will depend only on the individual preferences between $x$ and $y$. That is, $\forall x, y \in X, \forall (\succsim_1 \ldots, \succsim_n) \in \mathscr{A}, \forall (\succsim_1' \ldots, \succsim_n') \in \mathscr{A}$, with the property that

$$(x,y) \in \succsim_i \iff (x,y) \in \succsim_i' \text{ and } (y,x) \in \succsim_i \iff (y,x) \in \succsim_i' \quad \forall i \in N,$$

we have that

$$(x,y) \in W(\succsim_1, \ldots, \succsim_n) \iff (x,y) \in W(\succsim_1', \ldots, \succsim_n'); \text{ and}$$

$$(y,x) \in W(\succsim_1, \ldots, \succsim_n) \iff (y,x) \in W(\succsim_1', \ldots, \succsim_n').$$

*Example 2.48 (Pairwise Independence).* Consider the example as before and let

$$W_3(\succsim_{a_1}, \succsim_{a_2}) = W_3((x,y,z),(z,y,x)) = (x,y,z)$$

$$W_3(\succsim_{a_1}, \succsim_{b_2}) = W_3((x,y,z),(y,x,z)) = (y,z,x).$$

Here agent 1 prefers $x$ to $y$ in both the profiles while agent 2 prefers $y$ to $x$ in both the profiles. However in the first case, $x$ is socially preferred to $y$ while in the second case $y$ is socially preferred to $x$. Thus the social preference between $x$ and $y$ is not exclusively dependent on the individual preferences between $x$ and $y$. This shows that $W_1$ is not pairwise independent. On the other hand, consider $W_3$ given by

$$W_4((x,y,z),(z,y,x)) = (x,y,z)$$

$$W_4((x,y,z),(y,x,z)) = (z,x,y).$$

Now this social welfare functional satisfies pairwise independence.

The pairwise independence property is a very appealing property since it ensures that the social ranking between any pair of alternatives $x$ and $y$ does not in any way depend on other alternatives or the relative positions of these other alternatives in the individual preferences. Secondly, the pairwise independence property has a close connection to a property called the weak preference reversal property, which is quite crucial for ensuring dominant strategy incentive compatibility of social choice functions. Further, this property leads to a nice decomposition of the problem of social ranking. For instance, if we wish to determine a social ranking on the outcomes of a subset $Y$ of $X$, we do not need to worry about individual preferences on the set $X \backslash Y$.

**Definition 2.38 (Dictatorship).** A social welfare functional $W : \mathscr{A} \longrightarrow \mathscr{R}$ is called a dictatorship if there exists an agent, $d \in N$, called the dictator such that $\forall x, y \in X$ and $\forall (\succsim_1, \ldots, \succsim_n) \in \mathscr{A}$, we have

$$(x, y) \in \succsim_d \Rightarrow (x, y) \in W_p(\succsim_1, \ldots, \succsim_n).$$

This means that whenever the dictator prefers $x$ to $y$, then $x$ is also socially preferred to $y$, irrespective of the preferences of the other agents. A social welfare functional that does not have a dictator is said to be nondictatorial.

*Example 2.49 (Dictatorship).* Consider the social welfare functional

$$W_5((x, y, z), (z, y, x)) = (x, y, z)$$

$$W_5((x, y, z), (y, x, z)) = (x, y, z).$$

It is clear that agent 1 is a dictator here. On the other hand, the social welfare functional

$$W_3((x, y, z), (z, y, x)) = (x, y, z)$$

$$W_3((x, y, z), (y, x, z)) = (y, z, x)$$

is not dictatorial.

Ideally, a social planner would like to implement a social welfare functional that is unanimous, satisfies the pairwise independence property, and is nondictatorial. Unfortunately, this belongs to the realm of impossible situations when the preference profiles of the agents are *rich*. This is the essence of the Arrow's Impossibility Theorem, which is stated next.

**Theorem 2.6 (Arrow's Impossibility Theorem).** *Suppose*

*1. $|X| \geq 3$,*
*2. $\mathscr{A} = \mathscr{R}^n$ or $\mathscr{A} = \mathscr{P}^n$.*

*Then every social welfare functional $W : \mathscr{A} \longrightarrow \mathscr{R}$ that is unanimous and satisfies pairwise independence is dictatorial.*

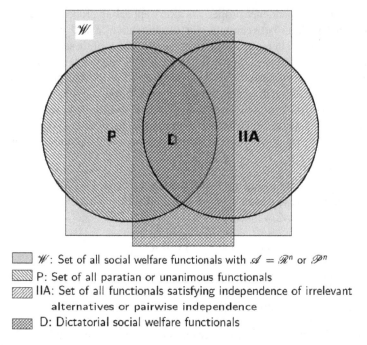

$\mathscr{W}$: Set of all social welfare functionals with $\mathscr{A} = \mathscr{R}^n$ or $\mathscr{P}^n$

P: Set of all paratian or unanimous functionals

IIA: Set of all functionals satisfying independence of irrelevant
alternatives or pairwise independence

D: Dictatorial social welfare functionals

**Fig. 2.11** An illustration of the Arrow's impossibility theorem

For a proof of this theorem, we refer the reader to proposition 21.C.1 of Mas-Colell, Whinston, and Green [6]. Arrow's Impossibility Theorem is pictorially depicted in Figure 2.11. The set $P$ denotes the set of all Paretian or unanimous social welfare functionals. The set IIA denotes the set of all social welfare functionals that satisfy independence of irrelevant alternatives (or pairwise independence). The diagram shows that the intersection of $P$ and IIA is necessarily a subset of $D$, the class of all dictatorial social welfare functionals.

The Gibbard–Satterthwaite theorem has close connections to Arrow's Impossibility Theorem. The property of unanimity of social welfare functionals is related to ex-post efficiency of social choice functions. The notions of dictatorship of social welfare functionals and social choice functions are closely related. The pairwise independence property of social welfare functionals has intimate connections with the DSIC property of social choice functions through the weak preference reversal property and monotonicity. We do not delve deep into this here; interested readers are referred to the book of Mas-Colell, Whinston, and Green [6] (Chapters 21 and 23).

## 2.13 The Quasilinear Environment

This is the most extensively studied special class of environments where the Gibbard–Satterthwaite theorem does not hold. In fact, the rest of this chapter assumes this environment most of the time. In the quasilinear environment, an alternative $x \in X$ is a vector of the form $x = (k, t_1, \dots, t_n)$, where $k$ is an element of a set $K$, which is called the set of project choices or set of allocations. The set $K$ is usually assumed to be finite. The term $t_i \in \mathbb{R}$ represents the monetary transfer to agent $i$. If $t_i > 0$ then agent $i$ will receive the money and if $t_i < 0$ then agent $i$ will pay the money. We assume that we are dealing with a system in which the $n$ agents have no external source of funding, i.e., $\sum_{i=1}^{n} t_i \leq 0$. This condition is known as the *weak budget balance* condition. The set of alternatives $X$ is therefore

$$X = \left\{ (k, t_1, \dots, t_n) : k \in K;\ t_i \in \mathbb{R}\ \forall\, i \in N;\ \sum_i t_i \leq 0 \right\}.$$

A social choice function in this quasilinear environment takes the form $f(\theta) = (k(\theta), t_1(\theta), \dots, t_n(\theta))$ where, for every $\theta \in \Theta$, we have $k(\theta) \in K$ and $\sum_i t_i(\theta) \leq 0$. Note that here we are using the symbol $k$ both as an element of the set $K$ and as a function going from $\Theta$ to $K$. It should be clear from the context as to which of these two we are referring. For a direct revelation mechanism $\mathscr{D} = ((\Theta_i)_{i \in N}, f(\cdot))$ in this environment, the agent $i$'s utility function takes the quasilinear form

$$u_i(x, \theta_i) = u_i((k, t_1, \dots, t_n), \theta_i) = v_i(k, \theta_i) + m_i + t_i$$

where $m_i$ is agent $i$'s initial endowment of the money and the function $v_i(\cdot)$ is known as agent $i$'s valuation function. Recall from our discussion of mechanism design environment (Section 2.6) that the utility functions $u_i(\cdot)$ are common knowledge. In the context of a quasilinear environment, this implies that for any given type $\theta_i$ of any agent $i$, the social planner and every other agent $j$ have a way to know the function $v_i(., \theta_i)$. In many cases, the set $\Theta_i$ of the direct revelation mechanism $\mathscr{D} = ((\Theta_i)_{i \in N}, f(\cdot))$ is actually the set of all feasible valuation functions $v_i$ of agent $i$. That is, each possible function represents the possible types of agent $i$. Therefore, in such settings, reporting a type is the same as reporting a valuation function.

Immediate examples of quasilinear environment include many of the previously discussed examples, such as the first price and second price auctions (Example 2.30), the public project problem (Example 2.31), the network formation problem (Example 2.33), bilateral trade (Example 2.32), etc. In the quasilinear environment, we can define two important properties of a social choice function, namely, allocative efficiency and budget balance.

**Definition 2.39 (Allocative Efficiency (AE)).** We say that a social choice function $f(\cdot) = (k(\cdot), t_1(\cdot), \dots, t_n(\cdot))$ is allocatively efficient if for each $\theta \in \Theta$, $k(\theta)$ satisfies the following condition[1]

---

[1] *We will be using the symbol $k^*(\cdot)$ for a function $k(\cdot)$ that satisfies Equation (2.17).*

$$k(\theta) \in \frac{\arg\max}{k \in K} \sum_{i=1}^{n} v_i(k, \theta_i). \tag{2.17}$$

Equivalently,

$$\sum_{i=1}^{n} v_i(k(\theta), \theta_i) = \frac{\max}{k \in K} \sum_{i=1}^{n} v_i(k, \theta_i).$$

The above definition implies that for every $\theta \in \Theta$, the allocation $k(\theta)$ will maximize the sum of the values of the players. In other words, every allocation is a value maximizing allocation, or the objects are allocated to the players who value the objects most. This is an extremely desirable property to have for any social choice function. The above definition implicitly assumes that for any given $\theta$, the function $\sum_{i=1}^{n} v_i(., \theta_i) : K \rightarrow \mathbb{R}$ attains a maximum over the set $K$.

*Example 2.50 (Public Project Problem).* Consider the public project problem with two agents $N = \{1,2\}$. Let the cost of the public project be 50 units of money. Let the type sets of the two players be given by

$$\Theta_1 = \Theta_2 = \{20, 60\}.$$

Each agent either has a low willingness to pay, 20, or a high willingness to pay, 60. Let the set of project choices be

$$K = \{0, 1\}$$

with 1 indicating that the project is taken up and 0 indicating that the project is dropped.

Assume that if $k = 1$, then the two agents will equally share the cost of the project by paying 25 each. If $k = 0$, the agents do not pay anything. A reasonable way of defining the valuation function would be

$$v_i(k, \theta_i) = k(\theta_i - 25).$$

This means, if $k = 0$, the agents derive zero value while if $k = 1$, the value derived is willingness to pay minus 25.

Define the following allocation function:

$$k(\theta_1, \theta_2) = 0 \text{ if } \theta_1 = \theta_2 = 20$$
$$= 1 \text{ otherwise.}$$

This means, the project is taken up only when at least one of the agents has a high willingness to pay. We can see that this function is allocatively efficient. This may be easily inferred from Table 2.12, which shows the values derived by the agents for different type profiles. The second column gives the actual value of $k$.

*Example 2.51 (A Non-Allocatively Efficient SCF).* Let the $v$ function be defined as under:

| $(\theta_1,\theta_2)$ | $k$ | $v_1(k,\theta_1)$ when $k=0$ | $v_2(k,\theta_2)$ when $k=0$ | $v_1(k,\theta_1)$ when $k=1$ | $v_2(k,\theta_2)$ when $k=1$ |
|---|---|---|---|---|---|
| $(20,20)$ | 0 | 0 | 0 | -5 | -5 |
| $(20,60)$ | 1 | 0 | 0 | -5 | 35 |
| $(60,20)$ | 1 | 0 | 0 | 35 | -5 |
| $(60,60)$ | 1 | 0 | 0 | 35 | 35 |

**Table 2.12** Values for different type profiles when $v_i(k,\theta_i) = k(\theta_i - 25)$

$$v_i(k,\theta_i) = k\theta_i \qquad i = 1,2.$$

With respect to the above function, the allocation function $k$ defined in the previous example can be seen to be not allocatively efficient. The values for different type profiles are shown in Table 2.13. If the type profile is $(20,20)$, the allocation is $k = 0$ and the total value of allocation is 0. However, the total value is 40 if the allocation were $k = 1$.

| $(\theta_1,\theta_2)$ | $k$ | $v_1(k,\theta_1)$ when $k=0$ | $v_2(k,\theta_2)$ when $k=0$ | $v_1(k,\theta_1)$ when $k=1$ | $v_2(k,\theta_2)$ when $k=1$ |
|---|---|---|---|---|---|
| $(20,20)$ | 0 | 0 | 0 | 20 | 20 |
| $(20,60)$ | 1 | 0 | 0 | 20 | 60 |
| $(60,20)$ | 1 | 0 | 0 | 60 | 20 |
| $(60,60)$ | 1 | 0 | 0 | 60 | 60 |

**Table 2.13** Values for different type profiles when $v_i(k,\theta_i) = k\theta_i$

**Definition 2.40 (Budget Balance (BB)).** We say that a social choice function $f(\cdot) = (k(\cdot),t_1(\cdot),\ldots,t_n(\cdot))$ is budget balanced if for each $\theta \in \Theta$, $t_1(\theta),\ldots,t_n(\theta)$ satisfy the following condition:

$$\sum_{i=1}^{n} t_i(\theta) = 0. \qquad (2.18)$$

Many authors prefer to call this property *strong budget balance*, and they refer to the property of having $\sum_{i=1}^{n} t_i(\theta) \leq 0$ as *weak budget balance*. In this monograph, we will use the term budget balance to refer to strong budget balance.

Budget balance ensures that the total receipts are equal to total payments. This means that the system is a closed one, with no surplus and no deficit. The weak budget balance property means that the total payments are greater than or equal to total receipts.

The following lemma establishes an important relationship of these two properties of an SCF with the ex-post efficiency of the SCF.

**Lemma 2.1.** *A social choice function* $f(\cdot) = (k(\cdot), t_1(\cdot), \ldots, t_n(\cdot))$ *is ex-post efficient in quasilinear environment if and only if it is allocatively efficient and budget balanced.*

**Proof:** Let us assume that $f(\cdot) = (k(\cdot), t_1(\cdot), \ldots, t_n(\cdot))$ is allocatively efficient and budget balanced. This implies that for any $\theta \in \Theta$, we have

$$\sum_{i=1}^{n} u_i(f(\theta), \theta_i) = \sum_{i=1}^{n} v_i(k(\theta), \theta_i) + \sum_{i=1}^{n} t_i(\theta)$$

$$= \sum_{i=1}^{n} v_i(k(\theta), \theta_i) + 0$$

$$\geq \sum_{i=1}^{n} v_i(k, \theta_i) + \sum_{i=1}^{n} t_i; \ \forall x = (k, t_1, \ldots, t_n)$$

$$= \sum_{i=1}^{n} u_i(x, \theta_i); \ \forall (k, t_1, \ldots, t_n) \in X.$$

That is if the SCF is allocatively efficient and budget balanced then for any type profile $\theta$ of the agent, the outcome chosen by the social choice function will be such that it maximizes the total utility derived by all the agents. This will automatically imply that the SCF is ex-post efficient.

To prove the other part, we will first show that if $f(\cdot)$ is not allocatively efficient, then, it cannot be ex-post efficient and next we will show that if $f(\cdot)$ is not budget balanced then it cannot be ex-post efficient. These two facts together will imply that if $f(\cdot)$ is ex-post efficient then it will have to be allocatively efficient and budget balanced, thus completing the proof of the lemma.

To start with, let us assume that $f(\cdot)$ is not allocatively efficient. This means that $\exists \, \theta \in \Theta$, and $k \in K$ such that

$$\sum_{i=1}^{n} v_i(k, \theta_i) > \sum_{i=1}^{n} v_i(k(\theta), \theta_i).$$

This implies that there exists at least one agent $j$ for whom $v_j(k, \theta_i) > v_j(k(\theta), \theta_i)$. Now consider the following alternative $x$

$$x = \left( k, (t_i = t_i(\theta) + v_i(k(\theta), \theta_i) - v_i(k, \theta_i))_{i \neq j}, t_j = t_j(\theta) \right).$$

It is easy to verify that $u_i(x, \theta_i) = u_i(f(\theta), \theta_i) \ \forall \, i \neq j$ and $u_j(x, \theta_i) > u_j(f(\theta), \theta_i)$, implying that $f(\cdot)$ is not ex-post efficient.

Next, we assume that $f(\cdot)$ is not budget balanced. This means that there exists at least one agent $j$ for whom $t_j(\theta) < 0$. Let us consider the following alternative $x$

$$x = \left( k, (t_i = t_i(\theta))_{i \neq j}, t_j = 0 \right).$$

It is easy to verify that for the above alternative $x$, we have $u_i(x, \theta_i) = u_i(f(\theta), \theta_i) \; \forall \, i \neq j$ and $u_j(x, \theta_i) > u_j(f(\theta), \theta_i)$ implying that $f(\cdot)$ is not ex-post efficient.

Q.E.D.

The next lemma summarizes another fact about social choice functions in quasilinear environment.

**Lemma 2.2.** *All social choice functions in quasilinear environments are nondictatorial.*

**Proof:** If possible, assume that a social choice function, $f(\cdot)$, is dictatorial in the quasilinear environment. This means that there exists an agent called the dictator, say $d \in N$, such that for each $\theta \in \Theta$, we have

$$u_d(f(\theta), \theta_d) \geq u_d(x, \theta_d) \; \forall \, x \in X.$$

However, because of the environment being quasilinear, we have $u_d(f(\theta), \theta_d) = v_d(k(\theta), \theta_d) + t_d(\theta)$. Now consider the following alternative $x \in X$ :

$$x = \begin{cases} \left( k(\theta), (t_i = t_i(\theta))_{i \neq d}, t_d = t_d(\theta) - \sum_{i=1}^n t_i(\theta) \right) & : \quad \sum_{i=1}^n t_i(\theta) < 0 \\ \left( k(\theta), (t_i = t_i(\theta))_{i \neq d, j}, t_d = t_d(\theta) + \varepsilon, t_j = t_j(\theta) - \varepsilon \right) & : \quad \sum_{i=1}^n t_i(\theta) = 0 \end{cases}$$

where $\varepsilon > 0$ is any arbitrary number, and $j$ is any agent other than $d$. It is easy to verify, for the above outcome $x$, that we have $u_d(x, \theta_d) > u_d(f(\theta), \theta_d)$, which contradicts the fact that $d$ is a dictator.

Q.E.D.

In view of Lemma 2.2, the social planner need not have to worry about the nondictatorial property of the social choice function in quasilinear environments and he can simply look for whether there exists any SCF that is both ex-post efficient and dominant strategy incentive compatible. Furthermore, in the light of Lemma 2.1, we can say that the social planner can look for an SCF that is allocatively efficient, budget balanced, and dominant strategy incentive compatible. Once again the question arises whether there could exist social choice functions which satisfy all these three properties — AE, BB, and DSIC. We explore this and other questions in the forthcoming sections.

## 2.14 Groves Mechanisms

The main result in this section is that in the quasilinear environment, there exist social choice functions that are both allocatively efficient and dominant strategy incentive compatible. These are in general called the VCG (Vickrey–Clarke–Groves) mechanisms.

## 2.14.1 VCG Mechanisms

The VCG mechanisms are named after their famous inventors William Vickrey, Edward Clarke, and Theodore Groves. It was Vickrey who introduced the famous Vickrey auction (second price sealed bid auction) in 1961 [15]. To this day, the Vickrey auction continues to enjoy a special place in the annals of mechanism design. Clarke [16] and Groves [17] came up with a generalization of the Vickrey mechanisms and helped define a broad class of dominant strategy incentive compatible mechanisms in the quasilinear environment. VCG mechanisms are by far the most extensively used among quasilinear mechanisms. They derive their popularity from their mathematical elegance and the strong properties they satisfy.

**William Vickrey** is the inventor of the famous *Vickrey Auction*, which is considered a major breakthrough in the design of auctions. He showed that the second price sealed bid auction enjoys the strong property of dominant strategy incentive compatibility, in his classic paper *Counterspeculation, Auctions, and Competitive Sealed Tenders* which appeared in the Journal of Finance in 1961. This work demonstrated for the first time the value of game theory in understanding auctions. Apart from this famous auction, Vickrey is known for an early version of revenue equivalence theorem, a key result in auction theory. He is also known for pioneering work in congestion pricing, where he introduced the idea of pricing roads and services as a natural means of regulating heavy demand. His ideas were subsequently put into practice in London city transportation. The Nobel prize in economic sciences in 1996 was jointly won by James A. Mirrlees and William Vickrey for their fundamental contributions to the economic theory of incentives under asymmetric information. However, just three days before the prize announcement, Vickrey passed away on October 11, 1996.

Vickrey was born on June 21, 1914 in Victoria, British Columbia. He earned a Ph.D. from Columbia University in 1948. His doctoral dissertation titled *Agenda for Progressive Taxation* is considered a pioneering piece of work. He taught at Columbia from 1946 until his retirement in 1982.

**Edward Clarke** distinguished himself as a senior economist with the Office of Management and Budget (Office of Information and Regulatory Affairs) involved in transportation regulatory affairs. He is a graduate of Princeton University and the University of Chicago, where he received an MBA and a Ph.D. (1978). He has worked in public policy at the city/regional (Chicago), state, federal, and international levels.

In public economics, he developed the demand revealing mechanism for public project selection, which was noted in the Nobel Committee's award of the 1996 Nobel Prize in Economics to William Vickrey. Clarke's 1971 paper *Multi-part Pricing of Public Goods* in the journal Public Choice in 1971 is a classic in mechanism design. Among VCG mechanisms, Clarke's mechanism is a natural and popular approach used in mechanism design problems. The website http://www.clarke.pair.com/clarke.html may be looked up for more details about Clarke's research and teaching.

 **Theodore Groves** is credited with the most general among the celebrated class of VCG mechanisms. In a classic paper entitled *Incentives in Teams* published in Econometrica in 1973, Groves proposed a general class of allocatively efficient, dominant strategy incentive compatible mechanisms. The Groves mechanism generalizes the Clarke mechanism (proposed in 1971), which in turn generalizes the Vickrey auction proposed in 1961. Groves earned a doctorate in economics at the University of California, Berkeley, and he is currently a Professor of Economics at the University of California, San Diego. The website http://weber.ucsd.edu/ tgroves/ may be looked up for more details about Groves's research and teaching.

## 2.14.2 The Groves' Theorem

The following theorem provides a sufficient condition for an allocatively efficient social function in quasilinear environment to be dominant strategy incentive compatible. We will refer to this theorem in the sequel as Groves theorem, rather than Groves' theorem.

**Theorem 2.7 (Groves Theorem).** *Let the SCF $f(\cdot) = (k^*(\cdot), t_1(\cdot), \ldots, t_n(\cdot))$ be allocatively efficient. Then $f(\cdot)$ is dominant strategy incentive compatible if it satisfies the following payment structure (popularly known as the Groves payment (incentive) scheme):*

$$t_i(\theta) = \left[ \sum_{j \neq i} v_j(k^*(\theta), \theta_j) \right] + h_i(\theta_{-i}) \ \forall \, i = 1, \ldots, n \tag{2.19}$$

*where $h_i : \Theta_{-i} \to \mathbb{R}$ is any arbitrary function that honors the feasibility condition $\sum_i t_i(\theta) \leq 0 \ \forall \, \theta \in \Theta$.*

**Proof**: The proof is by contradiction. Suppose $f(\cdot)$ satisfies both allocative efficiency and the Groves payment structure but is not DSIC. This implies that $f(\cdot)$ does not satisfy the following necessary and sufficient condition for DSIC: $\forall i \in N \ \forall \theta \in \Theta$,

$$u_i(f(\theta_i, \theta_{-i}), \theta_i) \geq u_i(f(\theta_i', \theta_{-i}), \theta_i) \ \forall \theta_i' \in \Theta_i \ \forall \theta_{-i} \in \Theta_{-i}.$$

This implies that there exists at least one agent $i$ for which the above is false. Let $i$ be one such agent. That is, for agent $i$,

$$u_i(f(\theta_i', \theta_{-i}), \theta_i) > u_i(f(\theta_i, \theta_{-i}), \theta_i)$$

for some $\theta_i \in \Theta_i$, for some $\theta_{-i} \in \Theta_{-i}$, and for some $\theta_i' \in \Theta_i$. Thus, for agent $i$, there would exist $\theta_i \in \Theta_i, \theta_i' \in \Theta_i, \theta_{-i} \in \Theta_{-i}$ such that

$$v_i(k^*(\theta_i', \theta_{-i}), \theta_i) + t_i(\theta_i', \theta_{-i}) + m_i > v_i(k^*(\theta_i, \theta_{-i}), \theta_i) + t_i(\theta_i, \theta_{-i}) + m_i.$$

Recall that

$$t_i(\theta_i, \theta_{-i}) = h_i(\theta_{-i}) + \sum_{j \neq i}(k^*(\theta_i, \theta_{-i}), \theta_j)$$

$$t_i(\theta_i', \theta_{-i}) = h_i(\theta_{-i}) + \sum_{j \neq i}(k^*(\theta_i', \theta_{-i}), \theta_j).$$

Substituting these, we get

$$v_i(k^*(\theta_i', \theta_{-i}), \theta_i) + \sum_{j \neq i} v_i(k^*(\theta_i', \theta_{-i}), \theta_j) > v_i(k^*(\theta_i, \theta_{-i}), \theta_i) + \sum_{j \neq i} v_i(k^*(\theta_i, \theta_{-i}), \theta_j),$$

which implies

$$\sum_{i=1}^{n} v_i(k^*(\theta_i', \theta_{-i}), \theta_i) > \sum_{i=1}^{n} v_i(k^*(\theta_i, \theta_{-i}), \theta_i).$$

The above contradicts the fact that $f(\cdot)$ is allocatively efficient. This completes the proof.

Q.E.D.

The following are a few interesting implications of the above theorem.

1. Given the announcements $\theta_{-i}$ of agents $j \neq i$, the monetary transfer to agent $i$ depends on his announced type only through effect of the announcement of agent $i$ on the project choice $k^*(\theta)$.
2. The change in the monetary transfer of agent $i$ when his type changes from $\theta_i$ to $\hat{\theta}_i$ is equal to the effect that the corresponding change in project choice has on total value of the rest of the agents. That is,

$$t_i(\theta_i, \theta_{-i}) - t_i(\hat{\theta}_i, \theta_{-i}) = \sum_{j \neq i}\left[v_j(k^*(\theta_i, \theta_{-i}), \theta_j) - v_j(k^*(\hat{\theta}_i, \theta_{-i}), \theta_j)\right].$$

Another way of describing this is to say that the change in monetary transfer to agent $i$ reflects exactly the externality he is imposing on the other agents.

After the famous result of Groves, a direct revelation mechanism in which the implemented SCF is allocatively efficient and satisfies the Groves payment scheme is called a *Groves Mechanism*.

**Definition 2.41 (Groves Mechanisms).** A direct mechanism, $\mathscr{D} = ((\Theta_i)_{i \in N}, f(\cdot))$ in which $f(\cdot) = (k(\cdot), t_1(\cdot), \ldots, t_n(\cdot))$ satisfies allocative efficiency (2.17) and Groves payment rule (2.19) is known as a Groves mechanism.

In mechanism design parlance, Groves mechanisms are popularly known as Vickrey–Clarke–Groves (VCG) mechanisms because the Clarke mechanism is a special case of Groves mechanism, and the Vickrey mechanism is a special case of Clarke mechanism. We will discuss this relationship later in this monograph.

The Groves theorem provides a sufficiency condition under which an allocatively efficient (AE) SCF will be DSIC. The following theorem due to Green and Laffont

[18] provides a set of conditions under which the condition of Groves Theorem also becomes a necessary condition for an AE SCF to be DSIC. In this theorem, we let $\mathscr{F}$ denote the set of all possible functions $f : K \to \mathbb{R}$.

**Theorem 2.8 (First Characterization Theorem of Green–Laffont).** *Suppose for each agent $i \in N$ that $\{v_i(., \theta_i) : \theta_i \in \Theta_i\} = \mathscr{F}$, that is, every possible valuation function from $K$ to $\mathbb{R}$ arises for some $\theta_i \in \Theta_i$. Then any allocatively efficient social choice function $f(\cdot)$ will be dominant strategy incentive compatible if and only if it satisfies the Groves payment scheme given by Equation (2.19).*

Note that in the above theorem, every possible valuation function from $K$ to $\mathbb{R}$ arises for any $\theta_i \in \Theta_i$. In the following characterization theorem, again due to Green and Laffont [18], $\mathscr{F}$ is replaced with with $\mathscr{F}_c$ where $\mathscr{F}_c$ denotes the set of all possible continuous functions $f : K \to \mathbb{R}$.

**Theorem 2.9 (Second Characterization Theorem of Green–Laffont).** *Suppose for each agent $i \in N$ that $\{v_i(., \theta_i) : \theta_i \in \Theta_i\} = \mathscr{F}_c$, that is, every possible continuous valuation function from $K$ to $\mathbb{R}$ arises for some $\theta_i \in \Theta_i$. Then any allocatively efficient social choice function $f(\cdot)$ will be dominant strategy incentive compatible if and only if it satisfies the Groves payment scheme given by Equation (2.19).*

## *2.14.3 Groves Mechanisms and Budget Balance*

Note that a Groves mechanism always satisfies the properties of AE and DSIC. Therefore, if a Groves mechanism is budget balanced, then it will solve the problem of the social planner because it will then be ex-post efficient and dominant strategy incentive compatible. By looking at the definition of the Groves mechanism, one can conclude that it is the functions $h_i(\cdot)$ that decide whether or not the Groves mechanism is budget balanced. The natural question that arises now is whether there exists a way of defining functions $h_i(\cdot)$ such that the Groves mechanism is budget balanced. In what follows, we present one possibility result and one impossibility result in this regard.

### 2.14.3.1 Possibility and Impossibility Results for Quasilinear Environments

Green and Laffont [18] showed that in a quasilinear environment, if the set of possible types for each agent is sufficiently rich then ex-post efficiency and DSIC cannot be achieved together. The precise statement is given in the form of the following theorem.

**Theorem 2.10 (Green–Laffont Impossibility Theorem).** *Suppose for each agent $i \in N$ that $\mathscr{F} = \{v_i(., \theta_i) : \theta_i \in \Theta_i\}$, that is, every possible valuation function from $K$ to $\mathbb{R}$ arises for some $\theta_i \in \Theta_i$. Then there is no social choice function that is ex-post efficient and DSIC.*

Thus, the above theorem says that if the set of possible types for each agent is sufficiently rich then there is no hope of finding a way to define the functions $h_i(\cdot)$ in Groves payment scheme so that we have $\sum_{i=1}^{n} t_i(\theta) = 0$. However, one special case in which a positive result arises is summarized in the form of following possibility result.

**Theorem 2.11 (A Possibility Result for Budget Balance of Groves Mechanisms).**
*If there is at least one agent whose preferences are known (that is, the type set is a singleton set) then it is possible to choose the functions $h_i(\cdot)$ so that $\sum_{i=1}^{n} t_i(\theta) = 0$.*

**Proof**: Let agent $i$ be such that his preferences are known, that is $\Theta_i = \{\theta_i\}$. In view of this condition, it is easy to see that for an allocatively efficient social choice function $f(\cdot) = (k^*(\cdot), t_1(\cdot), \ldots, t_n(\cdot))$, the allocation $k^*(\cdot)$ depends only on the types of the agents other than $i$. That is, the allocation $k^*(\cdot)$ is a mapping from $\Theta_{-i}$ to $K$. Let us define the functions $h_j(\cdot)$ in the following manner:

$$h_j(\theta_{-j}) = \begin{cases} h_j(\theta_{-j}) & : \quad j \neq i \\ -\sum_{r \neq i} h_r(\theta_{-r}) - (n-1) \sum_{r=1}^{n} v_r(k^*(\theta), \theta_r) & : \quad j = i. \end{cases}$$

It is easy to see that under the above definition of the functions $h_i(\cdot)$, we will have $t_i(\theta) = -\sum_{j \neq i} t_j(\theta)$.                                                                Q.E.D.

Figure 2.12 summarizes the main results of this section by showing what the space of social choice functions looks like in the quasilinear environment. The exhibit brings out various possibilities and impossibilities in the quasilinear environment, based on the results that we have discussed so far.

## 2.15 Clarke (Pivotal) Mechanisms

A special case of Groves mechanism was developed independently by Clarke in 1971 [16] and is known as the *Clarke*, or the *pivotal* mechanism. It is a special case of Groves mechanisms in the sense of using a natural special form for the function $h_i(\cdot)$. In the Clarke mechanism, the function $h_i(\cdot)$ is given by the following relation:

$$h_i(\theta_{-i}) = -\sum_{j \neq i} v_j(k^*_{-i}(\theta_{-i}), \theta_j) \ \forall \ \theta_{-i} \in \Theta_{-i}, \forall \ i = 1, \ldots, n \qquad (2.20)$$

where $k^*_{-i}(\theta_{-i}) \in K_{-i}$ is the choice of a project that is allocatively efficient if there were only the $n-1$ agents $j \neq i$. Formally, $k^*_{-i}(\theta_{-i})$ must satisfy the following condition.

$$\sum_{j \neq i} v_j(k^*_{-i}(\theta_{-i}), \theta_j) \geq \sum_{j \neq i} v_j(k, \theta_j) \ \forall \ k \in K_{-i} \qquad (2.21)$$

where the set $K_{-i}$ is the set of project choices available when agent $i$ is absent. Substituting the value of $h_i(\cdot)$ from Equation (2.20) in Equation (2.19), we get the following expression for agent $i$'s transfer in the Clarke mechanism:

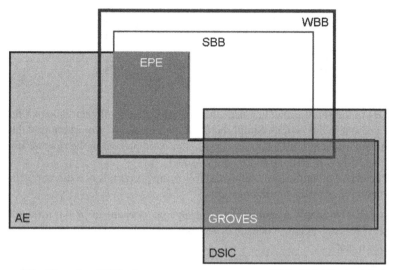

AE : Allocative Efficient                    SBB: Strict Budget Balanced
DSIC : Dominant strategy Incentive Compatible   EPE: Ex-post efficient
WBB : Weak Budget Balanced

**Fig. 2.12** Space of social choice functions in quasilinear environment

$$t_i(\theta) = \left[\sum_{j \neq i} v_j(k^*(\theta), \theta_j)\right] - \left[\sum_{j \neq i} v_j(k^*_{-i}(\theta_{-i}), \theta_j)\right]. \qquad (2.22)$$

The above payment rule has an appealing interpretation: Given a type profile $\theta = (\theta_1, \ldots, \theta_n)$, the monetary transfer to agent $i$ is given by the total value of all agents other than $i$ under an efficient allocation when agent $i$ is present in the system minus the total value of all agents other than $i$ under an efficient allocation when agent $i$ is absent in the system.

### 2.15.1 Clarke Mechanisms and Weak Budget Balance

Recall from the definition of Groves mechanisms that, for weak budget balance, we should choose the functions $h_i(\theta_{-i})$ in a way that the weak budget balance condition $\sum_{i=1}^{n} t_i(\theta) \leq 0$ is satisfied. In this sense, the Clarke mechanism is a useful special case because it achieves weak budget balance under fairly general settings. In order to understand these general sufficiency conditions, we define following quantities

$$B^*(\theta) = \left\{ k \in K : k \in \begin{array}{c} \arg\max \\ k \in K \end{array} \sum_{j=1}^{n} v_j(k, \theta_j) \right\}$$

$$B^*(\theta_{-i}) = \left\{ k \in K_{-i} : k \in \begin{array}{c} \arg\max \\ k \in K_{-i} \end{array} \sum_{j \neq i} v_j(k, \theta_j) \right\}$$

where $B^*(\theta)$ is the set of project choices that are allocatively efficient when all the agents are present in the system. Similarly, $B^*(\theta_{-i})$ is the set of project choices that are allocatively efficient if all agents except agent $i$ were present in the system. It is obvious that $k^*(\theta) \in B^*(\theta)$ and $k^*_{-i}(\theta_{-i}) \in B^*(\theta_{-i})$.

Using the above quantities, we define the following properties of a direct revelation mechanism in quasilinear environment.

**Definition 2.42 (No Single Agent Effect).** We say that mechanism $\mathcal{M}$ has no single agent effect if for each agent $i$, for each $\theta \in \Theta$, and for each $k^*(\theta) \in B^*(\theta)$, we have a $k \in K_{-i}$ such that

$$\sum_{j \neq i} v_j(k, \theta_j) \geq \sum_{j \neq i} v_j(k^*(\theta), \theta_j).$$

In view of the above properties, we have the following proposition that gives a sufficiency condition for Clarke mechanism to be weak budget balanced.

**Proposition 2.5.** *If the Clarke mechanism has no single agent effect, then the monetary transfer to each agent would be non-positive, that is, $t_i(\theta_i) \leq 0 \ \forall \ \theta \in \Theta; \ \forall \ i = 1, \ldots, n$. In such a situation, the Clarke mechanism would satisfy the weak budget balance property.*

**Proof:** Note that by virtue of no single agent effect, for each agent $i$, each $\theta \in \Theta$, and each $k^*(\theta) \in B^*(\theta)$, there exists a $k \in K_{-i}$ such that

$$\sum_{j \neq i} v_j(k, \theta_j) \geq \sum_{j \neq i} v_j(k^*(\theta), \theta_j).$$

However, by definition of $k^*_{-i}(\theta_{-i})$, given by Equation (2.21), we have

$$\sum_{j \neq i} v_j(k^*_{-i}(\theta_{-i}), \theta_j) \geq \sum_{j \neq i} v_j(k, \theta_j) \ \forall \ k \in K_{-i}.$$

Combining the above two facts, we get

$$\sum_{j \neq i} v_j(k^*_{-i}(\theta_{-i}), \theta_j) \geq \sum_{j \neq i} v_j(k^*(\theta), \theta_j)$$

$$\Rightarrow 0 \geq t_i(\theta)$$

$$\Rightarrow 0 \geq \sum_{i=1}^{n} t_i(\theta).$$

$$Q.E.D.$$

In what follows, we present an interesting corollary of the above proposition.

**Corollary 2.1.**

1. $t_i(\theta) = 0$ iff $k^*(\theta) \in B^*(\theta_{-i})$. *That is, agent i's monetary transfer is zero iff his announcement does not change the project decision relative to what would be allocatively efficient for agents $j \neq i$ in isolation.*
2. $t_i(\theta) < 0$ iff $k^*(\theta) \notin B^*(\theta_{-i})$. *That is, agent i's monetary transfer is negative iff his announcement changes the project decision relative to what would be allocatively efficient for agents $j \neq i$ in isolation. In such a situation, the agent i is known to be "pivotal" to the efficient project choice, and he pays a tax equal to his effect on the other agents.*

## 2.15.2 Clarke Mechanisms and Individual Rationality

We have studied individual rationality (also called voluntary participation) property in Section 2.10.3. The following proposition investigates the individual rationality of the Clarke mechanism. First, we provide two definitions.

**Definition 2.43 (Choice Set Monotonicity).** We say that a mechanism $\mathcal{M}$ is choice set monotone if the set of feasible outcomes $X$ (weakly) increases as additional agents are introduced into the system. An implication of this property is $K_{-i} \subset K \ \forall i = 1,\dots,n$.

**Definition 2.44 (No Negative Externality).** Consider a choice set monotone mechanism $\mathcal{M}$. We say that the mechanism $\mathcal{M}$ has no negative externality if for each agent $i$, each $\theta \in \Theta$, and each $k^*_{-i}(\theta_{-i}) \in B^*(\theta_{-i})$, we have

$$v_i(k^*_{-i}(\theta_{-i}), \theta_i) \geq 0.$$

We now state and prove a proposition which provides a sufficient condition for the ex-post individual rationality of the Clarke mechanism. Recall from Section 2.10.3 the notation $\overline{u}_i(\theta_i)$, which represents the utility that agent $i$ receives by withdrawing from the mechanism.

**Proposition 2.6 (Ex-Post Individual Rationality of Clarke Mechanism).** *Let us consider a Clarke mechanism in which*

1. $\overline{u}_i(\theta_i) = 0 \ \forall \theta_i \in \Theta_i; \ \forall i = 1,\dots,n,$
2. *The mechanism satisfies choice set monotonicity property,*
3. *The mechanism satisfies no negative externality property.*

*Then the Clarke mechanism is ex-post individual rational.*

**Proof:** Recall that utility $u_i(f(\theta), \theta_i)$ of an agent $i$ in Clarke mechanism is given by

$$u_i(f(\theta), \theta_i) = v_i(k^*(\theta), \theta_i) + \left[\sum_{j \neq i} v_j(k^*(\theta), \theta_j)\right] - \left[\sum_{j \neq i} v_j(k^*_{-i}(\theta_{-i}), \theta_j)\right]$$

$$= \left[\sum_j v_j(k^*(\theta), \theta_j)\right] - \left[\sum_{j \neq i} v_j(k^*_{-i}(\theta_{-i}), \theta_j)\right].$$

By virtue of choice set monotonicity, we know that $k^*_{-i}(\theta_{-i}) \in K$. Therefore, we have

$$u_i(f(\theta), \theta_i) \geq \left[\sum_j v_j(k^*_{-i}(\theta_{-i}), \theta_j)\right] - \left[\sum_{j \neq i} v_j(k^*_{-i}(\theta_{-i}), \theta_j)\right]$$

$$= v_i(k^*_{-i}(\theta_{-i}), \theta_i)$$

$$\geq 0 = \overline{u}_i(\theta_i).$$

The last step follows due to the fact that the mechanism has no negative externality.

$$Q.E.D.$$

*Example 2.52 (Individual Rationality in Sealed Bid Auction).* Let us consider the example of first-price sealed bid auction. If for each possible type $\theta_i$, the utility $\overline{u}_i(\theta_i)$ derived by the agents $i$ from not participating in the auction is 0, then it is easy to see that the SCF used in this example would be ex-post IR.

Let us next consider the example of a second-price sealed bid auction. If for each possible type $\theta_i$, the utility $\overline{u}_i(\theta_i)$ derived by the agents $i$ from not participating in the auction is 0, then it is easy to see that the SCF used in this example would be ex-post IR. Moreover, the ex post IR of this example also follows directly from Proposition 2.6 because this is a special case of the Clarke mechanism satisfying all the required conditions in the proposition.

## 2.16 Examples of VCG Mechanisms

VCG mechanisms derive their popularity on account of the elegant mathematical and economic properties that they have and the revealing first level insights they provide during the process of designing mechanisms for a game theoretic problem. For this reason, invariably, mechanism design researchers try out VCG mechanisms first. However, VCG mechanisms do have many limitations. The virtues and limitations of VCG mechanisms are captured by Ausubel and Milgrom [19], whereas a recent paper by Rothkopf [20] summarizes the practical limitations of applying VCG mechanisms. In this section, we provide a number of examples to illustrate some interesting nuances of VCG mechanisms.

*Example 2.53 (Vickrey Auction for a Single Indivisible Item).* Consider 5 bidders $\{1, 2, 3, 4, 5\}$, with valuations $v_1 = 20; v_2 = 15; v_3 = 12; v_4 = 10; v_5 = 6$, participating in a sealed bid auction for a single indivisible item. If Vickrey auction is the mechanism used, then it is a dominant strategy for the agents to bid their valuations. Agent 1 with valuation 20 will be the winner, and the monetary transfer to agent 1

$$= \sum_{j \neq 1} v_j(k^*(\theta), \theta_j) - \sum_{j \neq 1} v_j(k^*(\theta_{-1}), \theta_j)$$

$$= 0 - 15 = -15.$$

This means agent 1 would pay an amount equal to 15, which happens to be the second highest bid (in this case the second highest valuation). Note that 15 is the change in the total value of agents other than agent 1 when agent 1 drops out of the system. This is the externality that agent 1 imposes on the rest of the agents. This externality becomes the payment of agent 1 when he wins the auction.

Another way of determining the payment by agent 1 is to compute his marginal contribution to the system. The total value in the presence of agent 1 is 20, while the total value in the absence of agent 1 is 15. Thus the marginal contribution of agent 1 is 5. The above marginal contribution is given as a discount to agent 1 by the Vickrey payment mechanism, and agent 1 pays $20 - 5 = 15$. Such a discount is known as the *Vickrey discount*.

If we use the Clarke mechanism, we have

$$t_i(\theta_i, \theta_{-i}) = \sum_{j \neq i} v_j(k^*(\theta), \theta_j) - \sum_{j \neq i} v_j(k^*_{-i}(\theta_{-i}), \theta_j)$$

$$= \sum_{j \in N} v_j(k^*(\theta), \theta_j) - v_i(k^*(\theta), \theta_i) - \sum_{j \neq i} v_j(k^*_{-i}(\theta_{-i}), \theta_j)$$

$$= \sum_{j \in N} v_j(k^*(\theta), \theta_j) - \sum_{j \neq i} v_j(k^*_{-i}(\theta_{-i}), \theta_j) - v_i(k^*(\theta), \theta_i).$$

The difference in the first two terms represents the marginal contribution of agent $i$ to the system while the term $v_i(k^*(\theta), \theta_i)$ is the value received by agent $i$.

*Example 2.54 (Vickrey Auction for Multiple Identical Items).* Consider the same set of bidders as above but with the difference that there are 3 identical items available for auction. Each bidder wants only one item. If we apply the Clarke mechanism for this situation, bidders 1, 2, and 3 become the winners. The payment by bidder 1

$$= \sum_{j \neq 1} v_j(k^*(\theta), \theta_j) - \sum_{j \neq 1} v_j(k^*_{-1}(\theta_{-1}), \theta_j)$$

$$= (15 + 12) - (15 + 12 + 10)$$

$$= -10.$$

Thus bidder 1 pays an amount equal to the highest nonwinning bid. Similarly, one can verify that the payment to be made by the other two winners (namely agent 2 and agent 3) is also equal to 10. This payment is consistent with their respective marginal contributions.

Marginal contribution of agent $1 = (20 + 15 + 12) - (15 + 12 + 10) = 10$

Marginal contribution of agent $2 = (20 + 15 + 12) - (20 + 12 + 10) = 5$

Marginal contribution of agent $3 = (20 + 15 + 12) - (20 + 15 + 10) = 2.$

In the above example, let the demand by agent 1 be 2 units with the rest of agents continuing to have unit demand. Now the allocation will allocate 2 units to agent 1 and 1 unit to agent 2.

Payment by agent $1 = 15 - (15 + 12 + 10) = -22$

Payment by agent $2 = 40 - (40 + 12) = -12.$

This is because the marginal contribution of agent 1 and agent 2 are given by: agent 1: $55 - 37 = 18$; agent 2: $55 - 52 = 3$.

*Example 2.55 (Generalized Vickrey Auction).* Generalized Vickrey auction (GVA) refers to an auction that results when the Clarke mechanism is applied to a combinatorial auction. A combinatorial auction is one where the bids correspond to bundles or combinations of different items. In a *forward combinatorial auction*, a bundle of different types of goods is available with the seller; the buyers are interested in purchasing certain subsets of the items. In a *reverse combinatorial auction*, a bundle of different types of goods is required by the buyer; several sellers are interested in selling subsets of the goods to the buyer. There is a rich body of literature on combinatorial auctions, for example see the edited volume [21]. We discuss a simple example here. Let a seller be interested in auctioning two items A and B. Let there be three buying agents $\{1,2,3\}$. Let us abuse the notation slightly and denote the subsets $\{A\}, \{B\}, \{A,B\}$ by A, B, and AB, respectively. These are called combinations or bundles. Assume that the agents have *valuations* for the bundles as shown in Table 2.14. In the above table, a "*" indicates that the agent is not interested in

|         | A | B | AB |
|---------|---|---|----|
| Agent 1 | * | * | 10 |
| Agent 2 | 5 | * | *  |
| Agent 3 | * | 5 | *  |

**Table 2.14** Valuations of agents for bundles in scenario 1

that bundle. Note from Table 2.14 that agent 1 values bundle AB at 10 and does not have any valuation for bundle A and bundle B. Agent 2 is only interested in bundle A and has a valuation of 5 for this bundle. Agent 3 is only interested in bundle B and has a valuation of 5 for this bundle. If we apply the Clarke mechanism to this situation, the bids from the agents will be identical to the valuations because of the DSIC property of the Clarke mechanism. There are two allocatively efficient allocations, namely: (1) Allocate bundle AB to agent 1; (2) Allocate bundle A to agent

2 and bundle B to agent 3. Each of these allocations has a total value of 10. Suppose we choose allocation (2), which awards bundle A to agent 2 and bundle B to agent 3. To compute the payments to be made by agents 2 and 3, we have to use the Clarke payment rule. For this, we analyze what would happen in the absence of agent 2 and agent 3 separately. If agent 2 is absent, the allocation will award the bundle AB to agent 1 resulting in a total value of 10. Therefore, the Vickrey discount to agent 2 is $10 - 10 = 0$, which means payment to be made by agent 2 is $5 + 0 = 5$. Similarly the Vickrey discount to agent 3 is also 0 and the payment to be made by agent 3 is also equal to 5. The total revenue to the seller is $5 + 5 = 10$. Even if allocation (1) is chosen (that is, award bundle AB to agent 1), the total revenue to the seller remains as 10. This is a situation where the seller is able to capture the entire consumer surplus.

A contrasting situation will result if the valuations are as shown in Table 2.15. In

|        | A  | B  | AB |
|--------|----|----|----|
| Agent 1 | *  | *  | 10 |
| Agent 2 | 10 | *  | *  |
| Agent 3 | *  | 10 | *  |

**Table 2.15** Valuations of agents for bundles in scenario 2

this case, the winning allocation is: award bundle A to agent 2 and bundle B to agent 3, resulting in a total value of 20. If agent 2 is not present, the allocation will be to award bundle AB to agent 1, thus resulting in a total value of 10. Similarly, if agent 3 were not present, the allocation would be to award bundle AB to agent 1, thus resulting in a total value of 10. This would mean a Vickrey discount of 10 each to agent 2 and agent 3, which in turn means that the the payment to be made by agent 2 and agent 3 is 0 each! This represents a situation where the seller will end up with a zero revenue in the process of guaranteeing allocative efficiency and dominant strategy incentive compatibility. Worse still, if agent 2 and agent 3 are both the false names of a single agent, then the auction itself is seriously manipulated!

We now study a third scenario where the valuations are as described in Table 2.16. Here, the allocation is to award bundle AB to agent 1, resulting in a total value

|        | A | B | AB |
|--------|---|---|----|
| Agent 1 | * | * | 10 |
| Agent 2 | 2 | * | *  |
| Agent 3 | * | 2 | *  |

**Table 2.16** Valuations of agents for bundles in scenario 3

of 10. If agent 1 were absent, the allocation would be to award bundle A to agent 2 and bundle B to agent 3, which leads to a total value of 4. The Vickrey discount to agent 1 is therefore $10 - 4 = 6$, and the payment to be made by agent 1 is 4.

The revenue to the seller is also 4. Contrast this scenario with scenario 2, where the valuations of bidders 2 and 3 were higher, but they were able to win the bundles by paying nothing. This shows that the GVA mechanism is not foolproof against bidder collusion (in this case, bidders 2 and 3 can collude and deny the bundle to agent 1 and also seriously reduce the revenue to the seller.

*Example 2.56 (Strategy Proof Mechanism for the Public Project Problem).* Consider the public project problem discussed in Example 2.31. We shall compute the Clarke payments by each agent for each type of profile. We will also compute the utilities. First consider the type profile (20,20). Since $k = 0$, the values derived by either agent is zero. Hence the Clarke payment by each agent is zero, and the utilities are also zero.

Next consider the type profile (60, 20). Note that $k(60,20) = 1$. Agent 1 derives a value 35 and agent 2 derives a value $-5$. If agent 1 is not present, then agent 2 is left alone and the allocation will be 0 since its willingness to pay is 20. Thus the value to agent 2 when agent 1 is not present is 0. This means

$$t_1(60,20) = -5 - 0 = -5.$$

This implies agent 1 would pay an amount of 5 units in addition to 25 units, which is its contribution to the cost of the project. The above payment is consistent with the marginal contribution of agent 1, which is equal to $(60 - 25) + (20 - 25) - 0 = 35 - 5 = 30$.

We can now determine the utility of agent 1, which will be

$$u_1(60,20) = v_1(60,20) + t_1(60,20)$$
$$= 35 - 5 = 30.$$

To compute $t_2(60,20)$, we first note that the value to the agent 1 when agent 2 is not present is $(60 - 50)$. Therefore

$$t_2(60,20) = 35 - 10 = 25.$$

This means agent 2 receives 25 units of money; of course, this is besides the 25 units of money it pays towards the cost of the project. Now

$$u_2(60,20) = v_2(60,20) + t_2(60,20)$$
$$= -5 + 25$$
$$= 20.$$

Likewise, we can compute the payments and utilities of the agents for all the type profiles. Table 2.17 provides these values. Note that this mechanism is ex-post individually rational assuming that the utility for not participating in the mechanism is zero.

*Example 2.57 (Strategy Proof Network Formation).* Consider the problem of forming a supply chain as depicted in Figure 2.13. The node S represents a start-

| $(\theta_1,\theta_2)$ | $t_1(\theta_1,\theta_2)$ | $t_2(\theta_1,\theta_2)$ | $u_1(\theta_1,\theta_2)$ | $u_2(\theta_1,\theta_2)$ |
|---|---|---|---|---|
| (20, 20) | 0 | 0 | 0 | 0 |
| (60, 20) | -5 | 25 | 30 | 20 |
| (20, 60) | 25 | -5 | 20 | 30 |
| (60, 60) | 25 | 25 | 60 | 60 |

**Table 2.17** Payments and utilities for different type profiles

ing state and T represents a target state; A and B are two intermediate states. $SP_1, SP_2, SP_3, SP_4$ are four different service providers. In the figure, the service

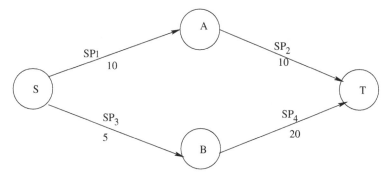

**Fig. 2.13** A network formation problem - case 1

providers are represented as owners of the respective edges. The cost of providing service (willingness to sell) is indicated on each edge. The problem is to procure a path from S to T having minimum cost. Let $(y_1, y_2, y_3, y_4)$ represent the allocation vector. The feasible allocation vectors are

$$K = \{(1,1,0,0), (0,0,1,1)\}.$$

Among these, the allocation $(1,1,0,0)$ is allocatively efficient since it minimizes the cost of allocation. We shall define the value as follows:

$$v_i((y_1, y_2, y_3, y_4); \theta_i) = -y_i \theta_i.$$

The above manner of defining the values reflects the fact that cost minimization is the same as value maximization. Applying Clarke's payment rule, we obtain

$$t_1(\theta) = -10 - (-25) = 15$$
$$t_2(\theta) = -10 - (-25) = 15.$$

Note that each agent gets a surplus of 5, being its marginal contribution. The utilities for these two agents are

$$u_1(\theta) = -10 + 15 = 5$$
$$u_2(\theta) = -10 + 15 = 5.$$

The payments and utilities for $SP_3$ and $SP_4$ are zero. Let us study the effect of changing the willingness to sell of $SP_4$. Let us make it as 15. Then, we find that both the allocations $(1,1,0,0)$ and $(0,0,1,1)$ are allocatively efficient. If we choose the allocation $(1,1,0,0)$, we get the payments as

$$t_1(\theta) = 10$$
$$t_2(\theta) = 10$$
$$u_1(\theta) = 0$$
$$u_2(\theta) = 0.$$

This means that the payments to the service providers are equal to the costs. There is no surplus payment to the winning agents. In this case, the mechanism is friendly to the buyer and unfriendly to the sellers.

If we make the willingness to sell of $SP_4$ as 95, the allocation $(1,1,0,0)$ is efficient and we get the payments as

$$t_1(\theta) = 90$$
$$t_2(\theta) = 90$$
$$u_1(\theta) = 80$$
$$u_2(\theta) = 80.$$

In this case, the mechanism is extremely unfriendly to the buyer but is very attractive to the sellers.

Let us introduce one more edge from B to A and see the effect. See Figure 2.14.

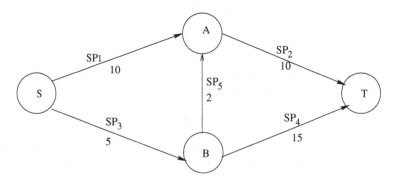

**Fig. 2.14** A network formation problem - case 2

The efficient allocation here is $(0, 1, 1, 0, 1)$. The payments are

$$t_2(\theta) = 13$$
$$t_3(\theta) = 8$$
$$t_5(\theta) = 5$$
$$u_2(\theta) = 3$$
$$u_3(\theta) = 3$$
$$u_2(\theta) = 3.$$

This shows that the total payments to be made by the buyer is 26 whereas the total payment if the service provider $SP_5$ were absent is 20. Thus in spite of an additional agent being available, the payment to the buyer is higher. This shows a kind of non-monotonicity exhibited by the Clarke payment rule.

*Example 2.58 (A Groves (but not Clarke) Mechanism).* Consider a sealed bid auction for a single indivisible item where the bidder with the highest bid is declared as the winner and the winner pays an amount equal to equal to twice the bid of the bidder with the lowest valuation among the rest of the agents. Such a payment rule is not a Clarke payment rule but does belong to the Groves payment scheme.

As an example consider 5 bidders with valuations 20, 15, 12, 10, 8. The bidder with valuation 20 is declared the winner and will pay an amount = 16. On the other hand, if there are only three bidders with values 20, 15, 12, the first bidder wins but has to pay 24. It is clear that this is not individually rational.

## 2.17 Bayesian Implementation: The dAGVA Mechanism

Recall that we mentioned two possible routes to get around the Gibbard–Satterthwaite Impossibility Theorem. The first was to focus on restricted environments like the quasilinear environment, and the second one was to weaken the implementation concept and look for an SCF which is ex-post efficient, nondictatorial, and Bayesian incentive compatible. In this section, our objective is to explore the second route.

Throughout this section, we will once again be working with the quasilinear environment. As we saw earlier, the quasilinear environments have a nice property that every social choice function in these environments is nondictatorial. Therefore, while working within a quasilinear environment, we do not have to worry about the nondictatorial part of the social choice function. We can just investigate whether there exists any SCF in quasilinear environment, which is both *ex-post efficient* and *BIC*, or equivalently, which has three properties — *AE, BB,* and *BIC*. Recall that in the previous section, we have already addressed the question whether there exists any SCF in quasilinear environments that is *AE, BB*, and *DSIC*, and we found that no function satisfies all these three properties. On the contrary, in this section, we will show that a wide range of SCFs in quasilinear environments satisfy three properties — AE, BB, and BIC.

### 2.17.1 The dAGVA Mechanism

The following theorem, due to d'Aspremont and Gérard-Varet [22] and Arrow [23] confirms that in quasilinear environments, there exist social choice functions that are both ex-post efficient and Bayesian incentive compatible. We refer to this theorem as the *dAGVA* Theorem.

**Theorem 2.12 (The dAGVA Theorem).** *Let the social choice function* $f(\cdot) = (k^*(\cdot), t_1(\cdot), \ldots, t_n(\cdot))$ *be allocatively efficient and the agents' types be statistically independent of each other (i.e. the density* $\phi(\cdot)$ *has the form* $\phi_1(\cdot) \times \ldots \times \phi_n(\cdot)$*). This function can be truthfully implemented in Bayesian Nash equilibrium if it satisfies the following payment structure, known as the dAGVA payment (incentive) scheme:*

$$t_i(\theta) = E_{\tilde{\theta}_{-i}}\left[\sum_{j \neq i} v_j(k^*(\theta_i, \tilde{\theta}_{-i}), \tilde{\theta}_j)\right] + h_i(\theta_{-i}) \ \forall \ i = 1, \ldots, n; \ \ \forall \theta \in \Theta \ (2.23)$$

*where* $h_i(\cdot)$ *is any arbitrary function of* $\theta_{-i}$.

**Proof:** Let the social choice function $f(\cdot) = (k^*(\cdot), t_1(\cdot), \ldots, t_n(\cdot))$ be allocatively efficient, i.e., it satisfies the condition (2.17) and also satisfies the dAGVA payment scheme (2.23). Consider

$$E_{\theta_{-i}}[u_i(f(\theta_i, \theta_{-i}), \theta_i)|\theta_i] = E_{\theta_{-i}}[v_i(k^*(\theta_i, \theta_{-i}), \theta_i) + t_i(\theta_i, \theta_{-i})|\theta_i].$$

Since $\theta_i$ and $\theta_{-i}$ are statistically independent, the expectation can be taken without conditioning on $\theta_i$. This will give us

$$E_{\theta_{-i}}[u_i(f(\theta_i, \theta_{-i}), \theta_i)|\theta_i] = E_{\theta_{-i}}\left[v_i(k^*(\theta_i, \theta_{-i}), \theta_i) + h_i(\theta_{-i}) + E_{\tilde{\theta}_{-i}}\left[\sum_{j \neq i} v_j(k^*(\theta_i, \tilde{\theta}_{-i}), \tilde{\theta}_j)\right]\right]$$

$$= E_{\theta_{-i}}\left[\sum_{j=1}^{n} v_j(k^*(\theta_i, \theta_{-i}), \theta_j)\right] + E_{\theta_{-i}}[h_i(\theta_{-i})].$$

Since $k^*(\cdot)$ satisfies the condition (2.17),

$$\sum_{j=1}^{n} v_j(k^*(\theta_i, \theta_{-i}), \theta_j) \geq \sum_{j=1}^{n} v_j(k^*(\hat{\theta}_i, \theta_{-i}), \theta_j) \ \forall \ \hat{\theta}_i \in \Theta_i.$$

Thus we get, $\forall \ \hat{\theta}_i \in \Theta_i$

$$E_{\theta_{-i}}\left[\sum_{j=1}^{n} v_j(k^*(\theta_i, \theta_{-i}), \theta_j)\right] + E_{\theta_{-i}}[h_i(\theta_{-i})] \geq E_{\theta_{-i}}\left[\sum_{j=1}^{n} v_j(k^*(\hat{\theta}_i, \theta_{-i}), \theta_j)\right] + E_{\theta_{-i}}[h_i(\theta_{-i})].$$

Again by making use of statistical independence we can rewrite the above inequality in the following form

$$E_{\theta_{-i}}[u_i(f(\theta_i, \theta_{-i}), \theta_i)|\theta_i] \geq E_{\theta_{-i}}[u_i(f(\hat{\theta}_i, \theta_{-i}), \theta_i)|\theta_i] \ \forall \ \hat{\theta}_i \in \Theta_i.$$

This shows that when agents $j \neq i$ announce their types truthfully, agent $i$ finds that truth revelation is his optimal strategy, thus proving that the SCF is BIC.

$$Q.E.D.$$

After the results of d'Aspremont and Gérard-Varet [22] and Arrow [23], a direct revelation mechanism in which the SCF is allocatively efficient and satisfies the dAGVA payment scheme is called as *dAGVA mechanism/expected externality mechanism/expected Groves mechanism*.

**Definition 2.45 (dAGVA/expected externality/expected Groves Mechanisms).** A direct revelation mechanism, $\mathscr{D} = ((\Theta_i)_{i \in N}, f(\cdot))$ in which $f(\cdot) = (k(\cdot), t_1(\cdot), \ldots, t_n(\cdot))$ satisfies (2.17) and (2.23) is known as dAGVA/expected externality/expected Groves Mechanism.[2]

### 2.17.2 The dAGVA Mechanism and Budget Balance

We now show that the functions $h_i(\cdot)$ above can be chosen to guarantee $\sum_{i=1}^{n} t_i(\theta) = 0$. Let us define,

$$\xi_i(\theta_i) = E_{\tilde{\theta}_{-i}} \left[ \sum_{j \neq i} v_j(k^*(\theta_i, \tilde{\theta}_{-i}), \tilde{\theta}_j) \right] \ \forall \, i = 1, \ldots, n$$

$$h_i(\theta_{-i}) = -\left(\frac{1}{n-1}\right) \sum_{j \neq i} \xi_j(\theta_j) \ \forall \, i = 1, \ldots, n.$$

In view of the above definitions, we can say that

$$t_i(\theta) = \xi_i(\theta_i) - \left(\frac{1}{n-1}\right) \sum_{j \neq i} \xi_j(\theta_j)$$

$$\Rightarrow \sum_{i=1}^{n} t_i(\theta) = \sum_{i=1}^{n} \xi_i(\theta_i) - \left(\frac{1}{n-1}\right) \sum_{i=1}^{n} \sum_{j \neq i} \xi_j(\theta_j)$$

$$\Rightarrow \sum_{i=1}^{n} t_i(\theta) = \sum_{i=1}^{n} \xi_i(\theta_i) - \left(\frac{1}{n-1}\right) \sum_{i=1}^{n} (n-1)\xi_j(\theta_j)$$

$$\Rightarrow \sum_{i=1}^{n} t_i(\theta) = 0.$$

The budget balanced payment structure of the agents in the above mechanism can be given a nice graph theoretic interpretation. Imagine a directed graph $G = (V, A)$ where $V$ is the set of $n+1$ vertices, numbered $0, 1, \ldots, n$, and $A$ is the set of $[n + n(n-1)]$ directed arcs. The vertices starting from 1 through $n$ correspond to the $n$

---

[2] *We will sometimes abuse the terminology and simply refer to a SCF $f(\cdot)$ satisfying (2.17) and (2.23) as dAGVA/expected externality/expected Groves Mechanism.*

agents involved in the system and the vertex number 0 corresponds to the social planner. The set $A$ consists of two types of the directed arcs:

1. Arcs $0 \rightarrow i$ $\forall i = 1, \ldots, n$,
2. Arcs $i \rightarrow j$ $\forall i, j \in \{1, 2, \ldots, n\}$; $i \neq j$.

Each of the arcs $0 \rightarrow i$ carries a flow of $t_i(\theta)$ and each of the arcs $i \rightarrow j$ carries a flow of $\frac{\xi_i(\theta_i)}{n-1}$. Thus the total outflow from a node $i \in \{1, 2, \ldots, n\}$ is $\xi_i(\theta_i)$ and total inflow to the node $i$ from nodes $j \in \{1, 2, \ldots, n\}$ is $-h_i(\theta_{-i}) = \left(\frac{1}{n-1}\right)\sum_{j \neq i}\xi_j(\theta_j)$ Thus for any node $i$, $t_i(\theta) + h_i(\theta_{-i})$ is the net outflow which it is receiving from node 0 in order to respect the flow conservation constraint. Thus, if $t_i(\cdot)$ is positive then the agent $i$ receives the money from the social planner and if it is negative, then the agent pays the money to the social planner. However, by looking at flow conservation equation for node 0, we can say that total payment received by the planner from the agents and total payment made by the planner to the agents will add up to zero. In graph theoretic terms, the flow from node $i$ to node $j$ can be justified as follows. Each agent $i$ first evaluates the expected total valuation that would be generated together by all his rival agents in his absence, which turns out to be $\xi_i(\theta_i)$. Now, agent $i$ divides it equally among the rival agents and pays to every rival agent an amount equivalent to this. The idea can be better understood with the help of Figure 2.15, which depicts the three agents case.

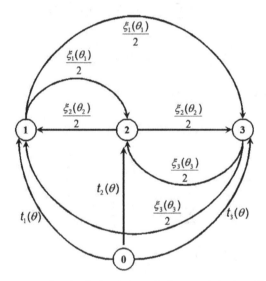

**Fig. 2.15** Payment structure showing budget balance in the expected externality mechanism

*Example 2.59 (dAGVA Mechanism for Sealed Bid Auction).* Consider a selling agent 0 and two buying agents 1, 2. The buying agents submit sealed bids to buy a single indivisible item. Let $\theta_1$ and $\theta_2$ be the willingness to pay of the buyers. Let us define the usual allocation function:

$$y_1(\theta_1, \theta_2) = 1 \text{ if } \theta_1 \geq \theta_2$$
$$= 0 \text{ else}$$
$$y_2(\theta_1, \theta_2) = 1 \text{ if } \theta_1 < \theta_2$$
$$= 0 \text{ else.}$$

Let $\Theta_1 = \Theta_2 = [0, 1]$ and assume that the bids from the bidders are i.i.d. uniform distributions on $[0, 1]$. Also assume that $\Theta_0 = \{0\}$. Assuming that the dAGVA mechanism is used, the payments can be computed as follows:

$$t_i(\theta_1, \theta_2) = E_{\theta_{-i}}\left[\sum_{j \neq i} v_j(k(\theta), \theta_j)\right] - \frac{1}{2}\left[\sum_{j \neq i} E_{\theta_{-i}}\left\{\sum_{l \neq j} v_l(k(\theta), \theta_j)\right\}\right].$$

It can be shown that

$$t_1(\theta) = -\left(\frac{1}{12} - \frac{\theta_1}{2} + \frac{\theta_2}{2}\right)y_1(\theta)$$
$$t_2(\theta) = -\left(\frac{1}{12} - \frac{\theta_2}{2} + \frac{\theta_1}{2}\right)y_2(\theta)$$
$$t_0(\theta) = -(t_1(\theta) + t_2(\theta))$$

This can be compared to the first price auction in which case

$$t_1(\theta) = -\frac{\theta_1}{2}y_1(\theta)$$
$$t_2(\theta) = -\frac{\theta_2}{2}y_2(\theta).$$

Also, one can compare with the second price auction, where

$$t_1(\theta) = -\theta_2 y_1(\theta)$$
$$t_2(\theta) = -\theta_1 y_2(\theta).$$

### 2.17.3 The Myerson–Satterthwaite Theorem

We have so far not seen a single example where we have all the desired properties in an SCF: AE, BB, BIC, and IR. This provides a motivation to study the feasibility of having all these properties in a social choice function.

The Myerson–Satterthwaite Theorem is a disappointing news in this direction, since it asserts that in a bilateral trade setting, whenever the gains from the trade are possible but not certain, then there is no SCF that satisfies AE, BB, BIC, and Interim IR all together. The precise statement of the theorem is as follows.

**Theorem 2.13 (Myerson–Satterthwaite Impossibility Theorem).** *Consider a bilateral trade setting in which the buyer and seller are risk neutral, the valuations $\theta_1$ and $\theta_2$ are drawn independently from the intervals $[\underline{\theta_1}, \overline{\theta_1}] \subset \mathbb{R}$ and $[\underline{\theta_2}, \overline{\theta_2}] \subset \mathbb{R}$ with strict positive densities, and $(\underline{\theta_1}, \overline{\theta_1}) \cap (\underline{\theta_2}, \overline{\theta_2}) \neq \emptyset$. Then there is no Bayesian incentive compatible social choice function that is ex-post efficient and gives every buyer type and every seller type nonnegative expected gains from participation.*

For a proof of the above theorem, refer to Proposition 23.E.1 of [6].

### 2.17.4 Mechanism Design Space in Quasilinear Environment

Figure 2.16 shows the space of mechanisms taking into account all the results we have studied so far. A careful look at the diagram suggests why designing a mechanism that satisfies a certain combination of properties is quite intricate.

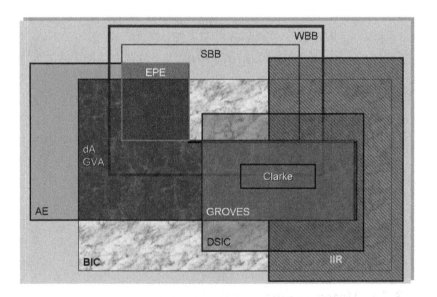

AE : Allocative Efficient                          SBB: Strict Budget Balanced
DSIC : Dominant strategy Incentive Compatible
WBB : Weak Budget Balanced                    BIC : Bayesian Incentive Compatible
IIR   : Interim Individually Rational             EPE: Ex-post efficient

**Fig. 2.16** Mechanism design space in quasilinear environment

## 2.18 Bayesian Incentive Compatibility in Linear Environment

The linear environment is a special, but often-studied, subclass of the quasilinear environment. This environment is a restricted version of the quasilinear environment in the following sense.

1. Each agent $i$'s type lies in an interval $\Theta_i = [\underline{\theta}_i, \overline{\theta}_i] \subset \mathbb{R}$ with $\underline{\theta}_i < \overline{\theta}_i$.
2. Agents' types are statistically independent, that is, the density $\phi(\cdot)$ has the form $\phi_1(\cdot) \times \ldots \times \phi_n(\cdot)$.
3. $\phi_i(\theta_i) > 0 \ \forall \ \theta_i \in [\underline{\theta}_i, \overline{\theta}_i] \ \forall \ i = 1, \ldots, n$.
4. Each agent $i$'s utility function takes the following form

$$u_i(x, \theta_i) = \theta_i v_i(k) + m_i + t_i.$$

The linear environment has very interesting properties in terms of being able to obtain a characterization of the class of BIC social choice functions. Before we present Myerson's Characterization Theorem for BIC social choice functions in a linear environment, we would like to define the following quantities with respect to any social choice function $f(\cdot) = (k(\cdot), t_1(\cdot), \ldots, t_n(\cdot))$ in this environment.

- Let $\overline{t}_i(\hat{\theta}_i) = E_{\theta_{-i}}[t_i(\hat{\theta}_i, \theta_{-i})]$ be agent $i$'s expected transfer given that he announces his type to be $\hat{\theta}_i$ and that all agents $j \neq i$ truthfully reveal their types.
- Let $\overline{v}_i(\hat{\theta}_i) = E_{\theta_{-i}}[v_i(\hat{\theta}_i, \theta_{-i})]$ be agent $i$'s expected "benefits" given that he announces his type to be $\hat{\theta}_i$ and that all agents $j \neq i$ truthfully reveal their types.
- Let $U_i(\hat{\theta}_i | \theta_i) = E_{\theta_{-i}}[u_i(f(\hat{\theta}_i, \theta_{-i}), \theta_i) | \theta_i]$ be agent $i$'s expected utility when his type is $\theta_i$, he announces his type to be $\hat{\theta}_i$, and that all agents $j \neq i$ truthfully reveal their types. It is easy to verify from the previous two definitions that

$$U_i(\hat{\theta}_i | \theta_i) = \theta_i \overline{v}_i(\hat{\theta}_i) + \overline{t}_i(\hat{\theta}_i).$$

- Let $U_i(\theta_i) = U_i(\theta_i | \theta_i)$ be the agent $i$'s expected utility conditional on his type being $\theta_i$ when he and all other agents report their true types. It is easy to verify that

$$U_i(\theta_i) = \theta_i \overline{v}_i(\theta_i) + \overline{t}_i(\theta_i).$$

With the above discussion as a backdrop, we now present Myerson's [24] theorem for characterizing the BIC social choice functions in this environment.

**Theorem 2.14 (Myerson's Characterization Theorem).** *In linear environment, a social choice function $f(\cdot) = (k(\cdot), t_1(\cdot), \ldots, t_n(\cdot))$ is BIC if and only if, for all $i = 1, \ldots, n$,*

*1. $\overline{v}_i(\cdot)$ is nondecreasing,*
*2. $U_i(\theta_i) = U_i(\underline{\theta}_i) + \int_{\underline{\theta}_i}^{\theta_i} \overline{v}_i(s) ds \ \forall \ \theta_i$.*

For a proof of the above theorem, refer to Proposition 23.D.2 of [6]. The above theorem shows that to identify all BIC social choice functions in a linear environment, we can proceed as follows: First identify which functions $k(\cdot)$ lead every agent $i$'s expected benefit function $\overline{v}_i(\cdot)$ to be nondecreasing. Then, for each such function identify transfer functions $\overline{t}_1(\cdot), \ldots, \overline{t}_n(\cdot)$ that satisfy the second condition of the above proposition. Substituting for $U_i(\cdot)$ in the second condition above, we get that expected transfer functions are precisely those that satisfy, for $i = 1, \ldots, n$,

$$\overline{t}_i(\theta_i) = \overline{t}_i(\underline{\theta_i}) + \underline{\theta_i}\overline{v}_i(\underline{\theta_i}) - \theta_i\overline{v}_i(\theta_i) + \int_{\underline{\theta_i}}^{\theta_i} \overline{v}_i(s)ds$$

for some constant $\overline{t}_i(\underline{\theta_i})$. Finally, choose any set of transfer functions $t_1(\cdot), \ldots, t_n(\cdot)$ such that $E_{\theta_{-i}}[t_i(\theta_i, \theta_{-i})] = \overline{t}_i(\theta_i)$ for all $\theta_i$. In general, there are many such functions, $t_i(\cdot, \cdot)$; one, for example, is simply $t_i(\theta_i, \theta_{-i}) = \overline{t}_i(\theta_i)$.

In what follows we discuss two examples where the environment is linear and analyze the BIC property of the social choice function by means of Myerson's Characterization Theorem.

*Example 2.60 (First-Price Sealed Bid Auction in Linear Environment).* Consider the first-price sealed bid auction. Let us assume that $S_i = \Theta_i = [\underline{\theta_i}, \overline{\theta_i}] \ \forall i \in N$. In such a case, the first-price auction becomes a direct revelation mechanism $\mathscr{D} = ((\Theta_i)_{i \in N}, f(\cdot))$, where $f(\cdot)$ is an SCF that is the same as the outcome rule of the first-price auction. Let us impose the additional conditions on the environment to make it linear. We assume that

1. Bidders' types are statistically independent, that is, the density $\phi(\cdot)$ has the form $\phi_1(\cdot) \times \ldots \times \phi_n(\cdot)$
2. Let each bidder draw his type from the set $[\underline{\theta_i}, \overline{\theta_i}]$ by means of a uniform distribution, that is $\phi_i(\theta_i) = 1/(\overline{\theta_i} - \underline{\theta_i}) \ \forall \ \theta_i \in [\underline{\theta_i}, \overline{\theta_i}] \ \forall i = 1, \ldots, n$.

Note that the utility functions of the agents in this example are given by

$$u_i(f(\theta), \theta_i) = \theta_i y_i(\theta) + t_i(\theta) \ \forall i = 1, \ldots, n.$$

Thus, observing $y_i(\theta) = v_i(k(\theta))$ will confirm that these utility functions also satisfy the fourth condition required for a linear environment. Now we can apply Myerson's Characterization Theorem to test the Bayesian incentive compatibility of the SCF involved here. It is easy to see that for any bidder $i$, we have

$$\begin{aligned}
\overline{v}_i(\theta_i) &= E_{\theta_{-i}}[v_i(\theta_i, \theta_{-i})] \\
&= E_{\theta_{-i}}[y_i(\theta_i, \theta_{-i})] \\
&= 1.P\{(\theta_{-i})_{(n-1)} \leq \theta_i\} + 0.\left(1 - P\{\theta_i < (\theta_{-i})_{(n-1)}\}\right) \\
&= P\{(\theta_{-i})_{(n-1)} \leq \theta_i\}
\end{aligned} \qquad (2.24)$$

where $P\{(\theta_{-i})_{(n-1)} \leq \theta_i\}$ is the probability that the given type $\theta_i$ of the bidder $i$ is the highest among all the bidders' types. This implies that in the presence of the independence assumptions made above, $\overline{v}_i(\theta_i)$ is a nondecreasing function.

We know that for a first-price sealed bid auction, $t_i(\theta) = -\theta_i y_i(\theta)$. Therefore, we can claim that for a first-price sealed bid auction, we have

$$\overline{t}_i(\theta_i) = -\theta_i \overline{v}_i(\theta_i) \ \forall \ \theta_i \in \Theta_i.$$

The above values of $\overline{v}_i(\theta_i)$ and $\overline{t}_i(\theta_i)$ can be used to compute $U_i(\theta_i)$ in the following manner:

$$U_i(\theta_i) = \theta_i \overline{v}_i(\theta_i) + \overline{t}_i(\theta_i) = 0 \ \forall \ \theta_i \in [\underline{\theta}_i, \overline{\theta}_i]. \tag{2.25}$$

The above equation can be used to test the second condition of Myerson's Theorem, which requires

$$U_i(\theta_i) = U_i(\underline{\theta}_i) + \int_{\underline{\theta}_i}^{\theta_i} \overline{v}_i(s)ds.$$

In view of Equations (2.24) and (2.25), it is easy to see that this second condition of Myerson's Characterization Theorem is not being met by the SCF used in the first-price sealed bid auction. Therefore, we can finally claim that a first-price sealed bid auction is not BIC in linear environment.

*Example 2.61 (Second-Price Sealed Bid Auction in a Linear Environment).* Consider the second-price sealed bid auction. Let us assume that $S_i = \Theta_i = [\underline{\theta}_i, \overline{\theta}_i] \ \forall \ i \in N$. In such a case, the second-price auction becomes a direct revelation mechanism $\mathscr{D} = ((\Theta_i)_{i \in N}, f(\cdot))$, where $f(\cdot)$ is an SCF that is the same as the outcome rule of the second-price auction. We have already seen that this SCF $f(\cdot)$ is DSIC in quasilinear environment, and a linear environment is a special case of a quasilinear environment; therefore, it is DSIC in the linear environment also. Moreover, we know that DSIC implies BIC. Therefore, we can directly claim that the SCF used in the second-price auction is BIC in a linear environment.

## 2.19 Revenue Equivalence Theorem

In this section, we prove two results that show the revenue equivalence of certain classes of auctions. The first theorem is a general result that shows the revenue equivalence of two auctions that satisfy certain conditions. The second result is a more specific result that shows the revenue equivalence of four different types of auctions (English auction, Dutch auction, first price auction, and second price auction) in the special context of an auction of a single indivisible item. The proof of the second result crucially uses the first result.

## 2.19.1 Revenue Equivalence of Two Auctions

Assume that $y_i(\theta)$ is the probability of agent $i$ getting the object when the vector of announced types is $\theta = (\theta_1, \ldots, \theta_n)$. The expected payoff to the buyer $i$ with a type profile $\theta = (\theta_1, \cdots, \theta_n)$ will be $y_i(\theta)\theta_i + t_i(\theta)$. The set of allocations is given by

$$K = \left\{ (y_1, \cdots, y_n) : y_i \in [0,1] \forall i = 1, \cdots, n; \sum_{i=1}^{n} y_i \leq 1 \right\}.$$

As earlier, let $\overline{y}_i(\hat{\theta}_i) = E_{\theta_{-i}}[y_i(\hat{\theta}_i, \theta_{-i})]$ be the probability that agent $i$ gets the object conditional to announcing his type as $\hat{\theta}_i$, with the rest of the agents announcing their types truthfully. Similarly, $\overline{t}_i(\hat{\theta}_i) = E_{\theta_{-i}}[t_i(\hat{\theta}_i, \theta_{-i})]$ denotes the expected payment received by agent $i$ conditional to announcing his type as $\hat{\theta}_i$, with the rest of the agents announcing their types truthfully. Let $\overline{v}_i(\hat{\theta}_i) = \overline{y}_i(\hat{\theta}_i)$. Then,

$$U_i(\theta_i) = \overline{y}_i(\theta_i)\theta_i + \overline{t}_i(\theta_i)$$

denotes the payoff to agent $i$ when all the buying agents announce their types truthfully. We now state and prove an important proposition.

**Theorem 2.15.** *Consider an auction scenario with:*

1. *$n$ risk-neutral bidders (buyers) $1, 2, \cdots, n$*
2. *The valuation of bidder $i$ $(i = 1, \cdots, n)$ is a real interval $[\underline{\theta}_i, \overline{\theta}_i] \subset \mathbb{R}$ with $\underline{\theta}_i < \overline{\theta}_i$.*
3. *The valuation of bidder $i$ $(i = 1, \cdots, n)$ is drawn from $[\underline{\theta}_i, \overline{\theta}_i]$ with a strictly positive density $\phi_i(.) > 0$. Let $\Phi_i(.)$ be the cumulative distribution function.*
4. *The bidders' types are statistically independent.*

*Suppose that a given pair of Bayesian Nash equilibria of two different auction procedures are such that:*

- *For every bidder $i$, for each possible realization of $(\theta_1, \cdots, \theta_n)$, bidder $i$ has an identical probability of getting the good in the two auctions.*
- *Every bidder $i$ has the same expected payoff in the two auctions when his valuation for the object is at its lowest possible level.*

*Then the two auctions generate the same expected revenue to the seller.*

Before proving the theorem, we elaborate on the first assumption above, namely risk neutrality. A bidder is said to be:

- *risk-averse* if his utility is a *concave* function of his wealth; that is, an increment in the wealth at a lower level of wealth leads to an increment in utility that is higher than the increase in utility due to an identical increment in wealth at a higher level of wealth;

- *risk-loving* if his utility is a *convex* function of his wealth; that is, an increment in the wealth at a lower level of wealth leads to an increment in utility that is lower than the increase in utility due to an identical increment in wealth at a higher level of wealth; and
- *risk-neutral* if his utility is a *linear* function of his wealth; that is, an increment in the wealth at a lower level of wealth leads to the same increment in the utility as an identical increment would yield at a higher level of wealth.

**Proof:** By the revelation principle, it is enough that we investigate two Bayesian incentive compatible social choice functions in this auction setting. It is enough that we show that two Bayesian incentive compatible social choice functions having (a) the same allocation functions $(y_1(\theta), \cdots, y_n(\theta)) \ \forall \theta \in \Theta$, and (b) the same values of $U_1(\theta_1), \cdots, U_n(\theta_n)$ will generate the same expected revenue to the seller.

We first derive an expression for the seller's expected revenue given any Bayesian incentive compatible mechanism. Expected revenue to the seller

$$= \sum_{i=1}^{n} E_\theta[-t_i(\theta)]. \tag{2.26}$$

Now, we have:

$$
\begin{aligned}
E_\theta[-t_i(\theta)] &= E_{\theta_i}[-E_{\theta_{-i}}[t_i(\theta)]] \\
&= \int_{\underline{\theta_i}}^{\overline{\theta_i}} [\overline{y_i}(\theta_i)\theta_i - U_i(\theta_i)]\phi_i(\theta_i)d\theta_i \\
&= \int_{\underline{\theta_i}}^{\overline{\theta_i}} \left[ [\overline{y_i}(\theta_i)\theta_i - U_i(\underline{\theta_i})] - \int_{\underline{\theta_i}}^{\theta_i} \overline{y_i}(s)ds \right] \phi_i(\theta_i)d\theta_i.
\end{aligned}
$$

The last step is an implication of Myerson's characterization of Bayesian incentive compatible functions in linear environment. The above expression is now equal to

$$= \left[ \int_{\underline{\theta_i}}^{\overline{\theta_i}} \left( \overline{y_i}(\theta_i)\theta_i - \int_{\underline{\theta_i}}^{\theta_i} \overline{y_i}(s)ds \right) \phi_i(\theta_i)d\theta_i \right] - U_i(\underline{\theta_i}).$$

Now, applying integration by parts with $\int_{\underline{\theta_i}}^{\theta_i} \overline{y_i}(s)ds$ as the first function, we get

$$\int_{\underline{\theta_i}}^{\overline{\theta_i}} \left( \int_{\underline{\theta_i}}^{\theta_i} \overline{y_i}(s)ds \right) \phi_i(\theta_i)d\theta_i$$

$$= \int_{\underline{\theta_i}}^{\overline{\theta_i}} \overline{y_i}(\theta_i)d\theta_i - \int_{\underline{\theta_i}}^{\overline{\theta_i}} \overline{y_i}(\theta_i)\Phi_i(\theta_i)d\theta_i$$

$$= \int_{\underline{\theta_i}}^{\overline{\theta_i}} \overline{y_i}(\theta_i)[1 - \Phi_i(\theta_i)]d\theta_i.$$

Therefore we get

$$E_{\theta_i}[-\bar{t}_i(\theta_i)] = -U_i(\underline{\theta_i}) + \left[ \int_{\underline{\theta_i}}^{\overline{\theta_i}} \bar{y}_i(\theta_i) \left\{ \theta_i - \frac{1 - \Phi_i(\theta_i)}{\phi_i(\theta_i)} \right\} \phi_i(\theta_i) d\theta_i \right]$$

$$= -U_i(\underline{\theta_i}) + \left[ \int_{\underline{\theta_1}}^{\overline{\theta_1}} \cdots \int_{\underline{\theta_n}}^{\overline{\theta_i}} y_i(\theta_1, \cdots, \theta_n) \right.$$

$$\left. \times \left( \theta_i - \frac{1 - \Phi_i(\theta_i)}{\phi_i(\theta_i)} \right) \left( \prod_{j=1}^{n} \phi_j(\theta_j) \right) d\theta_n \cdots d\theta_1 \right]$$

since

$$\bar{y}_i(\theta_i) = \int_{\underline{\theta_1}}^{\overline{\theta_1}} \cdots \int_{\underline{\theta_n}}^{\overline{\theta_n}} y_i(\theta_1, \cdots, \theta_n) \underbrace{d\theta_n \cdots d\theta_1}_{\text{without } d\theta_i}.$$

Therefore the expected revenue of the seller

$$= \left[ \int_{\underline{\theta_1}}^{\overline{\theta_1}} \cdots \int_{\underline{\theta_n}}^{\overline{\theta_n}} \sum_{i=1}^{n} y_i(\theta_1, \cdots, \theta_n) \left( \theta_i - \frac{1 - \Phi_i(\theta_i)}{\phi_i(\theta_i)} \right) \right] \left( \prod_{j=1}^{n} \phi_j(\theta_j) \right) d\theta_n \cdots d\theta_1$$

$$- \sum_{i=1}^{n} U_i(\underline{\theta_i}).$$

By looking at the above expression, we see that any two Bayesian incentive compatible social choice functions that generate the same functions $(y_1(\theta), \cdots, y_n(\theta))$ and the same values of $(U_1(\underline{\theta_1}), \cdots, U_n(\underline{\theta_n}))$ generate the same expected revenue to the seller.

As an application of the above theorem, we now state and prove a revenue equivalence theorem for a single indivisible item auction. The article by McAfee and McMillan [25] is an excellent reference for this topic and a part of the discussion in this section is inspired by this article.

### 2.19.2 Revenue Equivalence of Four Classical Auctions

There are four basic types of auctions when a single indivisible item is to be sold:

1. **English auction**: This is also called oral auction, open auction, open cry auction, and ascending bid auction. Here, the price starts at a low level and is successively raised until only one bidder remains in the fray. This can be done in several ways: (a) an auctioneer announces prices, (b) bidders call the bids themselves, or (b) bids are submitted electronically. At any point of time, each bidder knows the level of the current best bid. The winning bidder pays the latest going price.

2. **Dutch auction**: This is also called a descending bid auction. Here, the auctioneer announces an initial (high) price and then keeps lowering the price iteratively until one of the bidders accepts the current price. The winner pays the current price.

3. **First price sealed bid auction**: Recall that in this auction, potential buyers submit sealed bids and the highest bidder is awarded the item. The winning bidder pays the price that he has bid.
4. **Second price sealed bid auction**: This is the classic Vickrey auction. Recall that potential buyers submit sealed bids and the highest bidder is awarded the item. The winning bidder pays a price equal to the second highest bid (which is also the highest losing bid).

When a single indivisible item is to be bought or procured, the above four types of auctions can be used in a *reverse* way. These are called *reverse auctions* or *procurement auctions*. In this section, we would be discussing the Revenue Equivalence Theorem as it applies to selling. The procurement version can be analyzed on similar lines.

### 2.19.2.1 The Benchmark Model

There are four assumptions underlying the derivation of the Revenue Equivalence Theorem: (1) risk neutrality of bidders; (2) bidders have independent private values; (3) bidders are symmetric; (4) payments depend on bids alone. These are described below in more detail.

### (1) Risk Neutrality of Bidders

It is assumed in the benchmark model that all the bidders are risk neutral. This immediately implies that the utility functions are linear.

### (2) Independent Private Values Model

In the independent private values model, each bidder knows precisely how much he values the item. He has no doubt about the true value of the item to him. However, each bidder does not know anyone else's valuation of the item. Instead, he perceives any other bidder's valuation as a draw from some known probability distribution. Also, each bidder knows that the other bidders and the seller regard his own valuation as being drawn from some probability distribution. More formally, let $N = \{1, 2, \cdots, n\}$ be the set of bidders. There is a probability distribution $\Phi_i$ from which bidder $i$ draws his valuation $v_i$. Only bidder $i$ observes his own valuation $v_i$, but all other bidders and the seller know the distribution $\Phi_i$. Any one bidder's valuation is statistically independent from any other bidder's valuation.

An apt example of this assumption is provided by the auction of an antique in which the bidders are consumers buying for their own use and not for resale. Another example is government contract bidding when each bidder knows his own production cost if he wins the contract.

A contrasting model is the *common value model*. Here, if $V$ is the unobserved true value of the item, then the bidders' perceived values $v_i, i = 1, 2, \cdots, n$ are independent draws from some probability distribution $H(v_i|V)$. All the bidders know the distribution $H$. An example is provided by the sale of an antique that is being bid for by dealers who intend to resell it. The item has one single objective value, namely its market price. However, no one knows the true value. The bidders, perhaps having access to different information, have different guesses about how much the item is objectively worth. Another example is that of the sale of mineral rights to a particular tract of land. The objective value here is the amount of mineral actually lying beneath the ground. However no one knows its true value.

Suppose a bidder were somehow to learn another bidder's valuation. If the situation is described by the common value model, then the above provides useful information about the likely true value of the item, and the bidder would probably change his own valuation in light of this. If the situation is described by the independent private values model, the bidder knows his own mind, and learning about others' valuations will not cause him to change his own valuation (although he may, for strategic reasons, change his bid).

Real world auction situations are likely to contain aspects of both the independent private values model and the common value model. It is assumed in the benchmark model that the independent private values assumption holds.

### (3) Symmetry

This assumption implies that all the bidders have the same set of possible valuations, and further they draw their valuations using the same probability density $\phi$. That is, $\phi_1 = \phi_2 = \ldots = \phi_n = \phi$.

### (4) Dependence of Payments on Bids Alone

It is assumed that the payment to be made by the winner to the auctioneer is a function of bids alone.

**Theorem 2.16 (Revenue Equivalence Theorem for Single Indivisible Item Auctions).** *Consider a seller or an auctioneer trying to sell a single indivisible item in which n bidders are interested. For the benchmark model (bidders are risk neutral, bidders have independent private values, bidders are symmetric, and payments depend only on bids), all the four basic auction types (English auction, Dutch auction, first price auction, and second price auction) yield the same average revenue to the seller.*

The result looks counter intuitive: For example, it might seem that receiving the highest bid in a first price sealed bid auction must be better for the seller than receiving the second highest bid, as in second price auction. However, it is to be noted that bidders act differently in different auction situations. In particular, they bid more aggressively in a second price auction than in a first price auction.

**Proof**: The proof proceeds in three parts. In Part 1, we show that the first price auction and the second price auction yield the same expected revenue in their respective equilibria. In Part 2, we show that the Dutch auction and the first price auction produce the same outcome. In Part 3, we show that the English auction and the second price auction yield the same outcome.

## Part 1: Revenue Equivalence of First Price Auction and Second Price Auction

The first price auction and the second price auction satisfy the conditions of the theorem on revenue equivalence of two auctions.

- In both the auctions, the bidder with the highest valuation wins the auction.
- bidders' valuations are drawn from $[\theta_i, \overline{\theta_i}]$ and a bidder with valuation at the lower limit of the interval has a payoff of zero in both the auctions.

Thus the theorem can be applied to the equilibria of the two auctions: Note that in the case of the first price auction, it is a Bayesian Nash equilibrium while in the case of the second price auction, it is a weakly dominant strategy equilibrium. In fact, it can be shown in any *symmetric* auction setting (where the bidders' valuations are independently drawn from identical distributions) that the conditions of the above proposition will be satisfied by any Bayesian Nash equilibrium of the first price auction and the weakly dominant strategy equilibrium of the second price scaled bid auction.

## Part 2: Revenue Equivalence of Dutch Auction and First Price Auction

To see this, consider the situation facing a bidder in these two auctions. In each case, the bidder must choose how high to bid without knowing the other bidders' decisions. If he wins, the price he pays equals his own bid. This result is true irrespective of which of the assumptions in the benchmark model apply. Note that the equilibrium in the underlying Bayesian game in the two cases here is a Bayesian Nash equilibrium.

## Part 3: Revenue Equivalence of English Auction and Second Price Auction

First we analyze the English auction. Note that a bidder drops out as soon the going price exceeds his valuation. The second last bidder drops out as soon as the price exceeds his own valuation. This leaves only one bidder in the fray and he wins the auction. Note that the winning bidder's valuation is the highest among all the bidders and he earns some payoff in spite of the monopoly power of the seller. Only the winning bidder knows how much payoff he receives because only he knows his own valuation. Suppose the valuations of the $n$ bidders are $v_{(1)}, v_{(2)}, \cdots, v_{(n)}$. Since the bidders are symmetric, these valuations are draws from the same distribution and without loss of generality, assume that these are in descending order. The winning bidder gets a payoff of $v_{(1)} - v_{(2)}$.

Next we analyze the second price auction. In the second price auction, the bidder's choice of bid determines only whether or not he wins; the amount he pays if he wins is beyond his control. We have already shown that each bidder's equilibrium best response strategy is to bid his own valuation for the item. The payment here is equal to the actual valuation of the bidder with the second highest valuation. Thus the expected payment and payoff are the same in English auction and the second price auction. This establishes Part 3 and therefore proves the Revenue Equivalence Theorem.

Note that the outcomes of the English auction and the second price auction satisfy a weakly dominant strategy equilibrium. That is, each bidder has a well defined best response bid regardless of how high he believes his rivals will bid. In the second price auction, the weakly dominant strategy is to bid true valuation. In the English auction, the weakly dominant strategy is to remain in the bidding process until the price reaches the bidder's own valuation.

### 2.19.2.2 Some Observations

We now make a few important observations.

- The theorem does not imply that the outcomes of the four auction forms are always exactly the same. They are only equal on average.

  - Note that in the English auction or the second price auction, the price exactly equals the valuation of the bidder with the second highest valuation, $v_{(2)}$. In Dutch auction or the first price auction, the price is the expectation of the second highest valuation conditional on the winning bidder's own valuation. The above two prices will be equal only by accident; however, they are equal on average.

- Bidding strategy is very simple in the English auction and the second price auction. In the former, a bidder remains in bidding until the price reaches his valuation. In the latter, he submits a sealed bid equal to his own valuation.
- On the other hand, the bidding logic is quite complex in the Dutch auction and the first price auction. Here the bidder bids some amount less than his true valuation. Exactly how much less depends upon the probability distribution of the other bidders' valuations and the number of competing bidders. Finding the Nash equilibrium bid is a non-trivial computational problem.
- The Revenue Equivalence Theorem for the single indivisible item is devoid of empirical predictions about which type of auction will be chosen by the seller in any particular set of circumstances. However when the assumptions of the benchmark model are relaxed, particular auction forms emerge as being superior.

- The variance of revenue is lower in English auction or second price auction than in Dutch auction or first price auction. Hence if the seller were risk averse, he would choose English or second price rather than Dutch or first price.

For more details on the revenue equivalence theorems, the reader is referred to the papers by Myerson [24], McAfee and McMillan [25], Klemperer [26], and the books by Milgrom [27] and Krishna [28].

## 2.20 Myerson Optimal Auction

A key problem that faces a social planner is to decide which direct revelation mechanism (or equivalently, social choice function) is *optimal* for a given problem. We now attempt to formalize the notion of optimality of social choice functions and optimal mechanisms. For this, we first define the concept of a *social utility function*.

**Definition 2.46 (Social Utility Function).** A social utility function is a function $w : \mathbb{R}^n \to \mathbb{R}$ that aggregates the profile $(u_1, \ldots, u_n) \in \mathbb{R}^n$ of individual utility values of the agents into a social utility.

Consider a mechanism design problem and a direct revelation mechanism $\mathcal{D} = ((\Theta_i)_{i \in N}, f(\cdot))$ proposed for it. Let $(\theta_1, \ldots, \theta_n)$ be the actual type profile of the agents and assume for a moment that they will all reveal their true types when requested by the planner. In such a case, the social utility that would be realized by the social planner for a type profile $\theta$ of the agents is given by:

$$w(u_1(f(\theta), \theta_1), \ldots, u_n(f(\theta), \theta_n)). \tag{2.27}$$

However, recall the implicit assumption behind a mechanism design problem, namely, that the agents are autonomous, and they would report a type as dictated by their rational behavior. Therefore, the assumption that all the agents will report their true types is not true in general. In general, rationality implies that the agents report their types according to a strategy suggested by a Bayesian Nash equilibrium $s^*(\cdot) = (s_1^*(\cdot), \ldots, s_n^*(\cdot))$ of the underlying Bayesian game. In such a case, the social utility that would be realized by the social planner for a type profile $\theta$ of the agents is given by

$$w(u_1(f(s^*(\theta)), \theta_1), \ldots, u_n(f(s^*(\theta)), \theta_n)). \tag{2.28}$$

In some instances, the above Bayesian Nash equilibrium may turn out to be a dominant strategy equilibrium. Better still, truth revelation by all agents could turn out to be a Bayesian Nash equilibrium or a dominant strategy equilibrium.

## 2.20.1 Optimal Mechanism Design Problem

In view of the above notion of a social utility function, it is clear that the objective of
a social planner would be to look for a social choice function $f(\cdot)$ that would maximize the expected social utility for a given social utility function $w(\cdot)$. However,
being the social planner, it is always expected of him to be fair to all the agents.
Therefore, the social planner would first put a few desirable constraints on the set
of social choice functions from which he can probably choose. The desirable constraints may include any combination of all the previously studied properties of a
social choice function, such as ex-post efficiency, incentive compatibility, and individual rationality. This set of social choice functions is known as a *set of feasible
social choice functions* and is denoted by $F$. Thus, the problem of a social planner
can now be cast as an optimization problem where the objective is to maximize the
expected social utility, and the constraint is that the social choice function must be
chosen from the feasible set $F$. This problem is known as the *optimal mechanism
design* problem and the solution of the problem would be social choice function
$f^*(\cdot) \in F$, which is used to define the optimal mechanism $\mathscr{D}^* = ((\Theta_i)_{i \in N}, f^*(\cdot))$ for
the problem that is being studied.

Depending on whether the agents are loyal or autonomous rational entities, the
optimal mechanism design problem may take two different forms.

$$\underset{f(\cdot) \in F}{\text{maximize}} \; E_\theta \left[ w(u_1(f(\theta), \theta_1), \dots, u_n(f(\theta), \theta_n)) \right] \tag{2.29}$$

$$\underset{f(\cdot) \in F}{\text{maximize}} \; E_\theta \left[ w(u_1(f(s^*(\theta)), \theta_1), \dots, u_n(f(s^*(\theta)), \theta_n)) \right] \tag{2.30}$$

The problem (2.29) is relevant when the agents are loyal and always reveal their true
types whereas the problem (2.30) is relevant when the agents are rational. At this
point of time, one may ask how to define the set of feasible social choice functions
$F$. There is no unique definition of this set. The set of feasible social choice functions
is a subjective judgment of the social planner. The choice of the set $F$ depends on
the desirable properties the social planner would wish to have in the optimal social
choice function $f^*(\cdot)$. If we define

$$F_{\text{DSIC}} = \{ f : \Theta \to X | f(\cdot) \text{ is dominant strategy incentive compatible} \}$$
$$F_{\text{BIC}} = \{ f : \Theta \to X | f(\cdot) \text{ is Bayesian incentive compatible} \}$$
$$F_{\text{EPIR}} = \{ f : \Theta \to X | f(\cdot) \text{ is ex-post individual rational} \}$$
$$F_{\text{IIR}} = \{ f : \Theta \to X | f(\cdot) \text{ is interim individual rational} \}$$

$$F_{\text{EAIR}} = \{f : \Theta \to X \,|\, f(\cdot) \text{ is ex-ante individual rational}\}$$
$$F_{\text{EAE}} = \{f : \Theta \to X \,|\, f(\cdot) \text{ is ex-ante efficient}\}$$
$$F_{\text{IE}} = \{f : \Theta \to X \,|\, f(\cdot) \text{ is interim efficient}\}$$
$$F_{\text{EPE}} = \{f : \Theta \to X \,|\, f(\cdot) \text{ is ex post efficient}\}.$$

The set of feasible social choice functions $F$ may be either any one of the above sets or intersection of any combination of the above sets. For example, the social planner may choose $F = F_{BIC} \cap F_{IIR}$. In the literature, this particular feasible set is known as *incentive feasible set* due to Myerson [1]. Also, note that if the agents are loyal then the sets $F_{DSIC}$ and $F_{BIC}$ will be equal to the whole set of all the social choice functions.

### 2.20.2 Myerson's Optimal Reverse Auction

We now consider the problem of procuring a single indivisible item from among a pool of suppliers and present Myerson's optimal auction that minimizes the expected cost of procurement subject to Bayesian incentive compatibility and interim individual rationality of all the selling agents. The classical Myerson auction [24] is for maximizing the expected revenue of a selling agent who wishes to sell an indivisible item to a set of prospective buying agents. We present it here for the reverse auction case.

Each bidder $i$'s type lies in an interval $\Theta_i = [\underline{\theta_i}, \overline{\theta_i}]$. We impose the following additional conditions on the environment.

1. The auctioneer and the bidders are risk neutral.
2. Bidders' types are statistically independent, that is, the joint density $\phi(\cdot)$ has the form $\phi_1(\cdot) \times \ldots \times \phi_n(\cdot)$.
3. $\phi_i(\cdot) > 0 \,\forall\, i = 1, \ldots, n$.
4. We generalize the outcome set $X$ by allowing a random assignment of the good. Thus, we now take $y_i(\theta)$ to be seller $i$'s probability of selling the good when the vector of announced types is $\theta = (\theta_1, \ldots, \theta_n)$. Thus, the new outcome set is given by

$$X = \Big\{ (y_0, y_1 \ldots, y_n, t_0, t_1, \ldots, t_n) : y_0 \in [0,1], t_0 \leq 0, y_i \in [0,1], t_i \geq 0 \,\forall\, i = 1, \ldots, n,$$

$$\sum_{i=1}^{n} y_i \leq 1; \ \sum_{i=0}^{n} t_i = 0 \Big\}.$$

Recall that the utility functions of the agents in this example are given by, $\forall\, i = 1, \ldots, n$,

$$u_i(f(\theta), \theta_i) = u_i(y_0(\theta), \ldots, y_n(\theta), t_0(\theta), \ldots, t_n(\theta), \theta_i) = -\theta_i y_i(\theta) + t_i(\theta).$$

Thus, viewing $y_i(\theta) = v_i(k(\theta))$ in conjunction with the second and third conditions above, we can claim that the underlying environment here is linear.

In the above example, we assume that the auctioneer (buyer) is the social planner and he is looking for an optimal direct revelation mechanism to buy the good. Myerson's [24] idea was that the auctioneer must use a social choice function that is Bayesian incentive compatible and interim individual rational and at the same time minimizes the cost to the auctioneer. Thus, in this problem, the set of feasible social choice functions is given by $F = F_{BIC} \cap F_{IIR}$. The objective function in this case would be to minimize the total expected cost of the buyer, which would be given by

$$E_\theta \left[ w(u_1(f(\theta), \theta_1), \ldots, u_n(f(\theta), \theta_n)) \right] = E_\theta \left[ \sum_{i=1}^n t_i(\theta) \right].$$

Note that in the above objective function we have used $f(\theta)$ and not $f(s^*(\theta))$. This is because in the set of feasible social choice functions we are considering only BIC social choice functions, and for these functions we have $s^*(\theta) = \theta \ \forall \ \theta \in \Theta$. Thus, the Myerson's optimal auction design problem can be formulated as the following optimization problem:

$$\underset{f(\cdot) \in F}{\text{minimize}} \ E_\theta \left[ \sum_{i=1}^n t_i(\theta) \right] \tag{2.31}$$

where

$$F = \{ f(\cdot) = (y_0(\cdot), y_1(\cdot), \ldots, y_n(\cdot), t_0(\cdot), t_1(\cdot), \ldots, t_n(\cdot)) : f(\cdot) \text{ is BIC and interim IR} \}.$$

We have seen Myerson's Characterization Theorem (Theorem 2.12) for BIC SCFs in linear environment. Similarly, we can say that an SCF $f(\cdot)$ in the above context would be BIC iff it satisfies the following two conditions:

1. $\overline{y}_i(\cdot)$ is nonincreasing for all $i = 1, \ldots, n$.
2. $U_i(\theta_i) = U_i(\overline{\theta}_i) + \int_{\theta_i}^{\overline{\theta}_i} \overline{y}_i(s) ds \ \forall \ \theta_i \in \Theta_i; \ \forall \ i = 1, \ldots, n.$

Also, we can invoke the definition of interim individual rationality to claim that the an SCF $f(\cdot)$ in the above context would be interim IR iff it satisfies the following conditions:

$$U_i(\theta_i) \geq 0 \ \forall \theta_i \in \Theta_i; \ \forall \ i = 1, \ldots, n$$

where

- $\overline{t}_i(\hat{\theta}_i) = E_{\theta_{-i}}[t_i(\hat{\theta}_i, \theta_{-i})]$ is bidder $i$'s expected transfer given that he announces his type to be $\hat{\theta}_i$ and that all the bidders $j \neq i$ truthfully reveal their types.

- $\overline{y}_i(\hat{\theta}_i) = E_{\theta_{-i}}[y_i(\hat{\theta}_i, \theta_{-i})]$ is the probability that object will be procured from bidder $i$ given that he announces his type to be $\hat{\theta}_i$ and all bidders $j \neq i$ truthfully reveal their types.
- $U_i(\theta_i) = -\theta_i \overline{y}_i(\theta_i) + \overline{t}_i(\theta_i)$ ( we can take unconditional expectation because types are independent).

Based on the above, problem (2.31) can be rewritten as follows:

$$\underset{(y_i(\cdot), U_i(\cdot))_{i \in N}}{\text{minimize}} \sum_{i=1}^{n} \int_{\underline{\theta}_i}^{\overline{\theta}_i} (\theta_i \overline{y}_i(\theta_i) + U_i(\theta_i)) \, \phi_i(\theta_i) d\theta_i \qquad (2.32)$$

subject to

(i)   $\overline{y}_i(\cdot)$ is nonincreasing $\forall i = 1, \ldots, n$

(ii)  $y_i(\theta) \in [0,1], \sum_{i=1}^{n} y_i(\theta) \leq 1 \ \forall i = 1, \ldots, n, \forall \theta \in \Theta$

(iii) $U_i(\theta_i) = U_i(\overline{\theta}_i) + \int_{\theta_i}^{\overline{\theta}_i} \overline{y}_i(s) ds \ \forall \theta_i \in \Theta_i; \ \forall i = 1, \ldots, n$

(iv)  $U_i(\theta_i) \geq 0 \ \forall \theta_i \in \Theta_i; \ \forall i = 1, \ldots, n.$

We first note that if constraint (iii) is satisfied then constraint (iv) will be satisfied iff $U_i(\overline{\theta}_i) \geq 0 \ \forall i = 1, \ldots, n$. As a result, we can replace the constraint (iv) with

$$(iv') \quad U_i(\overline{\theta}_i) \geq 0 \ \forall i = 1, \ldots, n$$

Next, substituting for $U_i(\theta_i)$ in the objective function from constraint (iii), we get

$$\sum_{i=1}^{n} \int_{\underline{\theta}_i}^{\overline{\theta}_i} \left( \theta_i \overline{y}_i(\theta_i) + U_i(\overline{\theta}_i) + \int_{\theta_i}^{\overline{\theta}_i} \overline{y}_i(s) ds \right) \phi_i(\theta_i) d\theta_i.$$

Integrating by parts the above expression, the auctioneer's problem can be written as one of choosing the $y_i(\cdot)$ functions and the values $U_1(\overline{\theta}_1), \ldots, U_n(\overline{\theta}_n)$ to minimize

$$\int_{\underline{\theta}_1}^{\overline{\theta}_1} \cdots \int_{\underline{\theta}_n}^{\overline{\theta}_n} \left[ \sum_{i=1}^{n} y_i(\theta_i) J_i(\theta_i) \right] \left[ \prod_{i=1}^{n} \phi_i(\theta_i) \right] d\theta_n \ldots d\theta_1 + \sum_{i=1}^{n} U_i(\overline{\theta}_i)$$

subject to constraints (i), (ii), and (iv'), where

$$J_i(\theta_i) = \left( \theta_i + \frac{\Phi_i(\theta_i)}{\phi_i(\theta_i)} \right).$$

It is evident that the solution must have $U_i(\overline{\theta}_i) = 0$ for all $i = 1, \ldots, n$. Hence, the auctioneer's problem reduces to choosing functions $y_i(\cdot)$ to minimize

$$\int\limits_{\underline{\theta_1}}^{\overline{\theta_1}} \cdots \int\limits_{\underline{\theta_n}}^{\overline{\theta_n}} \left[ \sum_{i=1}^{n} y_i(\theta_i) J_i(\theta_i) \right] \left[ \prod_{i=1}^{n} \phi_i(\theta_i) \right] d\theta_n \dots d\theta_1$$

subject to constraints (i) and (ii).

Let us ignore constraint (i) for the moment. Then inspection of the above expression indicates that $y_i(\cdot)$ is a solution to this relaxed problem iff for all $i = 1, \dots, n$, we have

$$y_i(\theta) = \begin{cases} 0 & : \quad \text{if } J_i(\theta_i) > \min\left\{\overline{\theta_0}, \min_{h \neq i} J_h(\theta_h)\right\} \\ 1 & : \quad \text{if } J_i(\theta_i) < \min\left\{\overline{\theta_0}, \min_{h \neq i} J_h(\theta_h)\right\}. \end{cases} \tag{2.33}$$

Note that $J_i(\theta_i) = \min\left\{\overline{\theta_0}, \min_{h \neq i} J_h(\theta_h)\right\}$ is a zero probability event.

In other words, if we ignore the constraint (i) then $y_i(\cdot)$ is a solution to this relaxed problem iff the good is allocated to a bidder who has the lowest nonnegative value for $J_i(\theta_i)$. Now, recall the definition of $\overline{y_i}(\cdot)$. It is easy to write down the following expression:

$$\overline{y_i}(\theta_i) = E_{\theta_{-i}}[y_i(\theta_i, \theta_{-i})]. \tag{2.34}$$

Now, if we assume that $J_i(\cdot)$ is nondecreasing in $\theta_i$ then it is easy to see that the above solution $y_i(\cdot)$, given by (2.33), will be nonincreasing in $\theta_i$, which in turn implies, by looking at expression (2.34), that $\overline{y_i}(\cdot)$ is nonincreasing in $\theta_i$. Thus, the solution to this relaxed problem actually satisfies constraint (i) under the assumption that $J_i(\cdot)$ is nondecreasing. Assuming that $J_i(\cdot)$ is nondecreasing, the solution given by (2.33) seems to be the solution of the optimal mechanism design problem for single unit-single item procurement auction. The condition that $J_i(\cdot)$ is nondecreasing in $\theta_i$ is met by most of the distribution functions such as Uniform and Exponential.

So far we have computed the allocation rule for the optimal mechanism and now we turn our attention toward the payment rule. The optimal payment rule $t_i(\cdot)$ must be chosen in such a way that it satisfies

$$\overline{t_i}(\theta_i) = E_{\theta_{-i}}[t_i(\theta_i, \theta_{-i})] = U_i(\theta_i) + \theta_i \overline{y_i}(\theta_i) = \int\limits_{\underline{\theta_i}}^{\overline{\theta_i}} \overline{y_i}(s) ds + \theta_i \overline{y_i}(\theta_i). \tag{2.35}$$

Looking at the above formula, we can say that if the payment rule $t_i(\cdot)$ satisfies the following formula (2.36), then it would also satisfy the formula (2.35).

$$t_i(\theta_i, \theta_{-i}) = \int\limits_{\underline{\theta_i}}^{\overline{\theta_i}} y_i(s, \theta_{-i}) ds + \theta_i y_i(\theta_i, \theta_{-i}) \quad \forall \theta \in \Theta. \tag{2.36}$$

The above formula can be rewritten more intuitively as follows. For any vector $\theta_{-i}$, let us define

$$z_i(\theta_{-i}) = \sup\left\{\theta_i : J_i(\theta_i) < \overline{\theta_0} \text{ and } J_i(\theta_i) \leq J_j(\theta_j) \,\forall\, j \neq i\right\}.$$

Then $z_i(\theta_{-i})$ is the supremum of all winning bids for bidder $i$ against $\theta_{-i}$, so

$$y_i(\theta_i, \theta_{-i}) = \begin{cases} 1 & : \quad \text{if } \theta_i < z_i(\theta_{-i}) \\ 0 & : \quad \text{if } \theta_i > z_i(\theta_{-i}). \end{cases}$$

This gives us

$$\int_{\theta_i}^{\overline{\theta_i}} y_i(s, \theta_{-i})ds = \begin{cases} z_i(\theta_{-i}) - \theta_i & : \quad \text{if } \theta_i \leq z_i(\theta_{-i}) \\ 0 & : \quad \text{if } \theta_i > z_i(\theta_{-i}). \end{cases}$$

Finally, the formula (2.36) becomes

$$t_i(\theta_i, \theta_{-i}) = \begin{cases} z_i(\theta_{-i}) & : \quad \text{if } \theta_i \leq z_i(\theta_{-i}) \\ 0 & : \quad \text{if } \theta_i > z_i(\theta_{-i}). \end{cases}$$

That is, bidder $i$ will receive payment only when the good is procured from him, and then he receives an amount equal to his highest possible winning bid.

We make a few interesting observations:

1. When the various bidders have differing distribution function $\Phi_i(\cdot)$ then, the bidder who has the smallest value of $J_i(\theta_i)$ is *not* necessarily the bidder who has bid the lowest amount for the good. Thus Myerson's optimal auction need not be allocatively efficient, and therefore, need not be ex-post efficient.
2. If the bidders are symmetric, that is,

- $\Theta_1 = \ldots = \Theta_n = \Theta$
- $\Phi_1(\cdot) = \ldots = \Phi_n(\cdot) = \Phi(\cdot)$,

then the allocation rule would be precisely the same as that of first-price reverse auction and second-price reverse auction. In such a case the object would be allocated to the lowest bidder. In such a situation, the optimal auction would also become allocatively efficient, and the payment rule described above would coincide with the payment rules in second-price reverse auction. In other words, the second price reverse auction would be an optimal auction when the bidders are symmetric. Therefore, many times, the optimal auction is also known as *modified Vickrey auction*.

Riley and Samuelson [30] also have studied the problem of design of an optimal auction for selling a single unit of a single item. They assume the bidders to be symmetric. Their work is less general than that of Myerson [24].

## 2.21 Further Topics in Mechanism Design

Mechanism design is a rich area with a vast body of knowledge. So far in this chapter, we have seen essential aspects of game theory, followed by key results in mechanism design. We now provide a brief description of a few topics in mechanism design. The topics have been chosen, with an eye on their possible application to network economics problems of the kind discussed in the monograph. We have not followed any particular logical order while discussing the topics. We also caution the reader that the treatment is only expository. Pointers to the relevant literature are provided wherever appropriate.

### 2.21.1 Characterization of DSIC Mechanisms

We have seen that a direct revelation mechanism is specified as $\mathscr{D} = ((\Theta_i)_{i \in N}, f(.))$, where $f$ is the underlying social choice function and $\Theta_i$ is the type set of agent $i$. A valuation function of each agent $i$, $v_i(.)$, associates a value of the allocation chosen by $f$ to agent $i$, that is, $v_i : K \to \mathbb{R}$, where $K$ is the set of project choices.

In the case of an auction for selling a single indivisible item, suppose each agent $i$ has a valuation for the object $\theta_i \in [\underline{\theta_i}, \overline{\theta_i}]$. If agent $i$ gets the object, $v_i(., \theta_i) = \theta_i$. Otherwise the valuation is zero. Thus for the agent $i$, the set of valuation functions over the set of allocations $K$ can be written as $\Theta_i = [\underline{\theta_i}, \overline{\theta_i}]$. Thus $\Theta_i$ is single dimensional in this environment.

In a general setting, $\Theta_i$ may not be single dimensional. If we consider all real valued functions on $X$ and allow each agent to have a valuation function to be any of these functions, we say $\Theta_i$ is unconstrained. Suppose $|K| = m$, then $\theta_i \in \Theta_i$ is an $m$-dimensional vector:

$$\theta_i = (\theta_{i_1}, \ldots, \theta_{i_j}, \ldots, \theta_{i_m}).$$

Note that $\theta_{i_j}$ will be the valuation of agent $i$ if the $j^{th}$ allocation from $K$ is selected. In other words, $v_i(j) = \theta_{i_j}$. With such unconstrained type sets/valuation functions, an elegant characterization of DSIC social choice functions has been obtained by Roberts [31]. The work of Roberts generalizes the the Green–Laffont Theorem (Theorem 2.8) in the following way. Recall that the Green–Laffont Theorem asserts that an allocatively efficient and DSIC social choice function in the above unconstrained setting has to be necessarily a VCG mechanism. The result of Roberts asserts that all DSIC mechanisms are variations of the VCG mechanism. These variants are often referred to as the *weighted VCG mechanisms*. In a weighted VCG mechanism, weights are given to the agents and to the outcomes. The resulting social choice function is said to be an *affine maximizer*. The notion of an affine maximizer is defined below. Next we state the Roberts' Theorem.

**Definition 2.47.** A social choice function $f$ is called an affine maximizer if for some subrange $A' \subset X$, for some agent weights $w_1, w_2, \ldots, w_n \in \mathbb{R}^+$, and for some outcome weights $c_x \in \mathbb{R}$, and for every $x \in A'$, we have that

$$f(\theta_1, \theta_2, \ldots, \theta_n) \in \arg\max_{x \in A'}(c_x + \sum_i w_i v_i(x)).$$

**Theorem 2.17 (Roberts' Theorem).** *If $|X| \geq 3$ and for each agent $i \in N$, $\Theta_i$ is unconstrained, then any DSIC social choice function $f$ has nonnegative weights $w_1, w_2, \ldots, w_n$ (not all of them zero) and constants $\{c_x\}_{x \in X}$, such that for all $\theta \in \Theta$,*

$$f(\theta) \in \arg\max_{x \in X}\left\{\sum_{i=1}^{n} w_i v_i(x) + c_x\right\}.$$

For a proof of this important theorem, we refer the reader to the article by Roberts [31]. Lavi, Mu'alem, and Nisan have provided two more proofs for the theorem — interested readers might refer to their paper [32] as well.

## 2.21.2 Dominant Strategy Implementation of BIC Rules

Clearly, dominant strategy incentive compatibility is stronger and much more desirable than Bayesian incentive compatibility. A striking reason for this is any Bayesian implementation assumes that the private information structure is common knowledge. It also assumes that the social planner knows a common prior distribution. In many cases, this requirement might be quite demanding. Also, a slight misspecification of the common prior may lead the equilibrium to shift quite dramatically. This may result in unpredictable effects; for example it might cause an auction to to be highly nonoptimal.

A dominant strategy implementation overcomes these problems in a simple way since the equilibrium strategy does not depend upon the common prior distribution. We would therefore always wish to have a DSIC implementation. Since the class of BIC social choice functions is much richer than DSIC social choice functions, one would like to ask the question: Can we implement a BIC SCF as a DSIC rule with the same expected interim utilities to all the players? Mookherjee and Stefan [33] have answered this question by characterizing BIC rules that can be equivalently implemented in dominant strategies. When these sufficient conditions are satisfied, a BIC social choice function could be implemented without having to worry about a common prior. The article by Mookherjee and Stefan [33] may be consulted for further details.

## 2.21.3 Implementation in Ex-Post Nash Equilibrium

Dominant strategy implementation and Bayesian implementation are widely used for implementing a social choice function. There exists another notion of implementation, called ex-post Nash implementation, which is stronger than Bayesian implementation but weaker than dominant strategy implementation. This was formalized by Maskin [34]. Dasgupta, Hammond, and Maskin [35] generalized this to the Bayesian Nash implementation.

**Definition 2.48.** A profile of strategies $\left(s_1^*(.), s_2^*(.), \ldots, s_n^*(.)\right)$ is an ex-post Nash equilibrium if for every $\theta = (\theta_1, \ldots \theta_n) \in \Theta$, the profile $\left(s_1^*(\theta_1), \ldots, s_n^*(\theta_n)\right)$ is a Nash equilibrium of the complete information game defined by $(\theta_1, \ldots, \theta_n)$. That is, for all $i \in N$ and for all $\theta \in \Theta$, we have

$$u_i(s_i^*(\theta_i), s_{-i}^*(\theta_{-i}), \theta_i) \geq u_i(s_i'(\theta_i), s_{-i}^*(\theta_{-i}), \theta_i) \ \forall \ s_i' \in S_i.$$

In a Bayesian Nash equilibrium, the equilibrium strategy is played by the agents after observing their own private types and computing an expectation over others' types; it is an equilibrium only in the expected sense. On the other hand, in ex-post Nash equilibrium, even after the players are informed of the types of the other players, it is still a Nash equilibrium for each agent $i$ to play an action according to $s_i^*(\cdot)$. This is called the *lack of regret* feature. That is, even if agents come to know about the others' types, the agent need not regret playing this action. Bayesian Nash equilibrium may not have this feature since the agents may want to revise their strategies after knowing the types of the other agents.

For example, consider the first price sealed bid auction with two bidders. Let $\Theta_1 = \Theta_2 = [0,1]$ and $\theta_1$ denote the valuation of the first agent and $\theta_2$ that of the agent 2. It can be shown that it is a Bayesian Nash equilibrium for each bidder to bid according to the strategy $(b_1^*(\theta_1), b_2^*(\theta_2)) = (\frac{\theta_1}{2}, \frac{\theta_2}{2})$. Now suppose agent 1 is informed that the other agent values the object at 0.6. If agent 1 has a valuation of 0.8, say, it is not a Nash equilibrium for him to bid 0.4 even if agent 2 is still following a Bayesian Nash strategy.

**Definition 2.49.** We say that the mechanism $\mathscr{M} = ((S_i)_{i \in N}, g(\cdot))$ implements the social choice function $f(\cdot)$ in ex-post Nash equilibrium if there is a pure strategy ex-post Nash equilibrium $s^*(\cdot) = (s_1^*(\cdot), \ldots, s_n^*(\cdot))$ of the game $\Gamma^b$ induced by $\mathscr{M}$ such that

$$g\left(s_1^*(\theta_1), \ldots, s_n^*(\theta_n)\right) = f(\theta_1, \ldots, \theta_n) \ \forall (\theta_1, \ldots, \theta_n) \in \Theta.$$

Though ex-post implementation is stronger than Bayesian Nash implementation, it is still much weaker than dominant strategy implementation.

### 2.21.4 Interdependent Values

We have so far assumed that the private values or signals observed by the agents are independent of one another. This is a reasonable assumption in many situations. However, in the real world, there are environments where the valuation of agents might depend upon the information available or observed by the other agents. We will look at two examples.

*Example 2.62.* Consider an auction for an antique painting. There is no guarantee that the painting is an original one or a plagiarized version. If all the agents knew that the painting is not an original one, they would have a very low value for it independent of one another, whereas on the other hand, they would have a high value for it when it is a genuine piece of work. But suppose they have no knowledge about its authenticity. In such a case, if a certain bidder happens to get information about its genuineness, the valuations of all the other agents will naturally depend upon this signal (indicating the authenticity of the painting) observed by this agent.

*Example 2.63.* Consider an auction for oil drilling rights. At the time of the auction, buyers usually conduct geological tests and their private valuations depend upon the results of these tests. If a prospective bidder knew the results of the tests of the others, his own willingness to pay for the drilling rights would be modulated suitably based on the information available.

The interdependent private value models have been studied in the mechanism design literature. For example, there exists a popular model called the *common value model* (which we have already seen in Section 2.19.2). As another example, consider a situation when a seller is trying to sell an indivisible good or a fixed quantity of a divisible good. The value of the received good for the bidders depends upon each others' private signals. Also, the private signals observed by the agents are interdependent of specified properties. In such a scenario, Cremer and McLean [36] have designed an auction that extracts a revenue from the bidders, which is equal to what could have been extracted when the actual signals of the bidders are known. In this auction, it is an ex-post Nash equilibrium for the agents to report their true types. This auction is interim individually rational but may not be ex-post individually rational.

## 2.21.5 Implementation Theory

Dominant strategy incentive compatibility ensures that reporting true types is a weakly dominant strategy equilibrium. Bayesian incentive compatibility ensures that reporting true types is a Bayesian Nash equilibrium. Typically, the Bayesian game underlying a given mechanism may have multiple equilibria, in fact, could have infinitely many equilibria. These equilibria typically will produce different outcomes. Thus it is possible that nonoptimal outcomes are produced by truth revelation.

The *implementation problem* addresses the above difficulty caused by multiple equilibria. The implementation problem seeks to design mechanisms in which all the equilibrium outcomes are optimal. This property is called the *weak implementation property*. If it also happens that every optimum outcome is also an equilibrium, we call the property as *full implementation property*. Maskin [34] provided a general characterization of Nash implementable social choice functions using a monotonicity property, which is now called *Maskin Monotonicity*. This property has a striking similarity to the property of independence of irrelevant alternatives, which we encountered during our discussion on Arrow's impossibility theorem (Section 2.12). Maskin's work shows that Maskin monotonicity, in conjunction with another property called *no-veto-power* will guarantee that all Nash equilibria will produce an optimal outcome. His work has led to development of implementation theory. Dasgupta, Hammond, and Maskin [35] have summarized many important results in implementation theory, and they discuss incentive compatibility in detail. Maskin's results have now been generalized in many directions; for example, see the references in [37].

### 2.21.6 Computational Issues in Mechanism Design

We have seen several possibility and impossibility results in the context of mechanism design. While every possibility result is good news, there could be still be challenges involved in actually implementing a mechanism that is possible. For example, we have seen that the GVA mechanism (Example 2.55) is an allocatively efficient and dominant strategy incentive compatible mechanism for combinatorial auctions. A major difficulty with GVA is the computational complexity involved in determining the allocation and the payments. Both the allocation and payment determination problems are NP-hard, being instances of the weighted set packing problem (in the case of forward GVA) or the weighted set covering problem (in the case of reverse GVA). In fact, if there are $n$ agents, then in the worst case, the payment determination will involve solving as many as $n$ NP-hard problems, so overall, as many as $(n + 1)$ NP-hard problems will need to be solved for implementing the GVA mechanism. Moreover, approximately solving any one of these problems may compromise properties such as efficiency and/or incentive compatibility of the mechanism.

In mechanism design, computations are involved at two levels: first, at the agent level and secondly at the mechanism level [38, 39]. Complexity at the agent level involves strategic complexity (complexity of computing an optimal strategy) and valuation complexity (computation required to provide preference information within a mechanism). Complexity at the mechanism level includes communication complexity (how much communication is required between agents and the mechanism to compute an outcome) and winner determination complexity (computation required to determine an outcome given the strategies of the agents). Typically, insufficient computation leading to approximate solutions hinders mechanism design since properties such as incentive compatibility, allocative efficiency, individual rationality, etc., may be compromised. Novel algorithms and high computing power surely lead to better mechanisms.

For a detailed description of computational complexity issues in mechanism design, the reader is referred to the excellent survey articles [40, 41, 38, 39].

## 2.22 To Probe Further

For a microeconomics oriented treatment of mechanism design, the readers are requested to refer to textbooks, such as the ones by Mas-Colell, Whinston, and Green [6] (Chapter 23); Green and Laffont [18]; and Laffont [42]. There is an excellent recent survey article by Nisan [43], which targets a computer science audience. There are many other informative survey papers on mechanism design — for example by Myerson [44], Serrano [45], and Jackson [46, 47]. The Nobel Prize website has a scholarly technical summary of mechanism design theory [37]. The recent edited volume on *Algorithmic Game Theory* by Nisan, Roughgarden, Tardos, and Vazirani [48] also has valuable articles related to mechanism design.

This chapter is not to be treated as a survey on auctions in general. There are widely popular books (for example, by Milgrom [27], Krishna [28], and Klemperer [53]) and surveys on auctions (for example, [25, 49, 50, 51, 38]) that deal with auctions in a comprehensive way.

A related area where an extensive amount of work has been carried out in the past decade is combinatorial auctions. Exclusive surveys on combinatorial auctions include the articles by de Vries and Vohra [40, 41], Pekec and Rothkopf [52], and Narahari and Pankaj Dayama [54]. Cramton, Ausubel, and Steinberg [21] have brought out a comprehensive edited volume containing expository and survey articles on varied aspects of combinatorial auctions.

## References

1. R.B. Myerson. *Game Theory: Analysis of Conflict.* Harvard University Press, Cambridge, Massachusetts, 1997.
2. J. von Neumann and O. Morgenstern. *Theory of Games and Economic Behavior.* Princeton University Press, 1944.
3. R. Aumann. Agreeing to disagree. *The Annals of Statistics,* 4(6):1236–1239, 1976.
4. E. Tardos and V.V. Vazirani. Basic solution concepts and computational issues. In N. Nisan, T. Roughgarden, E. Tardos, and V.V. Vazirani, (eds.), *Algorithmic Game Theory,* pages 3–28. Cambridge University Press, New York, 2007.
5. J. Bertrand. Book review of theorie mathematique de la richesse sociale and of recherches sur les principles mathematiques de la theorie des richesses. *Journal de Savants,* 67:499–508, 1883.
6. A. Mas-Colell, M.D. Whinston, and J.R. Green. *Microeconomic Theory.* Oxford University Press, New York, 1995.
7. R.D. McKelvey and A. McLennan. Computation of equilibria in finite games. In J. Rust H.M. Amman, D.A. Keudrick, (eds.), *Handbook of Computational Economics, Handbooks in Economics (13), Volume 1,* pages 87–142. North-Holland, Amsterdam, 1996.

8. C.H. Papadimitriou. The complexity of finding nash equilibria. In N. Nisan, T. Roughgarden, E. Tardos, and V.V. Vazirani, (eds.), *Algorithmic Game Theory*, pages 29–52. Cambridge University Press, New York, 2007.

9. B.V. Stengel. Equilibrium computation for two-player games in strategic and extensive form. In N. Nisan, T. Roughgarden, E. Tardos, and V.V. Vazirani, (eds.), *Algorithmic Game Theory*, pages 53–78. Cambridge University Press, New York, 2007.

10. L. Hurwicz. On informationally decentralized systems. In C.B. McGuire and R. Radner, (eds.), *Decision and Organization*, North-Holland, Amsterdam, 1972.

11. L. Hurwicz. Optimality and informational efficiency in resource allocation processes. In K.J. Arrow, S. Karlin and P. Suppes (eds.), *Mathematical Methods in the Social Sciences*. Stanford University Press, Palo Alto, California, USA, 1960.

12. B. Holmstrom and R.B. Myerson. Efficient and durable decision rules with incomplete information. *Econometrica*, 51(6):1799–1819, 1983.

13. A. Gibbard. Manipulation of voting schemes. *Econometrica*, 41:587–601, 1973.

14. M.A. Satterthwaite. Strategy-proofness and Arrow's conditions: Existence and correspondence theorem for voting procedure and social welfare functions. *Journal of Economic Theory*, 10:187–217, 1975.

15. W. Vickrey. Counterspeculation, auctions, and competitive sealed tenders. *Journal of Finance*, 16(1):8–37, 1961.

16. E. Clarke. Multi-part pricing of public goods. *Public Choice*, 11:17–23, 1971.

17. T. Groves. Incentives in teams. *Econometrica*, 41:617–631, 1973.

18. J.R. Green and J.J. Laffont. *Incentives in Public Decision Making*. North-Holland, Amsterdam, 1979.

19. L.M. Ausubel and P. Milgrom. The lovely but lonely vickrey auction. In P. Cramton, Y. Shoham, and R. Steinberg, (eds.), *Combinatorial Auctions*, pages 17–40. The MIT Press, Cambridge, Massachusetts, 2006.

20. M. Rothkopf. Thirteen reasons why the Vickrey-Clarke-Groves process is not practical. *Operations Research*, 55(2):191–197, 2007.

21. C. Caplice and Y. Sheffi. Combinatorial auctions for truckload transportation. In P. Cramton, Y. Shoham, and R. Steinberg, (eds.), *Combinatorial Auctions*, pages 539–572. The MIT Press, Cambridge, Massachusetts, 2005.

22. C. d'Aspremont and L.A. Gérard-Varet. Incentives and incomplete information. *Journal of Public Economics*, 11:25–45, 1979.

23. K. Arrow. The property rights doctrine and demand revelation under incomplete information. In M. Boskin, (eds.), *Economics and Human Welfare*. Academic Press, New York, 1979.

24. R.B. Myerson. Optimal auction design. *Mathematics of Operations Research*, 6(1):58–73, 1981.

25. P.R. McAfee and J. McMillan. Auctions and bidding. *Journal of Economic Literature*, 25(2):699–738, 1987.

26. P. Klemperer. Why every economist should learn some auction theory. In M. Dewatripont, L. Hansen, and S. Turnovsky, (eds.), *Advances in Economics and Econometrics: Invited Lectures to 8th World Congress of the Econometric Society*. Cambridge University Press, Cambridge, UK, 2003.

27. P. Milgrom. *Putting Auction Theory to Work*. Cambridge University Press, Cambridge, UK, 2004.

28. V. Krishna. *Auction Theory*. Academic Press, San Diego, California, USA, 2002.

29. D. Garg, Y. Narahari, and S. Gujar. Foundations of mechanism design: A tutorial – Part 1: Key Concepts and Classical Results. *Sadhana — Indian Academy Proceedings in Engineering Sciences*, 33(2):83–130, 2008.

30. J.G. Riley and W.F. Samuelson. Optimal auctions. *American Economic Review*, 71(3):383–92, 1981.

31. K. Roberts. The characterization of implementable choice rules. In J.J. Laffont, (eds.), *Aggregation and Revelation of Preferences*, pages 321–349, Amsterdam, 1979.

32. R. Lavi, A. Mu'alem, and N. Nisan. Two simplified proofs for Roberts' theorem. Technical report, Working Paper, School of Computer Science and Engineering, The Hebrew University of Jerusalem, Israel, 2004.

33. D. Mookherjee and S. Reichelstein. Dominant strategy implementation of Bayesian incentive compatible allocation rules. *Journal of Economic Theory*, 56(2):378–399, 1992.

34. E. Maskin. Nash equilibrium and welfare optimality. *Review of Economic Studies*, 66:23–38, 1999.

35. P. Dasgupta, P. Hammond, and E. Maskin. The implementation of social choice rules: some general results on incentive compatibility. *Review of Economic Studies*, 46:181–216, 1979.

36. J. Cremer and R.P McLean. Optimal selling strategies under uncertainty for a discriminating monopolist when demands are interdependent. *Econometrica*, 53(2):345–61, 1985.

37. The Nobel Foundation. The Sveriges Riksbank Prize in Economic Sciences in Memory of Alfred Nobel 2007: Scientific Background. Technical report, The Nobel Foundation, Stockholm, Sweden, December 2007.

38. J.K. Kalagnanam and D.C. Parkes. Auctions, bidding, and exchange design. In D. Simchi-Levi, S.D. Wu, and Z.J. Shen, (eds.), *Handbook of Quantitative Supply Chain Analysis: Modeling in the E-Business Era*. Kluwer Academic Publishers, New York, 2005.

39. T. Sandholm. Computing in mechanism design. In S.N. Durlauf and L.E. Blume, (eds.), *The New Palgrave Dictionary of Economics*. Second Edition, Palgrave Macmillan, 2008.

40. S. de Vries and R.V. Vohra. Combinatorial auctions: A survey. *INFORMS Journal of Computing*, 15(1):284–309, 2003.

41. S. de Vries and R.V. Vohra. Design of combinatorial auctions. In *Handbook of Quantitative Supply Chain Analysis: Modeling in the E-Business Era*, pages 247–292. International Series in Operations Research and Management Science, Kluwer Academic Publishers, Norwell, MA, USA, 2005.

42. J.J. Laffont. *Fundamentals of Public Economics*. The MIT Press, Cambridge, Massachusetts, 1988.

43. N. Nisan. Introduction to mechanism design (for computer scientists). In N. Nisan, T. Roughgarden, E. Tardos, and V.V. Vazirani (eds.), *Algorithmic Game Theory*, pages 209–242. Cambridge University Press, New York, 2007.

44. R. Myerson. Mechanism design. In J. Eatwell, M. Milgate, and P. Newman (eds.), *The New Palgrave Dictionary of Economics*, pages 191–206. Norton, New York, 1989.

45. R. Serrano. The theory of implementation of social choice rules. *SIAM Review*, 46:377–414, 2004.

46. M.O. Jackson. A crash course in implementation theory. *Social Choice and Welfare*, 18:655–708, 2001.

47. M.O. Jackson. Mechanism theory. In U. Derigs (ed.), *Encyclopedia of Life Support Systems*. EOLSS Publishers, Oxford, U.K., 2003.

48. N. Nisan, T. Roughgarden, E. Tardos, and V.V. Vazirani (eds.). *Algorithmic Game Theory*. Cambridge University Press, New York, 2007.

49. P. Milgrom. Auctions and bidding: A primer. *Journal of Economic Perspectives*, 3(3):3–22, 1989.

50. P. Klemperer. *Auctions: Theory and Practice. The Toulouse Lectures in Economics*. Princeton University Press, Princeton, NJ, USA, 2004.

51. E. Wolfstetter. Auctions: An introduction. *Economic Surveys*, 10:367–421, 1996.

52. A. Pekec and M.H. Rothkopf. Combinatorial auction design. *Management Science*, 49:1485–1503, 2003.

53. P. Klemperer Auctions: Theory and Practice. Princeton University Press, Princeton, NJ, USA, 2004.

54. Y. Narahari and P. Dayama. Combinatorial auctions for electronic business. *Sadhana — Indian Academy Proceedings in Engineering Sciences*, 30(2-3):179–212, 2005.

# Chapter 3
# Mechanism Design for Sponsored Search Auctions

The sponsored search auction problem was introduced briefly as an example in Chapter 1. In this chapter, we study this problem in more detail to illustrate a compelling application of mechanism design. We first describe a framework to model this problem as a mechanism design problem under a reasonable set of assumptions. Using this framework, we describe three well known mechanisms for sponsored search auctions — *Generalized First Price (GFP), Generalized Second Price (GSP)*, and *Vickrey–Clarke–Groves (VCG)*. We then design an optimal auction mechanism by extending Myerson's optimal auction mechanism for a single indivisible good which was discussed in the previous chapter. For this, we impose the following well known requirements, which we feel are practical requirements for sponsored search auction, for any mechanism in this setting — *revenue maximization, individual rationality,* and *Bayesian incentive compatibility*. We call this mechanism the *Optimal (OPT)* mechanism. We then make a comparative study of three mechanisms, namely GSP, VCG, and OPT, along four different dimensions — *incentive compatibility, expected revenue earned by the search engine, individual rationality*, and *computational complexity*. This chapter is a detailed extension of the results presented in [22], [23].

## 3.1 Internet Advertising

The rapid growth of the Internet and the World Wide Web is transforming the way information is being accessed and used. Newer and innovative models for distributing, sharing, linking, and marketing the information are appearing. As with any major medium, a major way of financially supporting this growth has been advertising (popularly known as *Internet advertising* or *web advertising*). The advertisers-supported web site is one of the successful business models in the emerging web landscape.

The newspapers, magazines, radio, and TV channels built advertising revenue into their business models long ago. Television advertisements have long been accepted as a way of providing funding for programming and constitute a multi billion-dollar industry even today. After these traditional media of advertising, the World Wide Web seems to be the first new way of reaching a potentially worldwide audience in a personalized and meaningful way. Advertisers have realized that Internet

Y. Narahari et al., *Game Theoretic Problems in Network Economics and Mechanism Design Solutions*, Advanced Information and Knowledge Processing, DOI: 10.1007/978-1-84800-938-7_3, © Springer-Verlag London Limited 2009

advertising is a move from complex, expensive, multinational campaigns to the ones that are targeted at the individuals wherever they are. Moreover, the traditional media do not have an exact methodology to measure the success or failure of an advertising campaign, i.e., return on investment (ROI). Because of these artifacts, Internet advertising has become an obvious choice for advertisers today.

The rise of Internet advertising has witnessed a range of advertising formats and pricing models. In what follows, we provide a brief summary of the most common Internet advertising formats and the various pricing models that have emerged so far [44, 25, 2, 48]. The Interactive Advertising Bureau is another rich source of the information about various advertising formats (URL: http://www.iab.net/).

### 3.1.1 Internet Advertising Formats

#### Banner Ads or Display Ads

This is the oldest form of Internet advertising and also happens to be a popular format for advertising on the Internet. In this format, an ad is a long thin strip of information that may be either static or may include a hyperlink to the advertiser's web page. The advertiser pays an online company in return for space to display the banner ad on one or more of the online company's web pages.

#### Rich Media

Due to advancement in technology, the static and simple banner ads started becoming richer in terms of user interaction. In this format of advertising, the banner ads are integrated with some components of streaming video and/or audio and interactivity that can allow users to view and interact with products and services — for example, flash or Java script ads, a multimedia product description, a virtual test drive, etc. "Interstitials" have been consolidated within the rich media category and represent full- or partial-page text and image server-push advertisements that appear in the transition between two pages of content. Forms of interstitials can include splash screens, pop-up windows, and superstitials.

#### E-mail

In this format, an advertiser pays a fee to the commercial e-mail service providers, e.g., Rediff, Yahoo! etc., to include the banner or rich media ad in e-mail newsletters, e-mail marketing campaigns, and other commercial e-mail communication.

**Classifieds**

In this format, an advertiser pays on-line companies to list specific products or services. For example, online job boards and employment listings, real estate listings, automotive listings, yellow pages, etc.

**Referrals**

In this format, an advertiser pays online companies that refer qualified leads or purchase inquiries. For example, automobile dealers pay a fee in exchange for receiving a qualified purchase inquiry online, fees paid when users register or apply for credit card, contest or other service.

**Search**

In today's web advertising industry, this is the highest revenue generating format among all the other Internet advertising formats. In this format, advertisers pay on-line companies to list and/or link their company site domain names to a specific search word or phrase. The other popular terminologies used for this format are *paid placement, paid listing, paid links, sponsored links, and sponsored search links.* In this format, the text links appear at the top or side of the search results for specific keywords. The more the advertiser pays, the higher the position it gets. The paid placement format is extremely popular among those online companies that provide services based on the indexing-retrieval technologies, such as pure Web search engines (e.g., AskJeeves, Google, and LookSmart), information portals with search functionality (e.g., About, AOL, MSN, Rediff, and Yahoo!), meta search engines (e.g., Metacrawler), and comparison shopping engines (e.g., Amazon, eBay, MySimon, MakeMyTrip, and Shopping).

## *3.1.2 Pricing Models for Internet Advertising*

Each one of the previously mentioned advertising formats would use one or more of the following three pricing schemes for the purpose of charging the advertisement fee.

**Pay-Per-Impression (PPI) Models**

This is the earliest pricing model to sell the space for banner ads or display ads over the World Wide Web. In early 1994, Internet advertisements were largely sold on a per-impression basis. Advertisers used to pay flat fee to show their ads for a fixed number of times (typically, one thousand showings, or "impressions"). Contracts were negotiated on a case-by-case basis, minimum contracts for advertising purchases were large (typically, a few thousand dollars per month), and entry was slow. The commonly used term for this pricing model is CPM (Cost-Per-Million-Impressions) rather than PPI. Even today, PPI based pricing models are quite popular for major Internet portals, such as AOL, MSN, and Yahoo! to display the ads in the form of a banner.

### Pay-Per-Transaction (PPT) Models

Ideally, an advertiser would always prefer a pricing model in which the advertiser pays only when a customer actually completes a transaction. The PPT models were born out of this contention. A prominent example of PPT models is Amazon.com's Associates Program.[1] Under this program, a website that sends customers to Amazon.com receives a percentage of customers' purchases.

### Pay-Per-Click (PPC) Models

It is easy to see that an advertiser would always favor the PPT model to pay for its advertising spending, whereas a World Wide Web owner would always prefer the PPI model to charge the advertisers. The PPC models seem to be a compromise between these two. In these models, an advertiser pays only when a user clicks on the ad. These models were originally introduced by Overture in 1997, and today they have almost become a standard pricing model for the specific sector of Internet advertising, so called *search*. Today, all the major Internet search engine companies, including Google, MSN, and Yahoo!, use this pricing model to sell their advertising space [2]. The PPC models for the search engines basically rely on some or other form of the auction models. There are many terms currently used in practice to refer these auction models, e.g., *search auctions, Internet search auctions, sponsored search auctions, paid search auctions, paid placement auctions, AdWord auctions, slot auctions, etc.*

## 3.1.3 An Analysis of Internet Advertising

Internet advertisement is roughly 10 years old now. In just a relatively short time, advertising on the Internet has become a common activity embraced by advertisers and marketers across all industry sectors, and today there is no dearth of activity going on in the web advertising industry. This industry appears to be growing much

---

[1] See http://www.amazon.com/gp/browse.html?&node=3435371.

[2] The PPC models are susceptible to click fraud which is a major drawback of these models [27].

faster than traditional mass media advertising vehicles such as print, radio, and TV. The total Internet advertising revenue touched \$21.1 billion in the year of 2007. The following table gives a quick estimate of the size of the market dominated by Internet advertising and the pace with which it is growing (Source: Interactive Advertising Bureau. URL: `http://www.iab.net/resources/ad_revenue.asp` accessed on April 09, 2008) All the data in this table are in billion-dollars. The columns *Q1* through *Q4* represent the revenue generated from Internet advertising in all four quarters of each of the past 11 years. The *Annual Revenue* and *Year/Year* columns give the annual revenue generated and year-by-year growth of the Internet advertising industry. The last two columns are the most important in the sense that they give an estimate of the market share of two major formats of the Internet advertising — sponsored search and display ads.

| Year | Q1 | Q2 | Q3 | Q4 | Annual Revenue | Year/Year Growth | Market Share of Sponsored Search | Market Share of Display Ads |
|------|------|------|------|------|------|------|------|------|
| 2007 | 2.802 | 2.985 | 3.1 | 3.6 | 21.1 | +25% | - | - |
| 2006 | 2.802 | 2.985 | 3.1 | 3.6 | 16.879 | +35% | 40% | 22% |
| 2005 | 2.802 | 2.985 | 3.1 | 3.6 | 12.542 | +30.3% | 41% | 20% |
| 2004 | 2.230 | 2.369 | 2.333 | 2.694 | 9.626 | +33% | 40% | 19% |
| 2003 | 1.632 | 1.660 | 1.793 | 2.182 | 7.267 | +21% | 35% | 21% |
| 2002 | 1.520 | 1.458 | 1.451 | 1.580 | 6.010 | -16% | 15% | 29% |
| 2001 | 1.872 | 1.848 | 1.773 | 1.641 | 7.134 | -12% | 04% | 36% |
| 2000 | 1.922 | 2.091 | 1.951 | 2.123 | 8.087 | +75% | 01% | 48% |
| 1999 | 0.693 | 0.934 | 1.217 | 1.777 | 4.621 | +141% | - | 56% |
| 1998 | 0.351 | 0.423 | 0.491 | 0.656 | 1.920 | +112% | - | 56% |
| 1997 | 0.130 | 0.214 | 0.227 | 0.336 | 0.907 | +239% | - | 55% |

**Table 3.1** Historical revenue performance of the Internet advertising industry

It is evident from the Table 3.1 that sponsored search is a key factor in deciding the revenue performance of any search engine company. Therefore, our interest lies in studying the different kinds of mechanisms for sponsored search auction. In particular, we want to investigate

- Which mechanisms satisfy desirable economic properties, such as *incentive compatibility,* and/or *individual rationality?*
- Which mechanisms are the most favorable choice for advertisers, i.e., which mechanisms yield the minimum advertising expense for the advertisers?
- Which mechanisms are the most favorable choice for the search engines, i.e., which mechanisms yield the maximum revenue for the search engines?

With these objectives in mind, in the next section we first explain how a typical sponsored search auction works, and then we discuss different mechanisms for it.

## 3.2 Sponsored Search Auction

When an Internet user (which we will sometimes refer to as the user, searcher, or customer) enters a keyword (i.e., a search term) into a search engine, he gets back a page with results, containing both the links most relevant to the query and the sponsored links, i.e., paid advertisements. When a user clicks on a sponsored link, he is sent to the respective advertiser's web page. The advertiser then pays the search engine for sending the user to its web page. Figure 1 depicts the result of a search performed on Google using the keyword auctions. There are two different stacks — the left stack contains the links that are most relevant to the query term, and the right stack contains the sponsored links. Sometimes, a few sponsored links are placed on top of the search results as shown. These sponsored links are clearly distinguishable from the actual search results. However, the visibility of a sponsored search link depends on its location (slot) on the result page. Typically, a number of

**Fig. 3.1** Result of a search performed on Google

merchants (advertisers) are interested in advertising alongside the search results of a keyword. However, the number of slots available to display the sponsored links is limited. Therefore, against every search performed by the user, the search engine faces the problem of matching the advertisers to the slots. In addition, the search engine also needs to decide on a price to be charged to each advertiser. Note that each advertiser has different desirability for different slots on the search result page. The visibility of an ad shown at the top of the page is much better than an ad shown at the bottom, and, therefore, it is more likely to be clicked by the user. Therefore, an advertiser naturally prefers a slot with higher visibility. Hence, search engines need a system for allocating the slots to advertisers and deciding on a price to be charged to each advertiser. Due to increasing demands for advertising space, most search engines are currently using auction mechanisms for this purpose. These auctions are called sponsored search auctions. In a typical sponsored search auction, advertisers are invited to submit bids on keywords, i.e., the maximum amount they are willing to pay for an Internet user clicking on the advertisement. This is typically referred by the term Cost-Per-Click (CPC). Based on the bids submitted by the advertisers for a particular keyword, the search engine (which we will sometimes refer to as the auctioneer or the seller) picks a subset of advertisements along with the order in which to display them. The actual price charged also depends on the bids submitted by the advertisers. There are many terms currently used in practice to refer to these auctions models, e.g., search auctions, Internet search auctions, sponsored search auctions, paid search auctions, paid placement auctions, AdWord auctions, slot auctions, etc.

## 3.3 Sponsored Search Auction as a Mechanism Design Problem

Consider a search engine that has received a query from an Internet user, and it immediately faces the problem of conducting an auction for selling its advertising space among the available advertisers for this particular query word. Let us assume that

1. There are $n$ advertisers interested in this particular keyword, and $N = \{1,2,\ldots,n\}$ represents the set of these advertisers. Also, there are $m$ slots available with search engine to display the ads and $M = \{1,2,\ldots,m\}$ represents the set of these advertising slots.
2. $\alpha_{ij}$ is the probability that a user will click on the $i^{th}$ advertiser's ad if it is displayed in $j^{th}$ position (slot), where the first position refers to the top most position. We assume that $\alpha_{ij}$ satisfy the following condition:

$$1 \geq \alpha_{i1} \geq \alpha_{i2} \geq \ldots \geq \alpha_{im} \geq 0 \ \forall i \in N. \tag{3.1}$$

Note, here we are assuming that click probability $\alpha_{ij}$ does not depend on which other advertiser has been allocated to what other position. We refer to this assumption as *absence of allocative externality* among the advertisers.
3. Each advertiser precisely knows the value derived out of each click performed by the user on his ad[3] but does not know the value derived out of a single user-click by the other advertisers. Formally, this is modeled by supposing that advertiser $i$ observes a parameter, or signal $\theta_i$ that represents his value for each user click. The parameter $\theta_i$ is referred to as advertiser $i$'s *type*. The set of possible types of advertiser $i$ is denoted by $\Theta_i$.

---

[3] Note this value should be independent of the position of the ad and should only depend on whether a user clicks on the ad or not.

4. Each advertiser perceives any other advertiser's valuation as a draw from some probability distribution. Similarly, he knows that the other advertisers regard his own valuation as a draw from some probability distribution. More precisely, for advertiser $i$, $i = 1, 2, \ldots, n$, there is some probability distribution $\Phi_i(\cdot)$ from which he draws his valuation $\theta_i$. Let $\phi_i(\cdot)$ be the corresponding PDF. We assume that the $\theta_i$ takes values from a closed interval $\left[\underline{\theta_i}, \overline{\theta_i}\right]$ of the real line. That is, $\Theta_i = \left[\underline{\theta_i}, \overline{\theta_i}\right]$. We also assume that any advertiser's valuation is statistically independent from any other advertiser's valuation. That is, $\Phi_i(\cdot), i = 1, 2, \ldots, n$ are mutually independent. We refer to this assumption as *independent private values assumption*. Note that probability distribution $\Phi_i(\cdot)$ can be viewed as the distribution of a random variable that gives the profit earned by advertiser $i$ when a random customer clicks on advertiser's Ad.
5. Each advertiser $i$ is rational and intelligent in the sense of [38]. This fact is modeled by assuming that the advertisers always try to maximize a Bernoulli utility function $u_i : X \times \Theta_i \to \mathbb{R}$, where $X$ is the set of outcomes, which will be defined shortly.
6. The probability distribution functions $\Phi_i(\cdot)$, the type sets $\Theta_1, \ldots, \Theta_n$, and the utility functions $u_i(\cdot)$ are assumed to be common knowledge among the advertisers. Note that utility function $u_i(\cdot)$ of advertiser $i$ depends on both the outcome $x$ and the type $\theta_i$. Although the type $\theta_i$ is not common knowledge, by saying that $u_i(\cdot)$ is common knowledge we mean that for any given type $\theta_i$, the auctioneer (that is, search engine in this case) and every other advertiser can evaluate the utility function of advertiser $i$.

In view of the above modeling assumptions, the sponsored search auction problem can now be restated as follows. For any query word, each interested advertiser $i$, bids an amount $b_i \geq 0$, which depends on his actual type $\theta_i$. Now each time the search engine receives this query word, it first retrieves the information from its database of all the advertisers who are interested in displaying their ads against the search result of this query and their corresponding bid vector $b = (b_1, \ldots, b_n)$. The search engine then decides the winning advertisers along with the order in which their ads will be displayed against the search results and the amount that will be paid by each advertiser if the user clicks on his ad. These are called *allocation* and *payment rules*, respectively. Depending on what allocation and payment rules are employed by the search engine, it may take different forms. It is easy to verify that this is a perfect setting to formulate the sponsored search auction problem as a mechanism design problem. A sponsored search auction can be viewed as an *indirect mechanism* $\mathcal{M} = ((B_i)_{i \in N}, g(\cdot))$, where $B_i \subset \mathbb{R}^+$ is the set of bids that an advertiser $i$ can ever report to the search engine and $g(\cdot)$ is the allocation and payment rule. Note, if we assume that for each advertiser $i$, the set of bids $B_i$ is the same as type set $\Theta_i$, then indirect mechanism $\mathcal{M} = ((B_i)_{i \in N}, g(\cdot))$ becomes a direct revelation mechanism $\mathcal{D} = ((\Theta_i)_{i \in N}, f(\cdot))$, where $f(\cdot)$ becomes the allocation and payment rule. In the rest of this chapter, we will assume that $B_i = \Theta_i \ \forall \ i = 1, \ldots, n$. Thus, in view of this assumption, we can regard a sponsored search auction as a direct revelation mechanism.

The various components of a typical sponsored search mechanism design problem are listed below.

### 3.3.1 Outcome Set X

An outcome in the case of a sponsored search auction may be represented by a vector $x = (y_{ij}, p_i)_{i \in N, j \in M}$, where $y_{ij}$ is the probability that advertiser $i$ is allocated to the slot $j$, and $p_i$ denotes the price-per-click charged from advertiser $i$. The set of feasible alternatives is then

$$X = \left\{ (y_{ij}, p_i)_{i \in N, j \in M} \,\middle|\, y_{ij} \in [0,1] \; \forall i \in N, \; \forall j \in M, \sum_{i=1}^{n} y_{ij} \le 1 \; \forall j \in M, \sum_{j=1}^{m} y_{ij} \le 1 \; \forall i \in N, \right.$$
$$\left. p_i \ge 0 \; \forall i \in N \right\}.$$

Note that the randomized outcomes are also included in the above outcome set. This implies that the randomized mechanisms are also part of the design space.

### 3.3.2 Utility Function of Advertisers $u_i(\cdot)$

The utility function of advertiser $i$ can be given, for $x = (y_{ij}, p_i)_{i \in N, j \in M}$, by

$$u_i(x, \theta_i) = \left( \sum_{j=1}^{m} y_{ij} \alpha_{ij} \right) (\theta_i - p_i).$$

### 3.3.3 Social Choice Function $f(\cdot)$ (Allocation and Payment Rules)

The general structure of the allocation and payment rule for this case is

$$f(b) = (y_{ij}(b), p_i(b))_{i \in N, j \in M}$$

where $b = (b_1, \ldots, b_n)$ is a bid vector of the advertisers. The functions $y_{ij}(\cdot)$ form the allocation rule, and the functions $p_i(\cdot)$ form the payment rule.

### 3.3.4 Linear Environment

Through a slight modification in the definition of allocation rule, payment rule, and utility functions, we can show that a sponsored search auction is indeed a direct revelation mechanism in a linear environment. To transform the underlying environment to a linear one, we redefine the allocation and payment rule as below.

$$f(b) = (y(b), t_i(b))_{i \in N, j \in M}$$

where $y(b) = (y_{ij}(b))_{i \in N, j \in M}$ and $t_i(b) = \left( \sum_{j=1}^{m} y_{ij}(b) \alpha_{ij} \right) p_i(b)$. The quantity $t_i(b)$ can be viewed as the average payment made by the advertiser $i$ to the search engine against every search query received by the search engine, and when the bid vector of the advertisers is $b = (b_1, \ldots, b_n)$.

Now, we can rewrite the utility functions in following manner:

$$u_i(f(b), \theta_i) = \theta_i v_i(y(b)) - t_i(b)$$

where $v_i(y(b)) = \left( \sum_{j=1}^{m} y_{ij}(b) \alpha_{ij} \right)$. The quantity $v_i(y(b))$ can be interpreted as the probability that advertiser $i$ will receive a user click whenever there is a search query received by the search engine and when the bid vector of the advertisers is $b = (b_1, \ldots, b_n)$. Now, it is easy to verify that the underlying environment here is *linear*.

In what follows, we illustrate three basic mechanisms for sponsored search auctions with respect to the above model.

- Generalized First Price (GFP) Mechanism
- Generalized Second Price (GSP) Mechanism (also called Next Price Auction)
- Vickrey–Clarke–Groves (VCG) Mechanism

For each of these mechanisms, we describe the allocation rule $y_{ij}(\cdot)$ and payment rule $p_i(\cdot)$.

## 3.4 Generalized First Price (GFP) Mechanism

In 1997, Overture (formerly GoTo; recently acquired by Yahoo!) introduced the first auction mechanism ever used for sponsored search. The term *Generalized First Price Auction* was coined by Edelman, Ostrovsky, and Schwarz [15]. The allocation and payment rules under this mechanism are the following.

### 3.4.1 Allocation Rule

In this mechanism, the $m$ advertising slots are allocated to advertisers in *descending order of their bids* [15]. Let $b^{(k)}$ be the $k^{th}$ highest element in $(b_1, \ldots, b_n)$. Similarly, let $(b_{-i})^{(k)}$ be the $k^{th}$ highest element in $(b_1, \ldots, b_{i-1}, b_{i+1}, \ldots, b_n)$. In view of these definitions, we can say that if $b = (b_1, b_2, \ldots, b_n)$ is the profile of bids received from the $n$ advertisers, then the first slot is allocated to the advertiser whose bid is equal to $b^{(1)}$. Similarly, the second slot is allocated to the advertiser whose bid is equal to

$b^{(2)}$, and so on. [4] That is, for all $i \in N$ and all $j \in M$,

$$y_{ij}(b) = \begin{cases} 1 & : \quad \text{if } b_i = b^{(j)} \\ 0 & : \quad \text{otherwise.} \end{cases} \tag{3.2}$$

### 3.4.2 Payment Rule

Every time a user clicks on a sponsored link, an advertiser's account is automatically billed *the amount of the advertiser's bid*. That is, if $b = (b_1, b_2, \ldots, b_n)$ is the profile of bids received from the $n$ advertisers then, for all $i \in N$,

$$p_i(b) = \begin{cases} b_i & : \quad \text{if advertiser } i\text{'s Ad is displayed} \\ 0 & : \quad \text{otherwise.} \end{cases} \tag{3.3}$$

The ease of use, very low entry costs, and transparency of the mechanism quickly led to the success of Overture's paid search platform as the advertising provider for major search engines including MSN and Yahoo!.

## 3.5 Generalized Second Price (GSP) Mechanism

The primary motivation for this auction mechanism was instability of the GFP mechanism. In particular, it has been shown by Edelman, Ostrovsky, and Schwarz [15] that under the GFP mechanism, truth-telling is not an equilibrium bidding strategy for the advertisers, and this fact leads to instability in the system, which in turn leads to inefficient investments on behalf of the advertisers. The GFP mechanism also creates volatile prices, which in turn cause allocative inefficiencies. [5] Google realized these problems and tried fixing the problems by introducing its own new program, called *AdWord Select*, in February 2002.

Recognizing the tangible advantages, Yahoo!/Overture also switched to the GSP mechanism. The payment rules are the same in both Google's version of the GSP mechanism and the Yahoo!/Overture version of the GSP mechanism. However, the allocation rules are slightly different in these two versions of the GSP mechanisms. The Yahoo!/Overture version of the GSP mechanism follows the same allocation rule as the GFP mechanism but the allocation rule in Google's version of the GSP mechanism is more general. In what follows we describe the different allocation rules for the GSP mechanism and investigate their relationships.

---

[4] If two advertisers have the same bid, then the tie can be broken by an appropriate rule. It is easy to verify that two advertisers having the same bid value is a zero probability event.

[5] See [15] for a detailed discussion about demerits of the GFP mechanism.

## *3.5.1 Allocation Rule*

1. **Yahoo!/Overture's Allocation Rule:** This rule is the same as the allocation rule of GFP mechanisms, that is, the slots are allocated to the advertisers in descending order of their bids.
2. **Greedy Allocation Rule:** The primary motivation for this rule is *allocative efficiency*, which we will discuss later. In this rule, the first slot is allocated to the advertiser $i \in N$ for whom the quantity $\alpha_{i1} b_i$ is the maximum. If there is a tie then it is broken by appropriate rule. Now this advertiser is removed from the set $N$, and an advertiser among the remaining ones is chosen for whom $\alpha_{i2} b_i$ is the maximum. The second slot is allocated to this advertiser. In similar fashion, all the other slots are allocated.
3. **Google's Allocation Rule:** Google had also realized the fact that unappealing and poorly targeted ads attract relatively few clicks and thus provide less revenue for Google, and, therefore, the probability that the searcher will click on an ad link must be taken into account. In practice, Google uses some sort of stylized version of the *greedy* allocation rule. In Google's actual version of GSP mechanisms, for each advertiser Google computes its estimated *Click-Through-Rate (CTR)*, which is the ratio of the number of clicks received by the ad to the number of times the Ad was displayed against the search results — popularly known as number of *impressions*. Now the advertisers are ranked in decreasing order of the *ranking scores*, where ranking score of an advertiser is defined as the product of the advertiser's bid and estimated *CTR*. [6]

In order to understand the relationship among these three allocation rules, we need to first understand the relationship between click probability and CTR.

## *3.5.2 Relationship between Click Probability and CTR*

The notions of click probability and CTR seem to be quite similar in nature. The objective here is to develop a better understanding of these two quantities. Recall the following definitions that we presented earlier:

$\alpha_{ij}$ = Probability that user will click on $i^{th}$ advertiser's ad if it is displayed in $j^{th}$ position

$CTR_i$ = Probability that user will click on $i^{th}$ advertiser's ad if it is displayed

$y_{ij}$ = Probability that advertiser $i$'s ad is displayed in position $j$.

In view of above definitions, it is easy to verify that

---

[6] See Google's online AdWord demo on *bidding and ranking*. URL: `https://adwords.google.com/select/library/index.html` (accessed on November 21, 2005).

$$\text{CTR}_i = \sum_{j=1}^{m} y_{ij}\alpha_{ij} \; \forall \, i \in N$$

$$\text{CTR}_i \leq \sum_{j=1}^{m} \alpha_{ij} \; \forall \, i \in N.$$

In practice, the click probabilities ($\alpha_{ij}$) and CTR are learned by means of available data. In what follows, we present three different ways in which one can learn these quantities. This discussion is taken from [27].

### 3.5.2.1 Average over Fixed Time Window

In this method, we fix a time interval $T$ and consider an advertiser $i$. Let $I_i$ be the number of times advertiser $i$'s ad was displayed (irrespective of its position) against search results during the interval $T$. Further, let $I_{ij}$ impressions of the ad were made at position $j$ out of $I_i$ impressions. Let $C_i \leq I_i$ be the total number of clicks received on this ad during the interval $T$. Moreover, out of these $C_i$ clicks, $C_{ij}$ clicks were received when the ad was displayed in position $j$. Now we can define the click probabilities and CTR in following manner.

$$\alpha_{ij} = \frac{C_{ij}}{I_{ij}} \; \forall \, i \in N, \, \forall \, j \in M$$

$$\text{CTR}_i = \frac{C_i}{I_i} \; \forall \, i \in N.$$

### 3.5.2.2 Average over Fixed Impression Window

This method is the same as the previous one except that here we fix the number of impressions instead of time. Let us fix the total number of past impression equal to some positive integer constant, say $I$, and then count the total number of click $C_i$ that were received against these impressions. Similarly, we fix the total number of past impression made at position $j$ equal to same positive integer constant $I$ and then count the total number of click $C_{ij}$ that were received against these impressions. Now, the click probabilities and CTR can be defined in following manner:

$$\alpha_{ij} = \frac{C_{ij}}{I} \; \forall \, i \in N, \, \forall \, j \in M$$

$$\text{CTR}_i = \frac{C_i}{I} \; \forall \, i \in N.$$

### 3.5.2.3 Average over Fixed Click Window

In this method, we fix the total number of clicks received so far equal to some positive integer constant, say $C$, and then count the total number of impressions $I_i$ that were made in order to receive these clicks. Similarly, fix the total number of clicks received for position $j$ so far to the same constant $C$, and then count the total number of impressions $I_{ij}$ that were made in order to receive these clicks. Now, the click probabilities and CTR can be defined in following manner:

$$\alpha_{ij} = \frac{C}{I_{ij}} \ \forall\, i \in N, \forall\, j \in M$$

$$\mathrm{CTR}_i = \frac{C}{I_i} \ \forall\, i \in N.$$

## 3.5.3 Relationship Among Different Allocation Rules

Before moving on to the payment rule for GSP, we would like to explore the relationship among the allocation rules. In order to understand the relationship, let us assume that $b = (b_1, b_2, \ldots, b_n)$ is the profile of bids received from the $n$ advertisers. Consider the following optimization problem:

Maximize

$$\sum_{i=1}^{n} b_i v_i(y(b)) = \sum_{i=1}^{n} \sum_{j=1}^{m} (b_i \alpha_{ij}) y_{ij}(b)$$

subject to

$$\sum_{i=1}^{n} y_{ij}(b) \leq 1 \, \forall j \in M$$

$$\sum_{j=1}^{m} y_{ij}(b) \leq 1 \, \forall i \in N$$

$$y_{ij}(b) \leq 0 \, \forall i \in N, \forall j \in M.$$

It is easy to see that for given click probabilities $\alpha_{ij}$, where these probabilities satisfy the condition (3.1), the greedy allocation rule basically provides a solution to the above optimization problem. Such an allocation would be an *efficient* allocation. Yahoo!/Overture's allocation rule and Google's allocation rule become special cases of the greedy allocation rule under certain conditions that are summarized in the following propositions.

**Proposition 3.1.** *Given any bid vector $b = (b_1, \ldots, b_n)$, the greedy allocation rule and Yahoo!/Overture's allocation rule result in the same allocation if the following two conditions are satisfied:*

1. *The click probabilities satisfy the assumption of absence of allocative externality among the advertisers, that is,* $1 \geq \alpha_{i1} \geq \alpha_{i2} \geq \ldots \geq \alpha_{im} \geq 0 \ \forall i \in N$.
2. *The click probabilities depend only on the positions of the advertisements and are independent of the identities of the advertisers, that is,* $\alpha_{1j} = \alpha_{2j} = \ldots = \alpha_{nj} = \alpha_j \ \forall j \in M$.

**Proposition 3.2.** *Given any bid vector* $b = (b_1, \ldots, b_n)$, *both the greedy allocation rule and* Google*'s allocation rule result in the same allocation if the following two conditions are satisfied:*

1. *The click probabilities satisfy the assumption of absence of allocative externality among the advertisers, that is,* $1 \geq \alpha_{i1} \geq \alpha_{i2} \geq \ldots \geq \alpha_{im} \geq 0 \ \forall i \in N$.
2. *The click probabilities depend only on the identities of the advertisers and are independent of the positions of the advertisements, that is,* $\alpha_{i1} = \alpha_{i2} = \ldots = \alpha_{im} = \alpha_i = CTR_i \ \forall i \in N$.

In the rest of the chapter, we will work with the following assumptions:

1. Click probabilities depend only on the positions of the ads and are independent of the identities of the advertisers. That is, $\alpha_{1j} = \alpha_{2j} = \ldots = \alpha_{nj} = \alpha_j \ \forall j \in M$.
2. The allocation rule in a GSP mechanism is the same as the greedy allocation rule, which would be the same as Yahoo!/Overture's allocation rule because of the previous assumption.

### 3.5.4 Payment Rule

In this auction mechanism, every time a user clicks on a sponsored link, an advertiser's account is automatically billed *the amount of the advertiser's bid who is just below him in the ranking of the displayed ads plus a minimum increment (typically $0.01).* The advertiser whose ad appears at the bottom-most position is charged the amount of the highest bid among the disqualified bids plus the minimum increment. If there is no such bid then he is charged nothing. If $b = (b_1, b_2, \ldots, b_n)$ is the profile of bids received from the $n$ advertisers, then because of the assumptions we made earlier regarding the allocation rule in the GSP mechanism, the price per click that is charged to an advertiser $i$ would be given by

$$
p_i(b) = \begin{cases} \sum_{j=1}^{m} \left( b^{(j+1)} y_{ij}(b) \right) & : \quad \text{if either } m < n \text{ or } n \leq m \text{ but } b_i \neq b^{(n)} \\ 0 & : \quad \text{otherwise} \end{cases}
$$

where $b^{(j+1)}$ is the $(j+1)^{th}$ highest bid which is the same as the bid of an advertiser whose Ad is allocated to position $(j+1)$.[7]

---

[7] We have ignored the small increment $0.01 because all the future analysis and results are insensitive to this amount.

## 3.6 Vickrey–Clarke–Groves (VCG) Mechanism

On the face of it, the GSP mechanism appears to be a generalized version of the well known Vickrey auction, which is used for selling a single indivisible good. But as shown by Edelman, Ostrovsky, and Schwarz [15], and also shown in the later part of this chapter, the GSP mechanism is indeed not a generalization of the classical Vickrey auction to the setting where a set of ranked objects is being sold. The generalization of the Vickrey auction is the Clarke mechanism, which we introduced in the previous chapter. In this section, our objective is to develop the Clarke mechanism for the sponsored search auction. We refer to this as the VCG mechanism, following standard practice.

### 3.6.1 Allocation Rule

By definition, the VCG mechanism is allocatively efficient. Therefore, in the case of a sponsored search auction, the allocation rule $y^*(\cdot)$ in the VCG mechanism is

$$y^*(\cdot) = \frac{\arg\max}{y(\cdot)} \sum_{i=1}^{n} b_i v_i(y(b)) = \frac{\arg\max}{y_{ij}(\cdot)} \sum_{i=1}^{n} \sum_{j=1}^{m} (b_i \alpha_{ij}) y_{ij}(b). \tag{3.4}$$

In the previous section, we have already seen that the greedy allocation rule is a solution to (3.4). Moreover, under the assumption that click probabilities are independent of advertisers' identities, the allocation $y^*(\cdot)$ allocates the slots to the advertisers in the decreasing order of their bids. That is, if $b = (b_1, b_2, \ldots, b_n)$ is the profile of bids received from the $n$ advertisers then $y^*(\cdot)$ must satisfy the following condition:

$$y_{ij}^*(b) = \begin{cases} 1 & : \quad b_i = b^{(j)} \\ 0 & : \quad \text{otherwise.} \end{cases} \tag{3.5}$$

We state below an interesting observation regarding GFP and GSP mechanisms, which is based on the above observations.

**Proposition 3.3.** *If click probabilities depend only on the positions of the ads and are independent of the identities of the advertisers, then*

1. *The GFP mechanism is allocatively efficient.*
2. *The GSP mechanism is allocatively efficient if it uses the greedy allocation rule, which is the same as* Yahoo!/Overture's *allocation rule.*
3. *The allocation rule for the VCG mechanism, which is an efficient allocation, is given by (3.5). Moreover, this allocation rule is precisely the same as the GFP allocation rule and* Yahoo!/Overture's *allocation rule.*

### 3.6.2 Payment Rule

As per the definition of the VCG mechanism, the expected payment $t_i(b)$ made by an advertiser $i$, when the profile of the bids submitted by the advertisers is $b = (b_1, \ldots, b_n)$, must be calculated using the following Clarke's payment rule:

$$t_i(b) = \left[ \sum_{j \neq i} b_j v_j(y^*(b)) \right] - \left[ \sum_{j \neq i} b_j v_j(y^*_{-i}(b)) \right] \qquad (3.6)$$

where $y^*_{-i}(\cdot)$ is an efficient allocation of the slots among the advertisers when advertiser $i$ is removed from the scene. Substituting value of $y^*(\cdot)$ from Equation (3.5) and making use of the fact that $v_i(y^*(b)) = \sum_{j=i}^{m} y^*_{ij}(b)\alpha_j$, Equation (3.6) can be written as follows:

**Case 1 ($m < n$):**

$$t^{(j)}(b) = \alpha_j p^{(j)}(b)$$
$$= \begin{cases} \beta_j b^{(j+1)} + t^{(j+1)}(b) & : \quad \text{if } 1 \leq j \leq (m-1) \\ \alpha_m b^{(m+1)} & : \quad \text{if } j = m \\ 0 & : \quad \text{if } m < j \leq n \end{cases} \qquad (3.7)$$

where

- $t^{(j)}(b)$ is the expected payment made by the advertiser whose ad is displayed in $j^{th}$ position, for every search query received by the search engine and when the bid profile of the advertisers is $b = (b_1, \ldots, b_n)$,
- $p^{(j)}(b)$ is the payment made by the advertiser, whose ad is displayed in $j^{th}$ position, for every click made by the user and when the bid profile of the advertisers is $b = (b_1, \ldots, b_n)$,
- and $\beta_j = (\alpha_j - \alpha_{j+1})$,
- $b^{(j)}$ has its usual interpretation.

**Case 2 ($n \leq m$):**

$$t^{(j)}(b) = \alpha_j p^{(j)}(b)$$
$$= \begin{cases} \beta_j b^{(j+1)} + t^{(j+1)}(b) & : \quad \text{if } 1 \leq j \leq (n-1) \\ 0 & : \quad \text{if } j = n. \end{cases} \qquad (3.8)$$

Unfolding Equations (3.7) and (3.8) result in the following expressions:

**Case 1 ($m < n$):**

$$p^{(j)}(b) = \frac{1}{\alpha_j} t^{(j)}(b) = \begin{cases} \frac{1}{\alpha_j} \left[ \sum_{k=j}^{m-1} \beta_k b^{(k+1)} \right] + \frac{\alpha_m}{\alpha_j} b^{(m+1)} & : \quad \text{if } 1 \leq j \leq (m-1) \\ b^{(m+1)} & : \quad \text{if } j = m \\ 0 & : \quad \text{if } m < j \leq n. \end{cases} \qquad (3.9)$$

**Case 2** $(n \leq m)$:

$$p^{(j)}(b) = \frac{1}{\alpha_j} t^{(j)}(b) = \begin{cases} \frac{1}{\alpha_j} \sum_{k=j}^{n-1} \beta_k b^{(k+1)} & : \quad \text{if } 1 \leq j \leq (n-1) \\ 0 & : \quad \text{if } j = n. \end{cases} \quad (3.10)$$

Thus, we can say that Equation (3.5) describes the allocation rule for the VCG mechanism and Equations (3.9) and (3.10) describe the payment rule for the VCG mechanism.

So far we have discussed three basic mechanisms for sponsored search auction — GFP, GSP, and VCG. The GFP and GSP are the two mechanisms that have been used by the search engines in practice. The VCG mechanism, though not implemented so far by any search engine, is another mechanism that is worth studying because of its unique property, namely dominant strategy incentive compatibility. In fact, at least theoretically, there are infinitely many ways in which a sponsored search auction can be conducted where each mechanism has its own pros and cons. We will now focus on an another important mechanism, namely, Optimal (OPT) mechanism. This mechanism is interesting from a theoretical as well as a practical stand-point because it maximizes the search engine's revenue while ensuring individual rationality and incentive compatibility for the advertisers. Myerson first studied a similar auction mechanism in the context of selling a single indivisible good [37] (see Section 2.20). Myerson called such an auction mechanism an *optimal auction*. Following the same terminology, we would prefer to call a similar mechanism for the sponsored search auction an *optimal mechanism* for sponsored search auction (OPT mechanism for short). The readers are suggested to refer to the previous chapter (see Section 2.20) for details regarding this mechanism. It may be noted that the discussion in Section 2.20 is on optimal procurement auction (cost minimization) whereas here, it would be on optimal auctions for selling (revenue maximization).

## 3.7 Optimal (OPT) Mechanism

In this section, our goal is to compute the allocation and payment rule $f(\cdot)$ that results in an optimal mechanism for the sponsored search auction. This calls for extending Myerson's optimal auction to the case of the sponsored search auction. We follow a line of attack that is similar to that of Myerson [37]. Recall that we formulated the sponsored search auction as a direct revelation mechanism $\mathscr{D} = ((\Theta_i)_{i \in N}, f(\cdot))$ in a linear environment, where the Bernoulli utility function of an advertiser $i$ is given by

$$u_i(f(b), \theta_i) = \left( \sum_{j=1}^{m} y_{ij}(b) \alpha_j \right) (\theta_i - p_i(b))$$
$$= v_i(y(b))(\theta_i - p_i(b))$$
$$= \theta_i v_i(y(b)) - t_i(b)$$

where

- $v_i(y(b)) = \left( \sum_{j=1}^m y_{ij}(b)\alpha_j \right)$ is the valuation function of the advertiser $i$ and can be interpreted as the probability that advertiser $i$ will receive a user click whenever there is a search query received by the search engine and when the bid vector of the advertisers is $b$.
- $t_i(b) = v_i(y(b))p_i(b)$ can be viewed as the average payment made by advertiser $i$ to the search engine against every search query received by the search engine and when the bid vector of the advertisers is $b$.

### 3.7.1 Allocation Rule

It is convenient to define

- $\bar{t}_i(b_i) = E_{\theta_{-i}}[t_i(b_i, \theta_{-i})]$ is the expected payment made by advertiser $i$ when he bids an amount $b_i$ and all the advertisers $j \neq i$ bid their true types.
- $\bar{v}_i(b_i) = E_{\theta_{-i}}[v_i(y(b_i, \theta_{-i}))]$ is the probability that advertiser $i$ will receive a user click if he bids an amount $b_i$ and all the advertisers $j \neq i$ bid their true types.
- $U_i(\theta_i) = \theta_i \bar{v}_i(\theta_i) - \bar{t}_i(\theta_i)$ gives advertiser $i$'s expected utility from the mechanism conditional on his type being $\theta_i$ when he and all other advertisers bid their true types.

The problem of designing an optimal mechanism for the sponsored search auction can now be written as one of choosing functions $y_{ij}(\cdot)$ and $U_i(\cdot)$ to solve:

Maximize

$$\sum_{i=1}^n \int_{\underline{\theta_i}}^{\overline{\theta_i}} (\theta_i \bar{v}_i(\theta_i) - U_i(\theta_i)) \, \phi_i(\theta_i)d\theta_i \tag{3.11}$$

subject to

(i)   $\bar{v}_i(\cdot)$ is non-decreasing $\forall i \in N$

(ii)  $y_{ij}(\theta) \in [0,1], \sum_{j=1}^m y_{ij}(\theta) \leq 1, \sum_{i=1}^n y_{ij}(\theta) \leq 1 \ \forall i \in N, \ \forall j \in M, \ \forall \theta \in \Theta$

(iii) $U_i(\theta_i) = U_i(\underline{\theta_i}) + \int_{\underline{\theta_i}}^{\theta_i} \bar{v}_i(s)ds \ \forall i \in N, \ \forall \theta_i \in \Theta_i$

(iv)  $U_i(\theta_i) \geq 0 \ \forall i \in N, \ \forall \theta_i \in \Theta_i$.

In the above formulation, the objective function is the total expected payment received by the search engine from all the advertisers. Note that constraints (iv) are the advertisers' interim individual rationality constraints while constraint (ii) is the feasibility constraint. Constraints (i) and (iii) are the necessary and sufficient conditions for the allocation and payment rule $f(\cdot) = (y_{ij}(\cdot), t_i(\cdot))_{i \in N, j \in M}$ to be Bayesian incentive compatible. These constraints are taken from [37].

We have a critical observation to make here. Note that in the above optimization problem, we have replaced the bid $b_i$ by the actual type $\theta_i$. This is because we are imposing the Bayesian incentive compatibility constraints on the allocation and payment rule, and, hence, every advertiser will bid his true type. Thus, while dealing with the OPT mechanism, we can safely interchange $\theta_i$ and $b_i$ for any $i \in N$.

Note first that if constraint (iii) is satisfied, then constraint (iv) will be satisfied iff $U_i(\underline{\theta_i}) \geq 0 \; \forall i \in N$. As a result, we can replace the constraint (iv) with

$$(iv') \; U_i(\underline{\theta_i}) \geq 0 \; \forall i \in N.$$

Next, substituting for $U_i(\theta_i)$ in the objective function from constraint (iii), we get

$$\sum_{i=1}^{n} \int_{\underline{\theta_i}}^{\overline{\theta_i}} \left( \overline{v}_i(\theta_i)\theta_i - U_i(\underline{\theta_i}) - \int_{\underline{\theta_i}}^{\theta_i} \overline{v}_i(s)ds \right) \phi_i(\theta_i)d\theta_i.$$

Integrating by parts the above expression, the search engine's problem can be written as one of choosing the $y_{ij}(\cdot)$ functions and the values $U_1(\underline{\theta_1}), \ldots, U_n(\underline{\theta_n})$ to maximize

$$\int_{\underline{\theta_1}}^{\overline{\theta_1}} \cdots \int_{\underline{\theta_n}}^{\overline{\theta_n}} \left[ \sum_{i=1}^{n} v_i(y(\theta_i, \theta_{-i}))J_i(\theta_i) \right] \left[ \prod_{i=1}^{n} \phi_i(\theta_i) \right] d\theta_n \ldots d\theta_1 - \sum_{i=1}^{n} U_i(\underline{\theta_i})$$

subject to constraints (i), (ii), and (iv'), where

$$J_i(\theta_i) = \theta_i - \frac{1 - \Phi_i(\theta_i)}{\phi_i(\theta_i)}.$$

It is evident that the solution must have $U_i(\underline{\theta_i}) = 0$ for all $i = 1, 2, \ldots, n$. Hence, the search engine's problem reduces to choosing functions $y_{ij}(\cdot)$ to maximize

$$\int_{\underline{\theta_1}}^{\overline{\theta_1}} \cdots \int_{\underline{\theta_n}}^{\overline{\theta_n}} \left[ \sum_{i=1}^{n} v_i(y(\theta_i, \theta_{-i}))J_i(\theta_i) \right] \left[ \prod_{i=1}^{n} \phi_i(\theta_i) \right] d\theta_n \ldots d\theta_1$$

subject to constraints (i) and (ii).

Let us ignore the constraint (i) for the moment. Then an inspection of the above expression indicates that $y_{ij}(\cdot)$ is a solution to this relaxed problem iff for all $i = 1, 2, \ldots, n$ we have

$$y_{ij}(\theta) = \begin{cases} 0 & \forall j = 1, 2, \ldots, m & : \quad \text{if } J_i(\theta_i) < 0 \\ 1 & \forall j = 1, 2, \ldots, m < n & : \quad \text{if } J_i(\theta_i) = J^{(j)} \\ 1 & \forall j = 1, 2, \ldots, n \leq m & : \quad \text{if } J_i(\theta_i) = J^{(j)} \\ 0 & & : \quad \text{otherwise} \end{cases} \tag{3.12}$$

where $J^{(j)}$ is the $j^{th}$ highest value among $J_i(\theta_i)$s.

In other words, if we ignore the constraint (i) then $y_{ij}(\cdot)$ is a solution to this relaxed problem if and only if no slot is allocated to any advertiser having negative value $J_i(\theta_i)$, and the rest of the advertisers' ads are displayed in the same order as the values of $J_i(\theta_i)$. That is, the first slot is allocated to the advertiser who has the highest nonnegative value for $J_i(\theta_i)$, the second slot is allocated to the advertiser who has the second highest nonnegative value for $J_i(\theta_i)$, and so on.

Now, recall the definition of $\bar{v}_i(\cdot)$. It is easy to write down the following expression:

$$\bar{v}_i(\theta_i) = E_{\theta_{-i}}[v_i(y(\theta_i, \theta_{-i}))] \tag{3.13}$$

$$= E_{\theta_{-i}}\left[\sum_{j=1}^{m} y_{ij}(\theta_i, \theta_{-i})\alpha_j\right]. \tag{3.14}$$

Now if we assume that $J_i(\cdot)$ is nondecreasing in $\theta_i$, it is easy to see that the above solution $y_{ij}(\cdot)$, given by (3.12), will be nondecreasing in $\theta_i$, which in turn implies, by looking at expression (3.14), that $\bar{v}_i(\cdot)$ is nondecreasing in $\theta_i$. Thus, the solution to this relaxed problem actually satisfies constraint (i) under the assumption that $J_i(\cdot)$ is nondecreasing. Assuming that $J_i(\cdot)$ is nondecreasing, the solution given by (3.12) appears to be the solution of the optimal mechanism design problem for sponsored search auction. Note that in Equation (3.12), we have written the allocation rule $J_i(\cdot)$ as a function of actual type profile $\theta$ of the advertisers rather than the bid vector $b$. This is because in an OPT mechanism, each advertiser bids his true type, and we have $b_i = \theta_i \; \forall i = 1,\dots,n$.

The condition that $J_i(\cdot)$ is nondecreasing in $\theta_i$ is met by most distribution functions such as uniform and exponential. In the rest of this chapter, we will work with the assumption that for every advertiser $i$, $J_i(\cdot)$ is nondecreasing in $\theta_i$.

It is interesting to note that in the above allocation rule, the condition $J_i(.) > 0$ for qualifying an advertiser to display his/her ad can be expressed more conveniently in the form of reserve price in the following sense. For each advertiser $i$, we first compute the value of $\theta_i$ at which we have $J_i(\theta_i) = 0$. This value we call the reserve price of advertiser $i$. Now the allocation rule says that we first discard all those advertisers whose bid is less than their corresponding reserve price. Among the remaining advertisers, we allocate the advertisers in decreasing order of their $J_i(\theta_i)$ values. Further, if the advertisers are symmetric then the reserve price will be the same for all the advertisers, and moreover if $J_i(.)$ is nondecreasing in $\theta_i$ then among the qualified advertisers, the allocation rule will be the same as the GFP allocation rule. This interpretation of the allocation rule is the same as the allocation rule in [18] under the parallel case. This observation leads to the following proposition.

**Proposition 3.4.** *If the advertisers have nonidentical distribution functions* $\Phi_i(\cdot)$ *then the advertiser who has the $k^{th}$ largest value of $J_i(b_i)$ is* not *necessarily the advertiser who has bid the $k^{th}$ highest amount. Thus the OPT mechanism need not be allocatively efficient and, therefore, need not be ex post efficient.*

**Proposition 3.5.** *If the advertisers are symmetric in following sense*

- $\Theta_1 = \ldots = \Theta_n = \Theta$
- $\Phi_1(\cdot) = \ldots = \Phi_n(\cdot) = \Phi(\cdot)$

*and for every advertiser i, we have $J_i(\cdot) > 0$ and $J_i(\cdot)$ is nondecreasing, then*

- $J_i(\cdot) = \ldots = J_n(\cdot) = J(\cdot)$.
- *The rank of an advertiser in the decreasing order sequence of $J_1(b_1), \ldots, J_n(b_n)$ is precisely the same as the rank of the same advertiser in the decreasing order sequence of $b_1, \ldots, b_n$.*
- *For a given bid vector b, the OPT mechanism results in the same allocation as suggested by the GFP, the GSP, and the VCG mechanisms.*
- *The OPT mechanism is allocatively efficient.*

### 3.7.2 Payment Rule

Now we compute the payment rule. Again, following Myerson's line of attack, the optimal expected payment rule $t_i(\cdot)$ must be chosen in such a way that it satisfies

$$\bar{t}_i(\theta_i) = E_{\theta_{-i}}[t_i(\theta_i, \theta_{-i})] = \theta_i \bar{v}_i(\theta_i) - U_i(\theta_i) = \theta_i \bar{v}_i(\theta_i) - \int_{\underline{\theta_i}}^{\theta_i} \bar{v}_i(s) ds. \quad (3.15)$$

Looking at the above formula, we can say that if the payment rule $t_i(\cdot)$ satisfies the following formula (3.16) then it would also satisfy formula (3.15).

$$t_i(\theta_i, \theta_{-i}) = \theta_i v_i(y(\theta_i, \theta_{-i})) - \int_{\underline{\theta_i}}^{\theta_i} v_i(s, \theta_{-i}) ds \ \forall \ \theta \in \Theta \quad (3.16)$$

The above formula can be rewritten in a more intuitive way, for which we need to define the following quantities for any vector $\theta_{-i}$:

**Case 1** $(m < n)$:

$$z_{i1}(\theta_{-i}) = \inf\left\{\theta_i | J_i(\theta_i) > 0 \text{ and } J_i(\theta_i) \geq J_{-i}^{(1)}\right\}$$

$$z_{i2}(\theta_{-i}) = \inf\left\{\theta_i | J_i(\theta_i) > 0 \text{ and } J_{-i}^{(1)} > J_i(\theta_i) \geq J_{-i}^{(2)}\right\}$$

$$\vdots = \vdots$$

$$z_{im}(\theta_{-i}) = \inf\left\{\theta_i | J_i(\theta_i) > 0 \text{ and } J_{-i}^{(m-1)} > J_i(\theta_i) \geq J_{-i}^{(m)}\right\}.$$

**Case 2** $(n \leq m)$:

$$z_{i1}(\theta_{-i}) = \inf\left\{\theta_i | J_i(\theta_i) > 0 \text{ and } J_i(\theta_i) \geq J_{-i}^{(1)}\right\}$$

$$z_{i2}(\theta_{-i}) = \inf\left\{\theta_i | J_i(\theta_i) > 0 \text{ and } J_{-i}^{(1)} > J_i(\theta_i) \geq J_{-i}^{(2)}\right\}$$

$$\vdots = \vdots$$

$$z_{in}(\theta_{-i}) = \inf\left\{\theta_i | J_i(\theta_i) > 0 \text{ and } J_{-i}^{(n-1)} > J_i(\theta_i)\right\}.$$

In the above definitions, $J_{-i}^{(k)}$ is the $k^{th}$ highest value among the following $(n-1)$ values

$$J_1(\theta_1), \ldots, J_{i-1}(\theta_{i-1}), J_{i+1}(\theta_{i+1}), \ldots, J_n(\theta_n).$$

The quantity $z_{ik}(\theta_{-i})$ is the infimum of all the bids for advertisers $i$ that can make him win the $k^{th}$ slot against the bid vector $\theta_{-i}$ from the other advertisers. In view of the above definitions, we can write

**Case 1 ($m < n$):**

$$v_i(y(\theta_i, \theta_{-i})) = \begin{cases} \alpha_1 & : \quad \text{if } \theta_i \geq z_{i1}(\theta_{-i}) \\ \alpha_2 & : \quad \text{if } z_{i1}(\theta_{-i}) > \theta_i \geq z_{i2}(\theta_{-i}) \\ \vdots & : \quad \vdots \\ \alpha_m & : \quad \text{if } z_{i(m-1)}(\theta_{-i}) > \theta_i \geq z_{im}(\theta_{-i}) \\ 0 & : \quad \text{if } z_{im}(\theta_{-i}) > \theta_i. \end{cases}$$

**Case 2 ($n \leq m$):**

$$v_i(y(\theta_i, \theta_{-i})) = \begin{cases} \alpha_1 & : \quad \text{if } \theta_i \geq z_{i1}(\theta_{-i}) \\ \alpha_2 & : \quad \text{if } z_{i1}(\theta_{-i}) > \theta_i \geq z_{i2}(\theta_{-i}) \\ \vdots & : \quad \vdots \\ \alpha_n & : \quad \text{if } z_{i(n-1)}(\theta_{-i}) > \theta_i. \end{cases}$$

This gives us the following expressions for $\int_{\underline{\theta_i}}^{\theta_i} v_i(s, \theta_{-i}) ds$. In these expressions, $r$ is the position of advertiser $i$'s ad.

**Case 1 ($m < n$):**

$$\int_{\underline{\theta_i}}^{\theta_i} v_i(y(s, \theta_{-i})) ds = \begin{cases} \alpha_r(\theta_i - z_{ir}(\theta_{-i})) + \sum\limits_{j=(r+1)}^{m} \alpha_j(z_{i(j-1)}(\theta_{-i}) - z_{ij}(\theta_{-i})) & : \quad \text{if } 1 \leq r \leq (m-1) \\ \alpha_m(\theta_i - z_{im}(\theta_{-i})) & : \quad \text{if } r = m \\ 0 & : \quad \text{otherwise.} \end{cases}$$

**Case 2 ($n \leq m$):**

$$\int_{\underline{\theta_i}}^{\theta_i} v_i(y(s, \theta_{-i})) ds = \begin{cases} \alpha_r(\theta_i - z_{ir}(\theta_{-i})) + \sum\limits_{j=(r+1)}^{n} \alpha_j(z_{i(j-1)}(\theta_{-i}) - z_{ij}(\theta_{-i})) & : \quad \text{if } 1 \leq r \leq (n-1) \\ \alpha_n(\theta_i - z_{in}(\theta_{-i})) & : \quad \text{if } r = n. \end{cases}$$

Substituting the above value for $\int_{\underline{\theta_i}}^{\theta_i} v_i(y(s, \theta_{-i}))ds$ in formula (3.16), we get

**Case 1** $(m < n)$:

$$p_i(\theta_i, \theta_{-i}) = \frac{1}{\alpha_r}t_i(\theta_i, \theta_{-i}) = \begin{cases} \frac{\alpha_m}{\alpha_r}z_{im}(\theta_{-i}) + \frac{1}{\alpha_r}\sum_{j=r}^{m-1}\beta_j z_{ij}(\theta_{-i}) : \text{if } 1 \leq r \leq (m-1) \\ z_{im}(\theta_{-i}) \qquad\qquad\qquad\qquad\quad : \text{if } r = m \\ 0 \qquad\qquad\qquad\qquad\qquad\qquad\quad : \text{otherwise.} \end{cases} \quad (3.17)$$

**Case 2** $(n \leq m)$:

$$p_i(\theta_i, \theta_{-i}) = \frac{1}{\alpha_r}t_i(\theta_i, \theta_{-i}) = \begin{cases} \frac{\alpha_n}{\alpha_r}z_{in}(\theta_{-i}) + \frac{1}{\alpha_r}\sum_{j=r}^{n-1}\beta_j z_{ij}(\theta_{-i}) : \text{if } 1 \leq r \leq (n-1) \\ z_{in}(\theta_{-i}) \qquad\qquad\qquad\qquad\quad : \text{if } r = n \\ 0 \qquad\qquad\qquad\qquad\qquad\qquad\quad : \text{otherwise.} \end{cases} \quad (3.18)$$

The above relations say that an advertiser $i$ must pay only when his ad receives a click, and he pays the amount equal to $p_i(\theta)$. Note that in above relations, we have expressed the payment rule $p_i(\cdot)$ as a function of actual type profile $\theta$ of the advertisers rather than the bid vector $b$. This is because in an OPT mechanism, each advertiser bids his true type, and we have $b_i = \theta_i \; \forall i = 1, \ldots, n$.

Thus, we can say that Equation (3.12) describes the allocation rule for the OPT mechanism and Equations (3.17) and (3.18) describe the the payment rule for the OPT mechanism.

In what follows, we discuss an important special case of the OPT mechanism when the advertisers are symmetric.

### 3.7.3 OPT Mechanism and Symmetric Advertisers

Let us assume that advertisers are symmetric in the following sense:

- $\Theta_1 = \ldots = \Theta_n = \Theta = [L, U]$.
- $\Phi_1(\cdot) = \ldots = \Phi_n(\cdot) = \Phi(\cdot)$.

Also, we assume that

- $J(\cdot)$ is non-decreasing over the interval $[L, U]$.
- $J(x) > 0 \; \forall x \in [L, U]$.

Note that if $J(L) > 0$ then we must have $L > 0$.

Proposition 3.5 shows that if the advertisers are symmetric then the allocation rule under the OPT mechanism is the same as the GFP, the GSP, and the VCG mechanisms. Coming to the payment rule, it is easy to verify that if advertiser $i$ is allocated the slot $r$ for the bid vector $(\theta_i, \theta_{-i})$ then we should have

**Case 1 $(m < n)$:**

$$z_{ij}(\theta_{-i}) = \begin{cases} \theta^{(j)} & : \text{if } 1 \le j \le (r-1) \\ \theta^{(j+1)} & : \text{if } r \le j \le m. \end{cases} \tag{3.19}$$

**Case 2 $(n \le m)$:**

$$z_{ij}(\theta_{-i}) = \begin{cases} \theta^{(j)} & : \text{if } 1 \le j \le (r-1) \\ \theta^{(j+1)} & : \text{if } r \le j \le (n-1) \\ L & : \text{if } j = n. \end{cases} \tag{3.20}$$

If we substitute Equations (3.19) and (3.20) into Equations (3.17) and (3.18), then we get the following payment rule for the OPT mechanism when the advertisers are symmetric:

**Case 1 $(m < n)$:**

$$p_i(\theta_i, \theta_{-i}) = \frac{1}{\alpha_r} t_i(\theta_i, \theta_{-i}) = \begin{cases} \frac{\alpha_m}{\alpha_r} \theta^{(m+1)} + \frac{1}{\alpha_r} \sum_{j=r}^{m-1} \beta_j \theta^{(j+1)} & : \text{if } 1 \le r \le (m-1) \\ \theta^{(m+1)} & : \text{if } r = m \\ 0 & : \text{otherwise.} \end{cases} \tag{3.21}$$

**Case 2 $(n \le m)$:**

$$p_i(\theta_i, \theta_{-i}) = \frac{1}{\alpha_r} t_i(\theta_i, \theta_{-i}) = \begin{cases} \frac{\alpha_n}{\alpha_r} L + \frac{1}{\alpha_r} \sum_{j=r}^{n-1} \beta_j \theta^{(j+1)} & : \text{if } 1 \le r \le (n-1) \\ L & : \text{if } r = n \\ 0 & : \text{otherwise.} \end{cases} \tag{3.22}$$

Compare the above equations with the payment rule of the VCG mechanism given by Equations (3.9) and (3.10). This comparison leads to the following proposition:

**Proposition 3.6.** *If the advertisers are symmetric in following sense*

- $\Theta_1 = \ldots = \Theta_n = \Theta = [L, U]$
- $\Phi_1(\cdot) = \ldots = \Phi_n(\cdot) = \Phi(\cdot)$

*and for every advertiser i, we have $J_i(\cdot) > 0$ and $J_i(\cdot)$ is non-decreasing over the interval $[L,U]$, then*

- *the payment rule for Case 1 coincides with the corresponding payment rule in the VCG mechanism,*
- *and the payment rule for the Case 2 differs from the corresponding payment rule of the VCG mechanism just by a constant amount L.*

Note that $L$ cannot be zero because of the assumption that $J(L) > 0$.

## 3.8 Comparison of GSP, VCG, and OPT Mechanisms

We now compare the mechanisms GSP, VCG, and OPT along four dimensions:

1. Incentive compatibility
2. Expected revenue earned by the search engine
3. Individual rationality
4. Computational complexity

For the purpose of comparison we will make the following assumptions, which include the symmetry of advertisers:

- $\Theta_1 = \ldots = \Theta_n = \Theta = [L, U]$.
- $\Phi_1(\cdot) = \ldots = \Phi_n(\cdot) = \Phi(\cdot)$.
- $J(\cdot)$ is non-decreasing over the interval $[L, U]$.
- $J(x) > 0 \ \forall \, x \in [L, U]$.

### 3.8.1 Incentive Compatibility

Note that by design itself, the OPT mechanism is Bayesian incentive compatible, and the VCG mechanism is dominant strategy incentive compatible. In this section, we show that the GSP mechanism is not Bayesian incentive compatible. We will present two approaches to prove this.

#### 3.8.1.1 Approach 1 to Show That the GSP is not Bayesian IC

In this approach we will invoke Myerson's Theorem [37] for the Bayesian incentive compatibility of the social choice functions in a linear environment. According to this theorem, an allocation and payment rule $f(\cdot) = (y_{ij}(\cdot), t_i(\cdot))_{i \in N, j \in M}$ that is used by the GSP mechanism is Bayesian incentive compatible iff

(i) $\bar{v}_i(\cdot)$ is nondecreasing $\forall i \in N$,

(ii) $U_i(\theta_i) = U_i(\underline{\theta_i}) + \int_{\underline{\theta_i}}^{\theta_i} \bar{v}_i(s) ds \ \forall i \in N, \ \forall \, \theta_i \in \Theta_i$

where $\bar{v}_i(b_i) = E_{\theta_{-i}}\left[\sum_{j=1}^m y_{ij}(b_i, \theta_{-i}) \alpha_j\right]$ represents the probability that advertiser $i$ will receive a user click if he bids $b_i$, and all the advertisers $j \neq i$ truthfully bid their types. Note that by the definition of the allocation and payment rule for the GSP mechanism, it is easy to see that $\sum_{j=1}^m y_{ij}(b_i, \theta_{-i})$ is nondecreasing in $b_i$. This would essentially imply that $\bar{v}_i(\cdot)$ is nondecreasing. This implies that the allocation and the payment rule for the GSP mechanism satisfies the first condition of the above-mentioned two conditions. In order to investigate whether or not it satisfies the second condition, we need to compute the expressions for $\bar{v}_i(\theta_i)$ and $U_i(\theta_i)$. We consider two cases:

**Case 1** ($m < n$): Under this case, the probability that advertiser $i$, when he bids his true type $\theta_i$ and all the advertisers $j \neq i$ also truthfully bid their types, will be allocated a position $j$ ($j = 1, 2, \ldots, m$) is given by

$$\int_L^{\theta_i} \binom{n-1}{j-1} \binom{n-j}{1} (1 - \Phi(\theta_i))^{j-1} \Phi(x)^{n-j-1} \phi(x)dx.$$

Remember that while writing the above expression, we have invoked the assumption that advertisers are symmetric. It is now easy to see that under this case

$$\bar{v}_i(\theta_i) = \int_L^{\theta_i} \left( \sum_{j=1}^m \left[ \alpha_j j \binom{n-1}{j} (1 - \Phi(\theta_i))^{j-1} \Phi(x)^{n-j-1} \right] \right) \phi(x)dx \quad (3.23)$$

$$= \sum_{j=1}^m \left[ \alpha_j \binom{n-1}{j-1} (1 - \Phi(\theta_i))^{j-1} \Phi(\theta_i)^{n-j} \right] \quad (3.24)$$

$$\bar{t}_i(\theta_i) = \int_L^{\theta_i} \left( \sum_{j=1}^m \left[ \alpha_j j \binom{n-1}{j} (1 - \Phi(\theta_i))^{j-1} \Phi(x)^{n-j-1} \right] \right) x\phi(x)dx \quad (3.25)$$

$$= \theta_i \bar{v}_i(\theta_i) - \int_L^{\theta_i} \left( \sum_{j=1}^m \left[ \alpha_j \binom{n-1}{j-1} (1 - \Phi(\theta_i))^{j-1} \Phi(x)^{n-j} \right] \right) dx.$$

$$(3.26)$$

We already know that

$$U_i(\theta_i) = \theta_i \bar{v}_i(\theta_i) - \bar{t}_i(\theta_i). \quad (3.27)$$

Substituting Equation (3.26) in Equation (3.27), we get the following relation:

$$U_i(\theta_i) = \int_L^{\theta_i} \left( \sum_{j=1}^m \left[ \alpha_j \binom{n-1}{j-1} (1 - \Phi(\theta_i))^{j-1} \Phi(x)^{n-j} \right] \right) dx. \quad (3.28)$$

From Equation (3.28), we can conclude that due to symmetry of advertisers, we have

$$U_i(\underline{\theta_i}) = U_i(L) = 0. \quad (3.29)$$

Making use of Equations (3.24), (3.28), and (3.29), we can conclude that

$$U_i(\theta_i) < U_i(\underline{\theta_i}) + \int_{\underline{\theta_i}}^{\theta_i} \bar{v}_i(s)ds \ \forall i \in N, \theta_i \in [L, U].$$

This would imply that the allocation and payment rule for the GSP mechanism does not satisfy the second condition, and hence it is not Bayesian incentive compatible when $m < n$.

**Case 2** ($n \leq m$)**:** Under this case, the probability that an advertiser $i$, when he bids his true type $\theta_i$ and all the advertisers $j \neq i$ also truthfully bid their types, will be allocated a position $j$ ($j = 1, 2, \ldots, n-1$) is given by

$$\int_L^{\theta_i} \binom{n-1}{j-1} \binom{n-j}{1} (1 - \Phi(\theta_i))^{j-1} \Phi(x)^{n-j-1} \phi(x)dx,$$

and the probability that the advertiser will be allocated the $n^{th}$ position is given by

$$(1 - \Phi(\theta_i))^{n-1}.$$

It is easy to see under this case that

$$\bar{v}_i(\theta_i) = \alpha_n(1 - \Phi(\theta_i))^{n-1} + \int_L^{\theta_i} \left( \sum_{j=1}^{n-1} \left[ \alpha_j\, j \binom{n-1}{j} (1 - \Phi(\theta_i))^{j-1} \Phi(x)^{n-j-1} \right] \right) \phi(x)dx$$

$$= \sum_{j=1}^{n} \left[ \alpha_j \binom{n-1}{j-1} (1 - \Phi(\theta_i))^{j-1} \Phi(\theta_i)^{n-j} \right] \tag{3.30}$$

$$\bar{t}_i(\theta_i) = \int_L^{\theta_i} \left( \sum_{j=1}^{n-1} \left[ \alpha_j\, j \binom{n-1}{j} (1 - \Phi(\theta_i))^{j-1} \Phi(x)^{n-j-1} \right] \right) x\phi(x)dx \tag{3.31}$$

$$= \theta_i \left[ \bar{v}_i(\theta_i) - \alpha_n(1 - \Phi(\theta_i))^{n-1} \right]$$

$$\qquad - \int_L^{\theta_i} \left( \sum_{j=1}^{n-1} \left[ \alpha_j \binom{n-1}{j-1} (1 - \Phi(\theta_i))^{j-1} \Phi(x)^{n-j} \right] \right) dx. \tag{3.32}$$

Substituting Equation (3.32) in Equation (3.27), we get the following relation:

$$U_i(\theta_i) = \int_L^{\theta_i} \left( \sum_{j=1}^{n-1} \left[ \alpha_j \binom{n-1}{j-1} (1 - \Phi(\theta_i))^{j-1} \Phi(x)^{n-j} \right] \right) dx + \theta_i \alpha_n (1 - \Phi(\theta_i))^{n-1}$$

$$= \int_L^{\theta_i} \left( \sum_{j=1}^{n} \left[ \alpha_j \binom{n-1}{j-1} (1 - \Phi(\theta_i))^{j-1} \Phi(x)^{n-j} \right] \right) dx + L\alpha_n (1 - \Phi(\theta_i))^{n-1}. \tag{3.33}$$

We have already seen that $U_i(\underline{\theta_i}) = U_i(L) = 0$. Therefore, we can just compare $U_i(\theta_i)$ with $\int_{\underline{\theta_i}}^{\theta_i} \bar{v}_i(s)ds$. Making use of Equations (3.30) and (3.33), we can say that when

$\theta_i = U$, we must have

$$U_i(U) < U_i(\underline{\theta_i}) + \int\limits_{\underline{\theta_i}}^{U} \bar{v}_i(s)ds.$$

This would again imply that the allocation and payment rule for the GSP mechanism does not satisfy the second condition, and hence it is not Bayesian incentive compatible for $n \leq m$ also.

### 3.8.1.2 Approach 2 to Show That GSP is not Bayesian IC

This approach is quite similar to the one used by McAfee and McMillan [34] to compute the equilibrium bidding strategy of the buyers during the auction of a single indivisible good. Once again, we will assume that the advertisers are symmetric.

Consider an advertiser $i$, whose actual type is $\theta_i$. He conjectures that the other advertisers are following a bidding strategy $s(\cdot)$: That is, he predicts that any other advertiser $j$ will bid an amount $s(\theta_j)$ if his type is $\theta_j$ (although advertiser $i$ does not know this type). Assume that

1. $L \leq s(\theta_j) \leq \theta_j \ \forall \ \theta_j \in [L, U]$,
2. $s(\cdot)$ is a monotonically increasing function in $\theta_j$.

What is advertiser $i$'s best bid? Advertiser $i$ chooses his bid $b_i$ to maximize his expected utility, which in this case is given by

**Case 1 ($m < n$):**

$$\pi_i(\theta_i, b_i) = \int\limits_{L}^{\xi} \left( \sum_{j=1}^{m} \left[ \alpha_j \, j \binom{n-1}{j} [\bar{\Phi}(\xi)]^{j-1} \, \Phi(x)^{n-j-1} \right] \right) (\theta_i - s(x))\phi(x)dx.$$

$$(3.34)$$

**Case 2 ($n \leq m$):**

$$\pi_i(\theta_i, b_i) = \alpha_n \theta_i [\bar{\Phi}(\xi)]^{n-1} + \int\limits_{L}^{\xi} \left( \sum_{j=1}^{n-1} \left[ \alpha_j \, j \binom{n-1}{j} [\bar{\Phi}(\xi)]^{j-1} \Phi(x)^{n-j-1} \right] \right) (\theta_i - s(x))\phi(x)dx$$

$$(3.35)$$

where

- $\xi = s^{-1}(b_i)$,
- $\bar{\Phi}(\cdot) = 1 - \Phi(\cdot)$,
- The quantity

$$\int\limits_{L}^{\xi} \alpha_j \, j \binom{n-1}{j} [\bar{\Phi}(\xi)]^{j-1} [\Phi(x)]^{n-j-1} \phi(x)dx$$

gives the probability that advertiser $i$ will be allocated to slot $j$ if he bids $b_i$ and all the other advertisers bid according to the strategy $s(\cdot)$.

Thus, advertiser $i$ chooses bid $b_i$ such that

$$\frac{\partial \pi(\theta_i, b_i)}{\partial b_i} = 0. \tag{3.36}$$

Note that due to the Envelope Theorem, we can write

$$\frac{d\pi_i(\theta_i, b_i)}{d\theta_i} = \frac{\partial \pi_i(\theta_i, b_i)}{\partial b_i} \frac{db_i}{d\theta_i} + \frac{\partial \pi_i(\theta_i, b_i)}{\partial \theta_i}. \tag{3.37}$$

Thus, by substituting Equation (3.36) in Equation(3.37), we get the following condition that an optimally chosen bid $b_i$ must satisfy:

$$\frac{d\pi_i(\theta_i, b_i)}{d\theta_i} = \frac{\partial \pi_i(\theta_i, b_i)}{\partial \theta_i}. \tag{3.38}$$

By differentiating (3.34) and (3.35), we get

$$\frac{d\pi_i(\theta_i, b_i)}{d\theta_i} = \begin{cases} \sum\limits_{j=1}^{m} \left[ \alpha_j \binom{n-1}{j-1} [\overline{\Phi}(\xi)]^{j-1} \Phi(\xi)^{n-j} \right] &: \text{ if } m < n \\ \sum\limits_{j=1}^{n} \left[ \alpha_j \binom{n-1}{j-1} [\overline{\Phi}(\xi)]^{j-1} \Phi(\xi)^{n-j} \right] &: \text{ if } n \leq m. \end{cases} \tag{3.39}$$

So far, we have examined advertiser $i$'s best response to an arbitrary bidding strategy $s(\cdot)$ being used by his rivals. Now we impose the Nash requirement: The rivals' use of the bidding strategy $s(\cdot)$ must be consistent with the rivals themselves acting rationally. Together with an assumption of symmetry (any two advertisers with the same type will submit the same bid), this implies that advertiser $i$'s bid $b_i$, satisfying (3.38), must be the bid implied by the decision rule $s(\cdot)$ — in other words, at a Nash equilibrium, $b_i = s(\theta_i)$ or equivalently $\xi = \theta_i$. When we substitute this Nash condition into (3.39), we obtain the following equations

$$\frac{d\pi_i(\theta_i)}{d\theta_i} = \begin{cases} \sum\limits_{j=1}^{m} \left[ \alpha_j \binom{n-1}{j-1} [\overline{\Phi}(\theta_i)]^{j-1} [\Phi(\theta_i)]^{n-j} \right] &: \text{ if } m < n \\ \sum\limits_{j=1}^{n} \left[ \alpha_j \binom{n-1}{j-1} [\overline{\Phi}(\theta_i)]^{j-1} [\Phi(\theta_i)]^{n-j} \right] &: \text{ if } n \leq m. \end{cases} \tag{3.40}$$

We solve the above differential equations for $\pi_i$ simply by integrating in conjunction with the boundary condition $s(L) = L$. This results in the following expressions for $\pi_i$.

$$\pi_i(\theta_i) = \begin{cases} \int\limits_{L}^{\theta_i} \sum\limits_{j=1}^{m} \left[ \alpha_j \binom{n-1}{j-1} [\overline{\Phi}(x)]^{j-1} [\Phi(x)]^{n-j} \right] dx &: \text{ if } m < n \\ \alpha_n L + \int\limits_{L}^{\theta_i} \sum\limits_{j=1}^{n} \left[ \alpha_j \binom{n-1}{j-1} [\overline{\Phi}(x)]^{j-1} [\Phi(x)]^{n-j} \right] dx &: \text{ if } n \leq m. \end{cases} \tag{3.41}$$

We now use the definition of $\pi_i$ (Equations (3.34) and (3.35)) and Nash condition $s_i(\theta_i) = b_i$ or equivalently $\xi = \theta_i$ to obtain the following relations:

**Case 1** $(m < n)$:

$$\int_L^{\theta_i} \sum_{j=1}^{m} \left[ \alpha_j \binom{n-1}{j-1} [\overline{\Phi}(x)]^{j-1} [\Phi(x)]^{n-j} \right] dx =$$

$$\int_L^{\theta_i} \left( \sum_{j=1}^{m} \left[ \alpha_j \, j \binom{n-1}{j} [\overline{\Phi}(\theta_i)]^{j-1} [\Phi(x)]^{n-j-1} \right] \right) (\theta_i - s(x))\phi(x)dx.$$

**Case 2** $(n \leq m)$:

$$\alpha_n L + \int_L^{\theta_i} \sum_{j=1}^{n} \left[ \alpha_j \binom{n-1}{j-1} [\overline{\Phi}(x)]^{j-1} [\Phi(x)]^{n-j} \right] dx = \alpha_n \theta_i [\overline{\Phi}(\theta_i)]^{n-1} +$$

$$\int_L^{\theta_i} \left( \sum_{j=1}^{n-1} \left[ \alpha_j \, j \binom{n-1}{j} [\overline{\Phi}(\theta_i)]^{j-1} [\Phi(x)]^{n-j-1} \right] \right) (\theta_i - s(x))\phi(x)dx.$$

Differentiating the above equations with respect to $\theta_i$, we get each advertiser's bidding strategy $s(\cdot)$ as a solution of the following integral equations:

$$s(\theta_i) = \begin{cases} \theta_i - \frac{1}{g(\theta_i,m)} \int_L^{\theta_i} f(x,\theta_i,m)s'(x)dx & : \quad \text{if } m < n \\ \theta_i - \frac{1}{g(\theta_i,(n-1))} \int_L^{\theta_i} f(x,\theta_i,(n-1))s'(x)dx & : \quad \text{if } n \leq m \end{cases} \quad (3.42)$$

where

$$f(x,\theta_i,k) = \sum_{j=1}^{k} \alpha_j (j-1) \binom{n-1}{j-1} [\overline{\Phi}(\theta_i)]^{j-2} [\Phi(x)]^{n-j}$$

$$g(\theta_i,k) = \sum_{j=1}^{k-1} \left[ \beta_j \, j \binom{n-1}{j} [\overline{\Phi}(\theta_i)]^{j-1} [\Phi(\theta_i)]^{n-j-1} \right] + k\alpha_k \binom{n-1}{k} [\overline{\Phi}(\theta_i)]^{k-1} [\Phi(\theta_i)]^{n-k-1}.$$

It is easy to see from the above equations that truth-telling is not an equilibrium strategy of the advertisers, and, therefore, the allocation and payment rule for GSP mechanism is not Bayesian incentive compatible.

Observe that if $m = 1$ and $1 < n$, then this is precisely the scenario of auctioning a single indivisible good with $n$ bidders. For this scenario, the allocation and payment rules under GSP coincides precisely with the allocation and payment rules of a classical *Second Price (Vickrey) Auction*. It is a well known fact that in a Vickrey auction, truth-telling is a weakly dominant strategy equilibrium. This fact can be realized by substituting $m = 1$ in the above equation for the case $m < n$, which results in $s(x) = x$.

### 3.8.2 Revenue Equivalence of Mechanisms

Now we show that under some reasonable set of assumptions, the mechanisms we have discussed for sponsored search will produce the same expected revenue to the search engine. In what follows, we state and prove the *Revenue Equivalence Theorem* for sponsored search auctions. The Revenue Equivalence Theorem is a key result in the literature of single object auctions and has already been discussed in Section 2.19. Different variants of this theorem are discussed in [12, 34, 36]. The Revenue Equivalence Theorem presented below extends the classical Revenue Equivalence Theorem for single indivisible object auction setting to the sponsored search auction setting.

**Theorem 3.1.** *[A Revenue Equivalence of Mechanisms for Sponsored Search Auctions] Consider a sponsored search auction setting, in which*

*1. The advertisers are risk neutral,*
*2. The advertisers are symmetric, i.e.,*

- $\Theta_1 = \ldots = \Theta_n = \Theta = [L, U]$,
- $\Phi_1(\cdot) = \ldots = \Phi_n(\cdot) = \Phi(\cdot)$,

*3. For each advertiser i, we have $\phi_i(\cdot) > 0$; and*
*4. The advertisers draw their types independently.*

*Consider two different auction mechanisms, each having a symmetric and increasing Bayesian Nash equilibrium, such that*

*1. For each possible realization of $(\theta_1, \ldots, \theta_n)$, every advertiser i has an identical probability of getting slot j in the two mechanisms; and*
*2. Every advertiser i has the same expected utility level in the two mechanisms when his type $\theta_i$ is at its lowest possible level, i.e., L.*

*Then these equilibria of the two mechanisms generate the same expected revenue for the search engine against every search query.*

**Proof:** By the revelation principle, we know that any given indirect mechanism can be converted into a Bayesian incentive compatible direct revelation mechanism that results in the same outcome as the original mechanism for every type profile $\theta$ of the advertisers. This implies that the expected revenue earned by the search engine under both of these mechanisms will be the same. Therefore, we can establish the above theorem by showing that if two Bayesian incentive compatible direct revelation mechanisms have the same allocation rule $(y_{ij}(\theta))_{i \in N, j \in M}$ and the same value of $(U_i(L))_{i \in N}$ then they generate the same expected revenue for the search engine.

To show this, we derive an expression for the search engine's expected revenue from an arbitrary Bayesian incentive compatible direct revelation mechanism. Note, first, that the search engine's expected revenue from an arbitrary Bayesian incentive compatible direct revelation mechanism, under the assumption of risk neutral, symmetric, and independent advertisers, is equal to

$$R = n \int_{\theta_i=L}^{U} \bar{t}_i(\theta_i)\phi(\theta_i)d\theta_i$$

$$= n \int_{\theta_i=L}^{U} \left(\theta_i\bar{v}_i(\theta_i) - U_i(\theta_i)\right)\phi(\theta_i)d\theta_i. \tag{3.43}$$

We have already seen that according to Myerson's Theorem [37], a direct revelation mechanism is Bayesian incentive compatible iff

(i) $\bar{v}_i(\cdot)$ is nondecreasing $\forall i \in N$,

(ii) $U_i(\theta_i) = U_i(\underline{\theta_i}) + \int_{\underline{\theta_i}}^{\theta_i} \bar{v}_i(s)ds \ \forall i \in N, \ \forall \theta_i \in \Theta_i.$

Therefore, substituting for $U_i(\theta_i)$ Equation (3.43), we get

$$R = n \int_{\theta_i=L}^{U} \left(\bar{v}_i(\theta_i)\theta_i - U_i(\underline{\theta_i}) - \int_{s=L}^{\theta_i} \bar{v}_i(s)ds\right)\phi_i(\theta_i)d\theta_i.$$

Integrating by parts implies that

$$R = \int_{L}^{U} \cdots \int_{L}^{U} \left[\sum_{i=1}^{n} v_i(y(\theta_i,\theta_{-i}))J_i(\theta_i)\right]\left[\prod_{i=1}^{n}\phi_i(\theta_i)\right]d\theta_n\ldots d\theta_1 - \sum_{i=1}^{n}U_i(\underline{\theta_i})$$

$$= \int_{L}^{U} \cdots \int_{L}^{U} \left[\sum_{i=1}^{n}\left(\sum_{j=1}^{m}y_{ij}(\theta_i,\theta_{-i})\right)J_i(\theta_i)\right]\left[\prod_{i=1}^{n}\phi_i(\theta_i)\right]d\theta_n\ldots d\theta_1 - \sum_{i=1}^{n}U_i(\underline{\theta_i})$$

$$\tag{3.44}$$

where

$$J_i(\theta_i) = \theta_i - \frac{1-\Phi_i(\theta_i)}{\phi_i(\theta_i)}.$$

By inspection of (3.44), we see that any two Bayesian incentive compatible direct revelation mechanisms that generate the same allocation functions $(y_{ij}(\cdot))_{i\in N, j\in M}$ and the same vales of $(U_1(L),\ldots,U_n(L))$ generate the same expected revenue for the search engine.

*Q.E.D.*

We use the above revenue equivalence theorem to derive the following result stating the revenue comparison of GSP, VCG, and OPT mechanisms.

**Proposition 3.7.** *[Revenue Comparison of GSP, VCG, and OPT Mechanisms] Consider a sponsored search auction setting, in which*

*1. The advertisers are risk neutral,*
*2. The advertisers are symmetric, i.e.,*

- $\Theta_1 = \ldots = \Theta_n = \Theta = [L,U],$
- $\Phi_1(\cdot) = \ldots = \Phi_n(\cdot) = \Phi(\cdot),$

*3. For each advertiser i, we have $\phi_i(\cdot) > 0$; and*
*4. The advertisers draw their types independently,*
*5. For each advertiser i, we have $J_i(\cdot) > 0$ and $J_i(\cdot)$ is non-decreasing function.*

*If $R_{GSP}, R_{VCG}$, and $R_{OPT}$ are the expected revenue earned by the search engine, against every search query received by the search engine, under the GSP, the VCG, and the OPT mechanisms, respectively, then*

$$R_{GSP} = R_{VCG} = R_{OPT} \quad : \quad if\ m < n$$
$$R_{VCG} \leq R_{GSP} \leq R_{OPT} \quad : \quad if\ n \leq m.$$

**Proof:** Recall Proposition 3.5, which says that under the assumptions stated above, the VCG and the OPT mechanisms result in the same allocation for any given bid vector $b = (b_1, \dots, b_n)$. Also, recall that the VCG and the OPT mechanisms are incentive compatible, which implies that the advertisers bid their true types under both of these two mechanisms. Therefore, we can conclude that under the assumptions stated above, the VCG and the OPT mechanisms result in the same allocation for any given type profile $\theta = (\theta_1, \dots, \theta_n)$. Note that this result holds irrespective of whether $m < n$ or $n \leq m$. Now coming to the GSP mechanism, the Equation (3.42) shows that the GSP mechanism has a symmetric and increasing Bayesian Nash equilibrium. Therefore, if $\theta = (\theta_1, \dots, \theta_n)$ is the type profile of the advertisers then the bid profile would be $(s(\theta_1), \dots, s(\theta_n))$, where $s(\cdot)$ is given by Equation (3.42). Because $s(\cdot)$ is increasing, the ordering of the bids $s(\theta_1), \dots, s(\theta_n)$ is the same as ordering of the types $\theta_1, \dots, \theta_n$. Therefore, the GSP mechanism will also result in the same allocation as the VCG and the OPT. Once again this result holds irrespective of whether $m < n$ or $n \leq m$. Thus, we have shown that irrespective of whether $m < n$ or $n \leq m$, for any given type profile $\theta = (\theta_1, \dots, \theta_n)$ of the advertisers, advertiser $i$ has an identical probability of getting slot $j$ in all the three mechanisms, namely GSP, VCG, and OPT. This confirms the first condition required for the revenue equivalence theorem.

In order to show the second condition, we need to consider three scenarios separately. It is easy to see that if an advertiser $i$ has $\theta_i = L$ then under each one of these three mechanisms, the outcome of the mechanism will confirm any one of the following three scenarios.

1. **The advertiser $i$ does not get any slot:** Note that this scenario occurs only when $m < n$. In such a situation, irrespective of the auction mechanism, advertiser $i$ neither pays any amount to the search engine nor gets any click in return. Therefore, the utility of advertiser $i$ under this scenario is zero for all the three mechanisms.
2. **The advertiser $i$ gets the last slot:** This scenario may arise in both the cases — $m < n$ and $n \leq m$. Let us analyze these cases one by one.
   **The Case $m < n$:** In such a case, it is straightforward to verify that all the losing bids will be equal to $L$ under all the three mechanisms. This is because the VCG and the OPT mechanisms are incentive compatible. Hence, no bid can be smaller than $L$ for these mechanisms. Similarly, for the GSP mechanism, by virtue of Equation (3.42), we have $s(L) = L$, and moreover the function $s(\cdot)$ is increasing.

This again implies that all the losing bids will be equal to $L$ for the GSP mechanism as well. By invoking the respective payment rules for these three mechanisms, we can verify that under this case, advertiser $i$ needs to pay an amount $L$ to the search engine for every click received from a user under each one the three mechanisms. Thus, we see that for this case, advertiser $i$ pays an amount $L$ for each user click and gets a benefit of $L$ under each one of the three mechanisms. Therefore, the net utility of the advertiser under this case is zero for each one of the three mechanisms.

**The Case $n \leq m$:** Note that under this case, there is no losing advertiser. Therefore, by invoking the respective payment rule for the three mechanisms, we can say that under this case, the advertiser $i$ needs to pay an amount equal to $0, 0,$ and $L$ under the GSP, the VCG, and the OPT mechanisms, respectively. This implies that the advertiser's utility for every user click is $L, L,$ and $0$ for the GSP, the VCG, and the OPT mechanisms, respectively.

3. **The advertiser $i$ gets a slot other than the last slot:** Note that this scenario can arise under both the cases — $m < n$ and $n \leq m$. Let us analyze each case one by one.

**The Case $m < n$:** In this case, it is straightforward to verify that under all three mechanisms, the bid of an advertiser must be equal to $L$ if either the advertiser gets a slot that is below advertiser $i$ or the advertiser does not get any slot. Now by invoking the respective payment rule, we can claim that in this case, advertiser $i$ needs to pay an amount $L$ to the search engine for every click received from a user under each one the three mechanisms. Thus, we see that for this case, advertiser $i$ pays an amount $L$ for each user click and gets a benefit of $L$ under each one of the three mechanisms. Therefore, the net utility of the advertiser under this case is zero for each one of the three mechanisms.

**The Case $n \leq m$:** Similar to the previous case, in this case it is easy to verify that under all three mechanisms, the bid of an advertiser must be equal to $L$ if the advertiser gets a slot that is below advertiser $i$. By invoking the respective payment rule for the three mechanisms, we can say that under this case, advertiser $i$ needs to pay an amount equal to $L, L(1 - \frac{\alpha_n}{\alpha_j})$, and $L$ under the GSP, the VCG, and the OPT mechanisms, respectively. Here $\alpha_j$ is the click probability of the slot at which advertiser $i$'s ad is displayed. This implies that the utility of advertiser $i$ for every user click is $0, L\frac{\alpha_n}{\alpha_j}$, and $0$ for the GSP, the VCG, and the OPT mechanisms, respectively.

The above discussion implies that advertiser $i$ has zero expected utility level in all three mechanisms when his type $\theta_i$ is at its lowest possible level and when $m < n$. Thus, we can now invoke the Revenue Equivalence Theorem and get the first part of the desired result, that is

$$R_{\text{GSP}} = R_{\text{VCG}} = R_{\text{OPT}} \text{ if } m < n.$$

In order to get the second part, observe that in Equation (3.44), if the allocation rule is the same then the expected revenue of the search engine depends solely on the values of $U_i(\underline{\theta_i})$. In above discussion we have shown that for any advertiser $i$ we have

$$U_i^{\text{OPT}}(\underline{\theta_i}) \leq U_i^{\text{GSP}}(\underline{\theta_i}) \leq U_i^{\text{VCG}}(\underline{\theta_i}).$$

The above inequality can be used in conjunction with Equation (3.44) to conclude the second part of the desired result, that is

$$R_{\text{GSP}} \leq R_{\text{VCG}} \leq R_{\text{OPT}} \text{ if } n \leq m.$$

*Q.E.D.*

The proof of the above proposition depends on the assumption, namely, $J_i(.) > 0 \; \forall i \in N$. This is quite a restrictive assumption. We now look at what would happen when this assumption is relaxed.

**Proposition 3.8.** *[Revenue comparison when the condition $J_i(.) > 0$ for all $i$ may not be satisfied] Consider a sponsored search auction setting, in which*

1. *The advertisers are risk neutral,*
2. *The advertisers are symmetric, i.e.,*

   a. $\Theta_1 = \ldots = \Theta_n = \Theta = [L, U]$,
   b. $\Phi_1(\cdot) = \ldots = \Phi_n(\cdot) = \Phi(\cdot)$,

3. *For each advertiser i, we have $\phi_i(\cdot) > 0$,*
4. *The advertisers draw their types independently, and*
5. *For each advertiser i, we have $J_i(\cdot)$ is a non-decreasing function.*

*Then the following are satisfied:*

$$R_{VCG} \leq R_{GSP}$$

$$R_{VCG} = R_{GSP} \text{ if } m < n$$

$$R_{VCG} \leq R_{OPT}.$$

**Proof:**

1. The inequality $R_{\text{VCG}} \leq R_{\text{GSP}}$ holds even when assumption $J_i(.) > 0$ for all $i$ is not satisfied. This is because the comparison between the two (already discussed in the proof of Proposition 4.1) does not depend in any way on this assumption.
2. Also, from the proof of Proposition 4.1, observe that the above inequality becomes an equality when $m < n$.
3. Next, we note that the VCG mechanism in our case is individually rational. Of course, the VCG mechanism is dominant strategy incentive compatible (and hence Bayesian incentive compatible). Therefore, since the OPT mechanism maximizes the revenue over all the Bayesian incentive compatible and individually rational mechanisms, we have $R_{\text{VCG}} \leq R_{\text{OPT}}$. This result is again independent of the assumption $J_i(.) > 0$ for all $i$.

*Q.E.D.*

**Observation 1**: As an immediate corollary of Proposition 4.2, the following inequality holds when $m < n$:

$$R_{\text{VCG}} = R_{\text{GSP}} \leq R_{\text{OPT}}.$$

**Observation 2**: When $n \leq m$, it is clear that the following inequalities hold: $R_{\text{VCG}} \leq R_{\text{GSP}}$ and $R_{\text{VCG}} \leq R_{\text{OPT}}$. However, we do not have any concrete inequality to compare $R_{\text{GSP}}$ with $R_{\text{OPT}}$.

In what follows, we actually derive the exact expressions for the expected revenue earned by the search engine under these three different mechanisms.

### 3.8.3 Expected Revenue under GSP, VCG, and OPT

A natural way to compare these three mechanisms, from the point of view of expected revenue earned by the search engine, is the following:

1. Compute the equilibrium bidding strategies of the advertisers under each mechanism.
2. Compute the expected revenue earned by the search engine under each mechanism assuming that the advertisers will respond with corresponding equilibrium bidding strategies.
3. Compare these three auction mechanisms based on the calculated expected revenue.

We have already seen that

- Truth-telling is a dominant strategy equilibrium under the VCG mechanism,
- Truth-telling is a Bayesian Nash equilibrium under the OPT mechanism,
- Truth-telling is not an equilibrium under the GSP mechanism.

In view of this, we compute the expected revenue earned by the search engine in each of these three auction mechanisms and then present a comparison of these three auction mechanisms from the point of view of expected revenue earned by the search engine. Before we start, we would like to mention that we will continue to follow the assumptions made in the beginning of Section 3.8 throughout the rest of the discussion.

### 3.8.4 Expected Revenue under the VCG Mechanism

Under the assumption of the symmetric advertisers, it is easy to see that expected revenue, $R_{\text{VCG}}$, earned by the search engine can be computed by any one of the following two methods.

**Method 1:**

$$R_{\text{VCG}} = n \int_{\theta_i=L}^{U} \bar{t}_i(\theta_i) \phi(\theta_i) d\theta_i \tag{3.45}$$

where $\bar{t}_i(\theta_i)$ is the expected payment made by an advertiser $i$ to the search engine against every search query received by the search engine and when he announces his type to be $\theta_i$ and all the advertisers $j \neq i$ truthfully reveal their types. Recall that truth-telling is a dominant strategy equilibrium for the advertisers in the VCG mechanism. Therefore, the reported type (or bid) of advertiser $i$ is indeed his actual type.

**Method 2:**

$$R_{\text{VCG}} = E_\theta \left[ \sum_{j=1}^{\min(m,n)} \alpha_j p^{(j)}(\theta) \right] \tag{3.46}$$

where $p^{(j)}(\theta)$ is the payment made by the advertiser, whose ad is displayed in the $i^{th}$ position, to the search engine against every click made by the user and when the bid profile of the advertisers is $\theta = (\theta_1, \ldots, \theta_n)$. Recall that truth-telling is a dominant strategy equilibrium for the advertisers in the VCG mechanism. Therefore, the reported type (or bid) profile of the advertisers is indeed their actual type profile.

We will follow the second method in order to compute $R_{\text{VCG}}$. We consider two cases separately

**Case 1** ($m < n$)**:** Substituting Equation (3.9) in Equation (3.46), we get the following relation

$$R_{\text{VCG}} = E_\theta \left[ m\alpha_m \theta^{(m+1)} + \sum_{j=1}^{m-1} j\beta_j \theta^{(j+1)} \right]. \tag{3.47}$$

Now we need to compute the expectation of each term separately. For this, notice that the advertisers are assumed to be symmetric and they choose their bids independently; therefore, the probability that $(j+1)^{th}$ highest bid lies in an interval $[x, x+dx]$ can be given by

$$n \binom{n-1}{j} [1 - \Phi(x)]^j [\Phi(x)]^{n-j-1} \phi(x) dx$$

where $j = 0, \ldots, n-1$ and $x \in [L, U]$. Therefore, the expected value of the $(j+1)^{th}$ highest bid is given by

$$E_\theta \left[ \theta^{(j+1)} \right] = \int_L^U x\, n \binom{n-1}{j} [1 - \Phi(x)]^j [\Phi(x)]^{n-j-1} \phi(x) dx. \tag{3.48}$$

Substituting Equation (3.48) in Equation (3.47), we get the following relation for expected revenue earned by the search engine under this case

$$R_{\text{VCG}} = \int_L^U \left[ m\alpha_m \binom{n-1}{m} [\overline{\Phi}(x)]^m [\Phi(x)]^{n-m-1} + \sum_{j=1}^{m-1} j\beta_j \binom{n-1}{j} [\overline{\Phi}(x)]^j [\Phi(x)]^{n-j-1} \right] xn\phi(x)dx$$

(3.49)

where $\overline{\Phi}(\cdot) = 1 - \Phi(\cdot)$.

**Case 2** $(n \leq m)$**:** Substituting Equation (3.10) in Equation (3.46), we get the following relation:

$$R_{\text{VCG}} = E_\theta \left[ \sum_{j=1}^{n-1} j\beta_j \theta^{(j+1)} \right].$$

(3.50)

Following the same approach as for case 1, we get the following relation for expected revenue earned by the search engine under this case:

$$R_{\text{VCG}} = \int_L^U \left[ \sum_{j=1}^{n-1} j\beta_j \binom{n-1}{j} [\overline{\Phi}(x)]^j [\Phi(x)]^{n-j-1} \right] xn\phi(x)dx.$$

(3.51)

## 3.8.5 Expected Revenue under the OPT Mechanism

Because of the symmetric advertisers assumption and the fact that truth-telling is a Bayesian Nash equilibrium for the advertisers under the OPT mechanism, the expected revenue, $R_{\text{OPT}}$, earned by the search engine under the OPT mechanism can be computed either by method 1 (Equation 3.45) or method 2 (Equation 3.46), which were discussed earlier in the context of $R_{\text{VCG}}$.

Here also we will follow the second method (3.46) in order to compute $R_{\text{OPT}}$. Once again, We consider two cases separately.

**Case 1** $(m < n)$**:** Substituting Equation (3.21) in Equation (3.46), we get the following relation:

$$R_{\text{OPT}} = E_\theta \left[ m\alpha_m \theta^{(m+1)} + \sum_{j=1}^{m-1} j\beta_j \theta^{(j+1)} \right].$$

(3.52)

Following the same approach as for the case 1 of $R_{\text{VCG}}$, we get the following relation for expected revenue earned by the search engine under this case:

$$R_{\text{OPT}} = \int\limits_{L}^{U} \left[ m\alpha_m \binom{n-1}{m} [\overline{\Phi}(x)]^m [\Phi(x)]^{n-m-1} + \sum_{j=1}^{m-1} j\beta_j \binom{n-1}{j} [\overline{\Phi}(x)]^j [\Phi(x)]^{n-j-1} \right] x n\phi(x)dx.$$

$$(3.53)$$

It is easy to verify that $R_{\text{OPT}} = R_{\text{VCG}}$ for the case when $m < n$. This matches with the previous result about revenue equivalence of the OPT and the VCG mechanisms stated in the form of Proposition 3.7.

**Case 2** $(n \leq m)$**:** Substituting Equation (3.22) in Equation (3.46), we get the following relation:

$$R_{\text{OPT}} = E_\theta \left[ n\alpha_n L + \sum_{j=1}^{n-1} j\beta_j \theta^{(j+1)} \right]. \qquad (3.54)$$

Following the same approach as for the case 1, we get the following relation for expected revenue earned by the search engine under this case:

$$R_{\text{OPT}} = n\alpha_n L + \int\limits_{L}^{U} \left[ \sum_{j=1}^{n-1} j\beta_j \binom{n-1}{j} [\overline{\Phi}(x)]^j [\Phi(x)]^{n-j-1} \right] x n\phi(x)dx. \quad (3.55)$$

It is easy to verify that $R_{\text{VCG}} \leq R_{\text{OPT}}$ for the case when $n \leq m$. The equality holds if and only if $L = 0$. This matches with the previous result about revenue equivalence of the OPT and the VCG mechanisms stated in the form of Proposition 3.7.

### 3.8.6 Expected Revenue under the GSP Mechanism

Recall that truth-telling need not be a Bayesian Nash equilibrium for the advertisers under the GSP mechanism. Therefore, method 1 and method 2 for computing the expected revenue of the search engine under this auction mechanism can be modified in following manner:

#### 3.8.6.1 Method 1

$$R_{\text{GSP}} = n \int\limits_{\theta_i=L}^{U} \bar{t}_i(s(\theta_i))\phi(\theta_i)d\theta_i \qquad (3.56)$$

where $s(\cdot)$ is the symmetric equilibrium bidding strategy of advertiser $i$ and is given by Equation (3.42).

**3.8.6.2 Method 2**

$$R_{\text{GSP}} = E_\theta \left[ \sum_{j=1}^{\min(m,n)} \alpha_j p^{(j)}(s(\theta_1), \ldots, s(\theta_n)) \right] \qquad (3.57)$$

where $s(\cdot)$ is the symmetric equilibrium bidding strategy of advertiser $i$ and is given by Equation (3.42).

Note that computing the exact expression for expected revenue $R_{\text{GSP}}$ is a difficult problem because computing the exact expression for $s(\cdot)$ by solving Equation (3.42) is a hard problem. We therefore take a different approach here and instead of computing the exact expression for $R_{\text{GSP}}$, we appeal to Proposition 3.7, which says that

$$R_{\text{GSP}} = R_{\text{VCG}} = R_{\text{OPT}} \quad : \quad \text{if } m < n$$
$$R_{\text{VCG}} \leq R_{\text{GSP}} \leq R_{\text{OPT}} \quad : \quad \text{if } n \leq m.$$

Note that we have already computed $R_{\text{VCG}}$ and $R_{\text{OPT}}$ for both the cases — $m < n$ and $n \leq m$. Therefore, we can get the exact expression for $R_{\text{VCG}}$ when $m < n$, and an upper and a lower bound when $n \leq m$ by making use of the Equations (3.49), (3.53), (3.51), and (3.55).

## 3.9 Individual Rationality

We know that the OPT mechanism satisfies the interim individual rationality by definition.

In order to test whether the GSP mechanism satisfies it or not, we need to make the observation that under the GSP mechanism, an advertiser would never pay more than what he has has bid for each user click on his Ad. Therefore, as long as each advertiser $i$ uses a bidding strategy $s_i(\theta_i)$ such that $s_i(\theta_i) \leq \theta_i \; \forall \; \theta_i \in \Theta_i$, it will immediately imply that $U_i(\theta_i) \geq 0 \; \forall \; \theta_i \in \Theta_i$. This would satisfy the interim individual rationality constraints. It is easy to verify that under the symmetry assumption, no equilibrium of the GSP mechanism will ever have $s_i(\theta_i) > \theta_i$ for any advertiser $i$ and for any $\theta_i \in \Theta_i$. This proves that the GSP mechanism always satisfies interim individual rationality.

The VCG mechanism is also interim individually rational. This can be verified by observing that in the VCG mechanism, the payment made by an advertiser against each user click is always less than or equal to his bid amount, and the bid amount of each advertiser is always his true valuation. To show that the payment made by an advertiser per user click is less than or equal to his bid amount, we start with the payment rule of the VCG mechanism, which is given by Equations (3.9) and (3.10). We consider each case separately.

**Case 1** $(m < n)$: Notice that under this case,

- If an advertiser is not allocated any slot then by virtue of Equation (3.9), he pays nothing, which ensures interim IR.
- If an advertiser $i$, with his bid $\theta_i$, is allocated the last position, i.e., the $m^{th}$ position, then per Equation (3.9), he pays an amount $\theta^{(m+1)}$ for each user click. It is easy to see that $\theta^{(m+1)} \leq \theta_i$ because in the VCG mechanism, the advertisers are allocated the slots in decreasing order of their bids, and advertiser $i$ has received the $m^{th}$ slot. This again ensures interim IR.
- If an advertiser $i$, with his bid $\theta_i$, is allocated the position $r$, where $1 \leq r \leq (m-1)$, then according to Equation (3.9), he will be paying an amount

$$p_i(\theta_i, \theta_{-i}) = \frac{1}{\alpha_r} \left[ \sum_{j=r}^{m-1} \beta_j \theta^{(j+1)} \right] + \frac{\alpha_m}{\alpha_r} \theta^{(m+1)}$$

for every user click. Notice that because in the VCG mechanism, the advertisers are allocated the slots in decreasing order of their bids and advertiser $i$ has already received the $r^{th}$ slot, we must have

$$p_i(\theta_i, \theta_{-i}) \leq \frac{1}{\alpha_r} \left[ \sum_{j=r}^{m-1} \beta_j \theta_i \right] + \frac{\alpha_m}{\alpha_r} \theta_i$$

$$= \theta_i \left[ \sum_{j=r}^{m-1} \frac{\alpha_j - \alpha_{j+1}}{\alpha_r} + \frac{\alpha_m}{\alpha_r} \right]$$

$$= \theta_i.$$

This ensures interim IR even for this case.

**Case 2** $(n \leq m)$: For this case we make use of Equation (3.10) and go about applying arguments similar to those used in the previous case and show that the VCG mechanism is interim IR even under this case as well. Therefore, we can say that the VCG mechanism is interim individually rational.

## 3.10 Computational Complexity

Note that in each of the previously discussed auction schemes, after receiving the query word, the search engine needs to pull out of its database the bids of the advertisers who are interested in putting their ads against this particular query word. After getting these bid values, say $b_1, \ldots, b_n$, the search engine needs to sort them in decreasing order if it is either the GSP or the VCG mechanism. As is well known, the worst case complexity of sorting $n$ numbers is $O(n \log n)$. The sorted bids $b^{(1)}, \ldots, b^{(n)}$ can now be used for calculating the allocation and the payment of each advertiser. It is easy to verify that the allocation operation has a worst case complexity of $O(\min(m,n))$ for both the GSP and the VCG mechanisms. The payment operation has a worst case complexity of $O(\min(m,n))$ for the GSP mechanism and $O((\min(m,n))^2)$ for the VCG mechanism. Thus, we can claim that the computational complexity of the GSP is $O(n\log n + \min(m,n))$, which is the same as $O(n\log n)$, and the computational complexity of the VCG mechanism is $O(n\log n + (\min(m,n))^2)$.

The practical implementation of the OPT mechanism has its own challenges. Recall that the design of the OPT mechanism intrinsically assumes that the search engine precisely knows the type distribution $\Phi_i(\cdot)$ of each advertiser $i$. However, in practice this may not be true. The search engine typically has no information about an advertiser except his bid value and the history of click streams. However, the search engine can always learn the type distribution $\Phi_i(\cdot)$ of each advertiser $i$ from these given data. Assuming that the search engine knows the type distributions $\Phi_i(\cdot)$ for each advertiser $i$ and that $\phi_i(\cdot)$ is a positive function for each $i$, and $J_i(\cdot)$ is a non-decreasing function for each $i$, our objective here is to compute the computational complexity of the OPT mechanism. Note that after receiving the bid values (which are the same as actual types), say $\theta_1, \ldots, \theta_n$, from its database, the search engine needs to compute $J_1(\theta_1), \ldots, J_n(\theta_n)$. This is an $O(n)$ operation. Next, the search engine needs to sort $J_1(\theta_1), \ldots, J_n(\theta_n)$ in decreasing order, which is an $O(n\log n)$ operation. The search engine can use these sorted values to compute the assignment of the advertisers, which is an $O(\min(m,n))$ operation. Thus the complexity of the allocation operation under OPT mechanism is $O(n + n\log n + \min(m,n))$, which is the same as $O(n\log n)$. Coming to the payment determination, note that the search engine needs to compute the quantities $z_{ij}(\theta_{-i})$ for each advertiser $i$. Assuming that functions $J_i(\theta_i)$ are invertible, for example in the case of uniform type distribution, computing the quantity $z_{ij}(\theta_{-i})$ is a constant time operation. To illustrate this, suppose that advertiser $i$ is allocated the $r^{th}$ position, then we have

$$z_{ij} = \begin{cases} J_i^{-1}(J^{(j)}) & : \text{ if } j = 1, \ldots, r-1 \\ J_i^{-1}(J^{(j+1)}) & : \text{ if } j = r, \ldots, \min(m,n) \end{cases}$$

where $J^{(j)}$ is the value of the quantity $J_k(\theta_k)$ for an advertiser $k$ whose Ad is allocated to the $j^{th}$ position. In view of the assumption of invertibility of the functions $J_i(\cdot)$, we can say that computing the quantities $z_{ij}(\theta_{-i})$ is an $O((\min(m,n))^2)$ operation. After computing these quantities the payment for the advertisers can be computed in $O((\min(m,n))^2)$ time. Thus the complexity of the payment operation under the OPT mechanism is $O((\min(m,n))^2)$. Therefore, the computational complexity of the OPT mechanism, under the assumption that the function $J_i(\cdot)$ is invertible for every $i$, is $O(n\log n + (\min(m,n))^2)$, which is the same as the computational complexity of the VCG mechanism.

In what follows, we summarize the results of the comparative study made so far for three different sponsored search auction mechanisms, namely GSP, VCG, and OPT in the form of Table 3.2. The implicit assumptions here are

- The GSP mechanism uses the Yahoo!/Overture allocation rule,
- $\Theta_1 = \ldots = \Theta_n = \Theta = [L, U]$,

- $\Phi_1(\cdot) = \ldots = \Phi_n(\cdot) = \Phi(\cdot)$,
- $J(\cdot)$ is non-decreasing over the interval $[L, U]$,
- $J(x) > 0 \ \forall x \in [L, U]$,
- $J(\cdot)$ is invertible.

| Auction | AE | DSIC | BIC | Ex Post IR | Complexity |
|---------|----|------|-----|------------|------------|
| GSP | ✓ | × | × | ✓ | $O(n \log n)$ |
| VCG | ✓ | ✓ | ✓ | ✓ | $O(n \log n + (\min(m, n))^2)$ |
| OPT | ✓ | × | ✓ | ✓ | $O(n \log n + (\min(m, n))^2)$ |

**Table 3.2** Properties of various sponsored search auction mechanisms

## 3.11 Summary and Future Work

In this chapter, we explained how a typical sponsored search auction works and developed a modeling framework. We formulated the sponsored search auction as a mechanism design problem in a linear environment and then showed that three well-known mechanisms, GFP, GSP, and VCG, can be formulated in this framework. Next, we presented a mechanism, called the OPT mechanism. We compared the OPT mechanism with the GSP and the VCG from the point of view of incentive compatibility, expected revenue earned by the search engine, and individual rationality. We derived a symmetric equilibrium bidding strategy of the advertisers for the GSP mechanism, and this was instrumental in showing that the GSP is not a Bayesian incentive compatible mechanism. We extended the classical revenue equivalence theorem to the setting of a sponsored search auction and used it to show the revenue equivalence of the three mechanisms. Finally, we also computed expressions for the expected revenue earned by the search engine under the GSP, the VCG, and the OPT mechanisms.

The real world problem of conducting the sponsored search auction is far more complex than the one we have modeled here. One can study the above problem from the perspective of generalizing it in several directions. A few suggestive directions are discussed below.

- In the present formulation of the problem, we have focused on the problem of a sponsored search auction for a single query. However, in practice, the queries come in an online manner, and the auction needs to be conducted again and again. In this sense, a repeated game theoretic analysis of the problem will result in altogether a different insight.

- In a real world setting, the advertisers can leave and join at any point of time. Therefore, one can generalize the problem in terms of taking into account the probability that the advertiser will continue to participate in the next round of the auction also
- In practice, a given advertiser typically specifies a daily budget. The auctioneer does not allow a particular advertiser to participate in the auction after the advertiser's budget is exhausted. The budget constraint gives a new dimension to this problem, and the problem needs to be modeled from scratch. We have cited some recent work in this direction in next section.

## 3.12 Related Literature

The problem of the sponsored search auction is not more than 10 years old in the industry and 5 years old in the research community. The interest of the researchers and the practitioners has suddenly grown in the past few years toward addressing the problem of sponsored search auctions. The first workshop on sponsored search auctions was held in conjunction with the ACM Conference on Electronic Commerce (EC'05) in Vancouver, BC, Canada on June 5, 2005. The focus of the workshop was on mechanisms and behaviors in sponsored search. These workshops are now held annually. The other conferences where the related work has appeared includes International World Wide Web Conference (WWW), and Workshop on Internet and Network economics (WINE). In this section, we try to provide a glimpse of major research trends related to the sponsored search auctions problem. The literature can be broadly classified into three major categories:

1. Mechanism Design,
2. Bidder Behavior,
3. Click Through Rate and Click Fraud.

The relevant literature in the mechanism design category focuses primarily on the allocation and payment rules. The objective of these papers may include analyzing and comparing different auction mechanisms, designing new auction mechanisms, etc. The key contributions in this direction are as follows:

- Edelman, Ostrovsky, and Schwarz [15] investigate the Generalized Second Price (GSP) mechanism for sponsored search auction. They show that although GSP looks similar to VCG, its properties are very different, and equilibrium behavior is far from straightforward. In particular, they show that unlike the VCG mechanism, GSP generally does not have an equilibrium in dominant strategies, and truth telling is not an equilibrium of GSP.
- Aggarwal, Goel, and Motwani [4] design a simple truthful auction for a general class of ranking functions that includes direct ranking and revenue ranking. More specifically, the authors study the case where the merchants are assigned arbitrary

weights that do not depend on the bids, and then ranked in decreasing order of their weighted bids. They call such an auction a laddered auction, since the price for a merchant builds on the price of each merchant ranked below it. They show that this auction is truthful.

- Feng [18] studies the allocation mechanisms under a setting in which advertisers have a consistent ranking of advertising positions but different rates of decrease in absolute valuation.
- Varian [47] analyzes the equilibria of an assignment game that arises in the context of ad auctions. These equilibria are closely related to the equilibria of the assignment game studied by Shapley–Shubik [45], Demange–Gale–Sotomayer [13], and Roth–Sotomayer [41]. The author characterizes the symmetric Nash equilibria of such assignment games and uses it to derive an upper bound and a lower bound on the revenue generated by the search engine. Further, this revenue is also compared with the revenue in the VCG mechanism.
- Feng, Bhargava, and Pennock [16, 17] examine the paid-placement ranking strategies of the two dominant firms in this industry, and compare their revenue under different scenarios via computational simulation.
- Mehta, Saberi, Vazirani, and Vazirani [35] show that the problem of deciding what ads to display against each search query so as to maximize the search engine's revenue is a generalization of the online bipartite matching problem. They derive an optimal algorithm for this matching that has a competitive ratio of $1 - 1/e$.
- Lahaie [30] analyzes the incentive, efficiency, and revenue properties of the two popular slot auctions — first price and second price, under settings of incomplete and complete information.
- Iyengar and Kumar [28] formulate the general problem that allows the privately known valuation per click to depend both on the identity of merchant and the slot and present a compact characterization of the set of all deterministic incentive compatible direct mechanisms.
- Roughgarden and Sundararajan [42] study the simultaneous optimization of efficiency and revenue in pay-per-click keyword auctions in a Bayesian setting. Their main result is that the efficient keyword auction yields near-optimal revenue even under modest competition.
- Ganchev, Kulesza, Tan, Gabbard, Liu, and Kearns [21] present a characterization of empirical price data from sponsored search auctions. They show that simple models drawing bid values independently from a fixed distribution can be tuned to match empirical data on average, but still fail to account for deviations observed in individual auctions.
- Bu, Deng, and Qi [10] propose a new solution concept, the forward looking Nash Equilibrium, for the position auction by considering the strategic manipulations of an agent, that takes into consideration the effect of the existing strategies of other agents, as well as their future responses, to its own benefit. They prove that the forward-looking Nash equilibrium in its pricing and allocation scheme is equivalent to the VCG auction outcome.

- Feldman, Muthukrishnan, Pal, and Stein [19] consider the "Offline Ad Slot Scheduling" problem, where advertisers must be scheduled to "sponsored search" slots during a given period of time. Advertisers specify a budget constraint, as well as a maximum cost per click, and may not be assigned to more than one slot for a particular search. They give a truthful mechanism under the utility model where bidders try to maximize their clicks, subject to their personal constraints. In addition, they show that the revenue-maximizing mechanism is not truthful, but has a Nash equilibrium whose outcome is identical to their mechanism.
- Abrams, Mendelevitch, Mendelevitch, and Tomlin [1] propose an approach based on linear programming that takes bidder budgets into account, and uses them in conjunction with forecasting of query frequencies, and pricing and ranking schemes, to optimize ad delivery.
- Lahaie and Pennock [31] consider a family of ranking rules that contains those typically used to model Yahoo! and Google's auction designs as special cases. They find that in general neither of these is necessarily revenue-optimal in equilibrium. They propose a simple approach to determine a revenue-optimal ranking rule within their family, taking into account effects on advertiser satisfaction and user experience.
- Mahadian, Nazerzadeh, and Saberi [33] study the problem of optimally allocating online advertisement space to budget-constrained advertisers. Their objective is to find an algorithm that takes advantage of the given estimates of the frequencies of keywords to compute a near optimal solution when the estimates are accurate, while at the same time maintaining a good worst-case competitive ratio in case the estimates are totally incorrect.
- Bhargava and Feng [7] have formulated the search engine design problem as a tradeoff between placement revenue and user-based revenue.
- Borgs, Chayes, Immorlica, Mahdian, and Saberi [8] study a multi-unit (corresponds to a sequence of searches each with a single slot) auction with multiple agents, each of whom has a private valuation and budget.
- Aggarwal and Hartline [3] consider a special version of ad auction as the private value knapsack problem.
- Parkes and Sandholm [39] propose an architecture to improve the expressiveness of sponsored search auctions.
- Balcan, Blum, Hartline, and Mansour [6] use techniques from sample-complexity in machine learning theory to reduce the design of revenue maximizing incentive-compatible mechanisms to algorithmic pricing questions relevant to sponsored search.

The literature related to bidder behavior focuses on various kinds of challenges faced by the advertisers while participating in the sponsored search auction. These challenges include deciding the optimal bid amount, deciding the set of representative keywords, deciding the advertising budget, etc. The key contributions in this direction are as follows:

- Borgs, Chayes, Etesami, Immorlica, Jain, and Mahdian [9] propose a bidding heuristic to optimize the utility for bidders by equalizing the return-on-investment for each bidder across all keywords. They show that natural auction mechanisms combined with this heuristic can experience chaotic cycling (as is the case with many current advertisement auction systems), and therefore they propose a modified class of mechanisms with small random perturbations.

- Asdemir [5] analyzes how advertisers bid for search phrases in pay-per-click search engine auctions. The author develops an infinite horizon alternative-move game of advertiser bidding behavior.

- Szymanski and Lee [46] discuss how advertisers, by considering minimum return on investment (ROI), change their bids and, consequently the auctioneers' revenue in sponsored search advertisement auctions.

- Zhou and Lukose [49] study vindictive bidding, a strategic bidding behavior in keyword auctions where a bidder forces his competitor to pay more without affecting his own payment. The authors show that most Nash equilibria are vulnerable to vindictive bidding and are thus unstable.

- Sebastian, Bartz, and Murthi [43] address the advertisers' difficulty in discovering all the terms that are relevant to their products or services. They examine the performance of logistic regression and collaborative filtering models on two different data sources to predict terms relevant to a set of seed terms describing an advertisers product or service.

- Cray, Das, Edelman, Giotis, Heimerl, Karlin, Mathieu, and Schwarz [11] consider greedy bidding strategies for a repeated auction on a single keyword, where, in each round, each player chooses an optimal bid for the next round, assuming that the other players merely repeat their previous bid. They study the revenue, convergence and robustness properties of such strategies.

- Feldman, Muthukrishnan, Pal, and Stein [19] study the budget optimization problem. They show that simply randomizing between two uniform strategies that bid equally on all the keywords works well.

- Hosanagar and Cherepanov [26] study the problem of optimal bidding in stochastic budget constrained slot auctions.

The literature related to click through rate and click fraud focuses on how to estimate the click through rates of different advertisers, how to detect and prevent the fraudulent clicks, etc. The key contributions in this direction are as follows:

- Regelson and Fain [40] hypothesize that different keyword terms have an inherently different likelihood of receiving sponsored clicks. They seek to estimate a term level click-through rate (CTR) reflecting these inherent differences. They also propose the use of clusters of related terms for less frequent, or even completely novel, terms.

- Goodman [24] describes a simple method for selling advertising, pay-per-percentage of impressions, that is immune to both click fraud and impression fraud.

- Kitts, Laxminarayan, LeBlanc, and Meech [29] present a model of sponsored search auctions with very few assumptions. Their model provides analytical predictions on auction dynamics, bidder optima, and equilibria. The model makes a series of testable predictions that may be falsified through future observations.

Apart from the above literature, recently Lahaie, Pennock, Saberi, and Vohra [32] have put together a book chapter on sponsored search auctions that basically highlights the application of mechanism design in the context of sponsored search auctions.

# References

1. Z. Abrams, O. Mendelevitch, and J. Tomlin. Optimal delivery of sponsored search advertisements subject to budget constraints. In *ACM Conference on Electronic Commerce (EC'07)*, San Diego, California, 2007.
2. R. Adams. *www.Advertising:Advertising and Marketing on the World Wide Web (Design Directories)*. Watson-Guptill Publications, New York, 2003.
3. G. Aggarwal and J.D. Hartline. Knapsack auctions. In *1st Workshop on Sponsored Search Auctions in conjunction with the ACM Conference on Electronic Commerce (EC'05)*, pages 1083–1092, Vancouver, BC, Canada, 2005.
4. G. Aggarwal, A. Goel, and R. Motwani. Truthful auctions for pricing search keywords. In *7th ACM Conference on Electronic Commerce (EC'06)*, Ann Arbor, Michigan, USA, 2006.
5. K. Asdemir. Bidding patterns in search engine auctions In *Second Workshop on Sponsored Search Auctions in Conjunction with the ACM Conference on Electronic Commerce (EC'06)*, Ann Arbor, Michigan, 2006.
6. M.F. Balcan, A. Blum, J.D. Hartline, and Y. Mansour. Sponsored search auction design via machine learning. In *1st Workshop on Sponsored Search Auctions in conjunction with the ACM Conference on Electronic Commerce (EC'05)*, Vancouver, BC, Canada, 2005.
7. H.K. Bhargava and J. Feng. Paid placement strategies for Internet search engines. In *11th International Conference on World Wide Web*, Honolulu, Hawaii, USA, pages 117–123, 2002.
8. C. Borgs, J. Chayes, N. Immorlica, M. Mahdian, and A. Saberi. Multi-unit auctions with budget-constrained bidders. In *6th ACM Conference on Electronic Commerce (EC'05)*, pages 44–51, 2005.
9. C. Borgs, J. Chayes, O. Etesami, N. Immorlica, K. Jain, and M. Mahdian. Bid optimization in online advertisement auctions. In *International World Wide Web Conference (WWW'07)*, Banff, Alberta, Canada, 2007.
10. T.-M. Bu, X. Deng, and Q. Qi. Dynamics of strategic manipulation in ad-words auction. In *International World Wide Web Conference (WWW'07)*, Banff, Alberta, Canada, 2007.
11. M. Cary, A. Das, B. Edelman, I. Giotis, K. Heimerl, A. Karlin, C. Mathieu, and M. Schwarz. Greedy bidding strategies for keyword auctions. In *ACM Conference on Electronic Commerce (EC'07)*, San Diego, California, USA, 2007.
12. A. Mas-Colell, M.D. Whinston, and J.R. Green. *Microeconomic Theory*. Oxford University Press, New York, 1995.
13. G. Demange, D. Gale, and M. Sotomayor. Multi-item auctions. *Journal of Political Economy*, 94(4):863–72, 1986.
14. B. Edelman and M. Ostrovsky. Strategic bidder behavior in sponsored search auctions. In *1st Workshop on Sponsored Search Auctions in conjunction with the ACM Conference on Electronic Commerce (EC'05)*, Vancouver, BC, Canada, 2005.
15. B. Edelman, M. Ostrovsky, and M. Schwarz. Internet advertising and the generalized second price auction: Selling billions of dollars worth of keywords. In *2nd Workshop on Sponsored Search Auctions in Conjunction with the ACM Conference on Electronic Commerce (EC'06)*, Ann Arbor, MI, USA, 2006.
16. J. Feng, H.K. Bhargava, and D. Pennock. Comparison of allocation rules for paid placement advertising in search engines. In *5th International Conference on Electronic Commerce (ICEC'03)*, Pittsburgh, PA, USA, 2003.

17. J. Feng, H. Bhargava, and D. Pennock. Implementing sponsored search in web search engines: Computational evaluation of alternative mechanisms. INFORMS Journal on Computing, Forthcoming, http://ssrn.com/abstract=721262, 2005.
18. J. Feng. Optimal mechanism for selling a set of commonly ranked objects. *Marketing Sciencey*, To appear:2008.
19. J. Feldman, S. Muthukrishnan, M. Pal, and C. Stein. Budget optimization in search-based advertising auctions. In *ACM Conference on Electronic Commerce (EC'07)*, San Diego, California, USA, 2007.
20. J. Feldman, S. Muthukrishnan, E. Nikolova, and M. Pál. A truthful mechanism for offline ad slot scheduling. In *Symposium on Algorithmic Game Theory (SAGT)*, pages 182–193, Paderborn, Germany, 2008.
21. K. Ganchev, A. Kulesza, J. Tan, R. Gabbard, Q. Liu, and M. Kearns. Empirical price modeling for sponsored search. In *International World Wide Web Conference (WWW'07)*, Banff, Alberta, Canada, 2007.
22. D. Garg and Y. Narahari. Design of an optimal mechanism for sponsored search auctions on the web. To appear in: *IEEE Transactions on Automation Science and Engineering*, 2008.
23. D. Garg. Design for sponsored search auctions. Chapter 4 of the Ph.D. Dissertation "Innovative Mechanisms for Emerging Game Theoretic Problems in Electronic Commerce," Department of Computer Science and Automation, Indian Institute of Science, 2006.
24. J. Goodman. Pay-per-percentage of impressions: An advertising method that is highly robust to fraud. In *Workshop on Sponsored Search Auctions in Conjunction with the ACM Conference on Electronic Commerce (EC'05)*, Vancouver, BC, Canada, 2005.
25. D.L. Hoffman and T.P. Novak. Advertising pricing models for the world wide web. In B. Kahin and H.R. Varian (eds.), *Internet Publishing and Beyond: The Economics of Digital Information and Intellectual Property*. MIT Press, Cambridge, 2000.
26. K. Hosanagar and V. Cherepanov. Optimal bidding in stochastic budget constrained slot auctions. In *ACM Conference on Electronic Commerce (EC'08)*, Chicago, Illinois, USA, 2008.
27. N. Immorlica, K. Jain, M. Mahdian, and K. Talwar. Click fraud resistant methods for learning click-through rates. In *1st International Workshop on Internet and Network Economics (WINE'05)*, Hong Kong, 2005.
28. G. Iyengar and A. Kumar. Characterizing optimal keyword auctions. In *Second Workshop on Sponsored Search Auctions in Conjunction with the ACM Conference on Electronic Commerce (EC'06)*, Ann Arbor, Michigan, 2006.
29. B. Kitts, P. Laxminarayan, B. LeBlanc, and R. Meech. A formal analysis of search auctions including predictions on click fraud and bidding tactics. In *Workshop on Sponsored Search Auctions in Conjunction with the ACM Conference on Electronic Commerce (EC'05)*, Vancouver, BC, Canada, 2005.
30. S. Lahaie. An analysis of alternative slot auction designs for sponsored search. In *7th ACM Conference on Electronic Commerce (EC'06)*, Ann Arbor, Michigan, USA, 2006.
31. S. Lahaie and D. Pennock. Revenue analysis of a family of ranking rules for keyword auctions. In *ACM Conference on Electronic Commerce (EC'07)*, San Diego, California, USA, 2007.
32. S. Lahaie, D. Pennock, A. Saberi, and R.V. Vohra. Sponsored search auctions. In N. Nisan, T. Roughgarden, E. Tardos, and V.V. Vazirani (eds.), *Algorithmic Game Theory*. pages 699–716, Cambridge University Press, New York, 2007.
33. M. Mahdian, H. Nazerzadeh, and A. Saberi. Allocating online advertisement space with unreliable estimates. In *ACM Conference on Electronic Commerce (EC'07)*, San Diego, California, USA, 2007.
34. P.R. McAfee and J. McMillan. Auctions and bidding. *Journal of Economic Literature*, 25(2):699–738, 1987.
35. A. Mehta, A. Saberi, V. Vazirani, and U. Vazirani. Adwords and generalized online matching. In *46th Annual IEEE Symposium on Foundations of Computer Science (FOCS'05)*, Pittsburgh, PA, USA, 2005.

36. P.R. Milgrom. Auction Theory. In T. Bewley (eds.), *Advances in Economic Theory: Fifth World Congress*. Cambridge University Press, Cambridge, U.K., 1987.

37. R.B. Myerson. Optimal auction design. *Math. Operations Res.*, 6(1):58–73, 1981.

38. R.B. Myerson. *Game Theory: Analysis of Conflict*. Harvard University Press, Cambridge, Massachusetts, 1997.

39. D.C. Parkes and T. Sandholm. Optimize-and-dispatch architecture for expressive ad auctions. In *1st Workshop on Sponsored Search Auctions in Conjunction with the 6th ACM Conference on Electronic Commerce (EC'05)*, Vancouver, BC, Canada, 2005.

40. M. Regelson and D.C. Fain. Predicting clickthrough rate using keyword clusters. In *Second Workshop on Sponsored Search Auctions in Conjunction with the ACM Conference on Electronic Commerce (EC'06)*, Ann Arbor, Michigan, 2006.

41. A. Roth and M. Sotomayor. *Two-Sided Matching*. Cambridge University Press, Cambridge, U.K., 1990.

42. T. Roughgarden and M. Sundararajan. Is efficiency expensive? In *International World Wide Web Conference (WWW'07)*, Banff, Alberta, Canada, 2007.

43. S. Sebastian, K. Bartz, and V. Murthi. Logistic regression and collaborative filtering for sponsored search term recommendation. In *Second Workshop on Sponsored Search Auctions in Conjunction with the ACM Conference on Electronic Commerce (EC'06)*, Ann Arbor, Michigan, 2006.

44. C. Seda. *Search Engine Advertising: Buying Your Way to the Top to Increase Sale (Voices that Matter)*. New Riders Press, New York, 2004.

45. L. Shapley and M. Shubik. The assignment game I: The core. *International Journal of Game Theory*, 1:111–130, 1972.

46. B.K. Szymanski and J-S. Lee. Impact of roi on bidding and revenue in sponsored search advertisement auctions. In *Second Workshop on Sponsored Search Auctions in Conjunction with the ACM Conference on Electronic Commerce (EC'06)*, Ann Arbor, Michigan, 2006.

47. H.R. Varian. Position auctions. Technical report, University of California, Berkeley, USA, 2006.

48. R. L. Zeff and B. Aronson *Advertising on the Internet*. John Wiley & Sons, New York, USA, 1999.

49. Y. Zhou and R. Lukose. Vindictive bidding in keyword auctions. In *Second Workshop on Sponsored Search Auctions in conjunction with the ACM Conference on Electronic Commerce (EC'06)*, Ann Arbor, Michigan, 2006.

# Chapter 4
# Mechanism Design for Resource Procurement in Grid Computing

This chapter is written with the objective of getting the readers to appreciate the value of game theoretic modeling and mechanism design to the problem of resource allocation in grid computing. As we proceed to show later, the dynamics of resource sharing and selfishness involved in grid computing offers itself in a natural way for the use of game theoretic analysis. This chapter is by no means an exhaustive treatment of such an application but is an eye-opener for the reader to appreciate what could be possible if the chisel of mechanism design is used on the marble of grid computing. The chapter is an extended version of the results presented in [13].

## 4.1 Grid Computing

The growth of the Internet, along with the availability of powerful computers and high-speed networks as low-cost commodity components, is changing the way scientists and engineers do computing, and is also changing how society in general manages information and information services. These new technologies have enabled the clustering of a wide variety of geographically distributed resources, such as supercomputers, storage systems, data sources, instruments, and special devices and services, which can then be used as a unified resource. Furthermore, they have enabled seamless access to and interaction among these distributed resources, services, applications, and data. A recent paradigm that harnesses these advancements and is continuously evolving is popularly termed *grid computing*.

Grid computing can be defined as flexible, secure, coordinated resource sharing and problem solving in dynamic, multi-institutional virtual organizations [7]. Grids enable the sharing, selection, and aggregation of a wide variety of resources including powerful computers, storage systems, data sources, and specialized devices that are geographically distributed and owned by different organizations for solving large-scale computational and data intensive problems in science, engineering, and commerce.

Y. Narahari et al., *Game Theoretic Problems in Network Economics and Mechanism Design Solutions*, Advanced Information and Knowledge Processing,
DOI: 10.1007/978-1-84800-938-7_4, © Springer-Verlag London Limited 2009

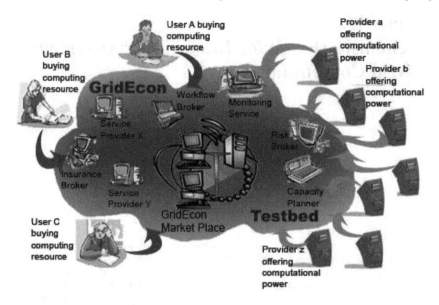

**Fig. 4.1** Overview of grid resource procurement (courtesy: http://www.gridecon.eu)

### 4.1.1 Resource Procurement Problem in Grid Computing

The resources in a grid include computational resources, storage systems, I/O devices, scientific instruments, computer networks, etc. These resources are geographically distributed and owned by different organizations. The sharing of these resources is subject to different administrative policies. Grid resource management is defined as the process of identifying requirements, matching resources to applications, allocating those resources, and scheduling and monitoring grid resources over time in order to run grid applications as efficiently as possible [22]. Resource procurement is an important aspect of grid resource management. Refer to Figure 4.1 for a pictorial representation of the grid resource procurement scenario (thanks:http://www.gridecon.eu).

In a computational grid, a typical user would want to submit a job that can be split into tasks that execute independently on various nodes on the grid. Specifically, the *parameter sweep type jobs* consist of a number of independent, identical tasks that can all be run in parallel. A user trying to procure the necessary resources for such a job would be ready to pay a usage fee for the resources that will be procured for the duration of execution of these tasks. This allocation of tasks to the resource providers is done on the basis of the cost and capacity bids of these resource providers.

Consider a grid user who has a *parameter sweep type* job to be submitted to the grid for computation. The grid user could conduct a procurement auction to procure resources for running the tasks contained in the job. While selecting the resource providers who are ready to offer their resources for running these tasks, the grid user would like to minimize the total cost of executing the overall job on hand. The grid user therefore has to come up with an allocation scheme that minimizes the total cost of using the available resources required for executing all the tasks. In order to solve this optimization problem, the grid user should know the true values of the costs of using the resources. However, the grid resource providers may not be willing to provide the information in a truthful way. This is because the resource providers are in general *autonomous*, *rational*, and *intelligent*.

- A grid resource provider is *autonomous* in the sense that neither the grid users nor the other resource providers have any direct control over the actions of the grid resource provider. There is no compulsion for the resource provider to comply with the dictates of a central authority.
- A grid resource provider is *rational* in the game theoretic sense of making decisions consistently in pursuit of his own objectives. Each resource provider's objective is to maximize the expected value of his own payoff measured in some utility scale. Recall that *selfishness* or *self-interest* is an important implication of rationality.
- Each resource provider is *intelligent* in the game theoretic sense of knowing everything about the underlying game that a game theorist knows, and he can make any inferences about the game that a game theorist can make. In particular, each resource provider is *strategic*, that is, he takes into account his knowledge or expectation of behavior of other agents. He is capable of doing the required computations.

In this chapter, we focus on the grid resource procurement problem when the resource providers are autonomous, rational, and intelligent. Henceforth, we will use the term *rational* to mean all three attributes: rationality, autonomy, and intelligence. In the presence of such resource providers, there is an overall degradation of the total efficiency as achieved by the computational grid when compared to the efficiency that can be achieved if the participating users were globally aligned under a single thread of control. This loss in efficiency might arise due to the resource providers' unwillingness to either perform at all or perform to the fullest capability, the computational jobs of other users in the grid. Hence, study of methods for grid resource procurement without analysis of the strategic behavior of the resource providers will not be useful.

## *4.1.2 Incentive Compatible Resource Procurement Mechanisms*

As we progress through the chapter, we design *procurement mechanisms* that a grid user can use for procuring resources for parameter sweep types of jobs in *computational grids with rational resource providers*. Specifically, we define the *incentive compatible resource procurement* problem in computational grids and focus on developing three elegant solutions to this problem.

The user announces on the grid information service, certain conditions and requirements of both the hardware and software resources required for executing the tasks making up his job. This information is received by all the prospective resource providers. Based on the requirements and conditions, resource providers decide whether or not to bid for executing the tasks. Once the user receives these bids, he aggregates them and, based on his requirements, attempts to find a cost minimizing allocation of resources for executing all the tasks.

As we had discussed, the resource providers might provide false bids about their cost and capacity values in an attempt to increase their pay-offs. With this as the setting, we attempt to design incentive compatible mechanisms that induce truth revelation by the resource providers.

In addition to the requirement of *incentive compatibility*, economic mechanisms should be designed in a way as to induce the resource providers to participate voluntarily in the scheme. For this to happen, the scheme must ensure that the resource providers are not worse off by participating than by not participating in the scheme, which is in essence *individual rationality*. As a grid user who is procuring resources for a job on hand, one would like to minimize the cost of procurement under the conditions of *incentive compatibility* and *individual rationality*. So, what we are looking for is an *optimal mechanism*.

### 4.1.3 Outline of the Chapter

We consider the resource procurement problem from the viewpoint of a grid user. By the end of this chapter we will have three elegant mechanisms as solutions to the resource procurement problem in computational grids with autonomous, rational, and intelligent resource providers. The mechanisms that are presented are:

- *G-DSIC* (Grid-Dominant Strategy Incentive Compatible) mechanism, which guarantees that truthful bidding is a best response for each resource provider, irrespective of what the other resource providers bid
- *G-BIC* (Grid-Bayesian Nash Incentive Compatible) mechanism, which only guarantees that truthful bidding is a best response for each resource provider whenever all other resource providers also bid truthfully
- *G-OPT* (Grid-Optimal) mechanism, which minimizes the cost to the grid user, satisfying at the same time (1) *Bayesian Incentive Compatibility* and (2) *Individual Rationality*.

Note that the mechanisms designed in this chapter are in the context of *parameter sweep* types of jobs, which consist of multiple homogeneous and independent tasks. However, the applicability of the mechanisms proposed transcends parameter sweep types of jobs, and in general, the proposed mechanisms could be extended to provide a robust way of procuring resources in a computational grid where the resource providers exhibit rational and strategic behavior.

The rest of this chapter is organized as follows. In Section 4.2 we describe a game theoretic model for the grid resource procurement problem and bring out the relevance of mechanism design. We first model the resource procurement problem as a noncooperative game with incomplete information. We use this model as the basis for all the three mechanisms developed in this chapter.

Section 4.3 explores the design of a dominant strategy incentive compatible mechanism, *G-DSIC*, for the resource procurement problem in computational grids with rational resource providers. We first design an allocation rule and pricing scheme based on the VCG mechanisms. This particular payment rule makes it an optimal strategy for the resource providers to reveal their true cost and capacity valuations, irrespective of the strategies of the other resource providers. We inspect certain properties of this mechanism.

The *G-DSIC* mechanism has very strong and desirable properties. However, the applicability of the *G-DSIC* scheme is limited due to two difficulties — lack of budget balance and another requirement, namely, absence of critical resource providers. Motivated by this, we develop, in Section 4.4, the *G-BIC* mechanism, which is a Bayesian incentive compatible resource procurement mechanism. The *G-BIC* protocol overcomes the difficulties identified in the design of the *G-DSIC* scheme and also satisfies a few additional desirable game theoretic properties.

Though the *G-DSIC* satisfies both *incentive compatibility* and *individual rationality*, it results in quite high procurement costs. The *G-BIC* mechanism overcomes this problem but does not ensure voluntary participation (*individual rationality*). Using this as the motivation, in Section 4.5, we design a mechanism that achieves both incentive compatibility and individual rationality while at the same time minimizes the cost of procurement for the grid user. We call this mechanism *G-OPT*. We discuss the properties of *G-OPT*.

## 4.2 The Model

Consider a grid user who has a job that he wishes to submit to the grid for computation. The nature of the job is of *parameter sweep type*, that is, the job consists of a number of tasks all of which can be run independently. We denote the number of tasks composing the main job by $m$. That is, the user can split the job into $m$ tasks. These tasks can run independently, and there is no dependency between any subset of these $m$ tasks. The user conducts a procurement auction to procure resources for running these $m$ tasks. With resource providers who are ready to offer their resources for running these tasks, the grid user would like to minimize the total cost of executing the overall job on hand. The grid user therefore has to come up with an allocation scheme that minimizes the total cost of using the available resources required for executing all the tasks. In order to solve this optimization problem, the grid user should know the true values of the costs of the resources.

The job information announced by the user on the grid information service is received by all the prospective resource providers. Based on the requirements and conditions, resource providers decide whether or not they want to bid for executing the tasks. Once the user receives these bids, he aggregates them and based on his requirements attempts to find a cost minimizing allocation of resources for executing all the tasks. He selects up to a maximum of $m$ resource providers (this case occurs when each resource provider is allocated only 1 task) out of $n$ bidders on whose resources the tasks will be executed. We represent the set of bidders (resource providers) by $N = \{1, 2, \ldots, n\}$.

In this setting, once the grid user provides the information about the tasks and calls for the bids, the resource providers respond with two-dimensional bids $(\hat{c}_1, \hat{q}_1)$, $\ldots$, $(\hat{c}_n, \hat{q}_n)$, where $\hat{c}_i$ ($i = 1, 2, \ldots, n$) represents the unit cost of provider $i$ for executing the tasks and $\hat{q}_i$ is the capacity of the resource provider $i$ (maximum number of tasks that the resource provider can handle). The grid user accrues a benefit $R(q)$ from completing $q$ tasks of the job. Also, in this model, we assume an all-or-nothing payoff to the grid user. That is, the grid user receives zero payoff if only a fraction of the $m$ tasks are completed and receives a certain payoff $R(q)$ only when all the $q$ tasks allocated to him are completed. This constant payoff is the *value* that the grid user attaches to the job. This reflects the real world quite accurately. Each resource provider $i$ has constant marginal execution cost $c_i \in [\underline{c}, \bar{c}] \subset [0, \infty]$ and also a maximum capacity $q_i \in [\underline{q}, \bar{q}] \subset [0, \infty]$. $\Phi_i$ denotes the joint distribution of the marginal cost $c_i$ and execution capacity $q_i$. We work with a joint distribution in order to allow for the cost and quantity values to be correlated. Note that the actual values of $(c_i, q_i)$ are known only to bidder $i$.

Since the resource providers are rational and intelligent, they will try to maximize their profit by possibly providing false bids of their valuations. The job of designing an incentive compatible mechanism addresses exactly this issue. The mechanism should make it optimal for the resource providers to bid their true valuations.

In our mechanism, the true type of each bidder is represented by $\theta_i = (c_i, q_i)$ and each bid is represented by $s_i = (\hat{c}_i, \hat{q}_i)$. The set of possible true types (and the bids) for resource provider $i$ is denoted by $\Theta_i$. Note that in our setting $\Theta_i = [\underline{c}, \bar{c}] \times [\underline{q}, \bar{q}]$. Throughout this chapter, we also adopt the notational convention of $\theta_{-i}$ for $(\theta_1, \ldots, \theta_{i-1}, \theta_{i+1}, \ldots, \theta_n)$ and $\Theta_{-i}$ for $\Theta_1 \times \ldots \times \Theta_{i-1} \times \Theta_{i+1} \times \ldots \times \Theta_n$. Thus a type profile $\theta$ is also represented as $(\theta_i, \theta_{-i})$. Table 4.1 summarizes the notation that we use in this chapter.

Now, one may recall from mechanism design theory in quasi-linear environments (Section 2.13) that a grid resource allocation mechanism, consists of:

1. an **allocation vector:** $k(\theta) = (k_1(\theta), k_2(\theta), \ldots, k_n(\theta))$, where $k_i(\theta)$ represents the number of tasks allocated to provider $i, (i = 1, 2, \ldots, n)$, and
2. a **payment vector:** $t(\theta) = (t_1(\theta), t_2(\theta), \ldots, t_n(\theta))$, where $t_i(\theta)$ represents the amount of monetary transferred from the grid user to the resource provider $i(i = 1, 2, \ldots, n)$.

The utility, $u_i((k(\theta), t(\theta)), \theta_i)$, of resource provider $i$ is given by

| $m$ | number of tasks contained in the job of the grid user |
|---|---|
| $n$ | number of resource providers |
| $N$ | set of resource providers in the auction $= \{1, \ldots, n\}$ |
| $i$ | index for resource providers |
| $R(q)$ | total revenue to the user by completing the $q$ tasks |
| $c_i$ | actual cost of executing a task for provider $i$ |
| $q_i$ | actual capacity of (maximum number of tasks that can be executed by) resource provider $i$ |
| $\theta_i$ | actual type of resource provider $i$, $\quad \theta_i = (c_i, q_i)$ |
| $\hat{c}_i$ | reported cost of executing a task for provider $i$ |
| $\hat{q}_i$ | reported capacity of the provider $i$ |
| $s_i$ | reported type of provider $i$, $\quad s_i = (\hat{c}_i, \hat{q}_i)$ |
| $\Theta_i$ | set of possible types of provider $i$ $\Theta_i = [\underline{c}, \bar{c}] \times [\underline{q}, \bar{q}]$ |
| $\theta$ | a profile of types of the providers, represented by $(\theta_1, \ldots, \theta_n)$ |
| $s$ | a profile of bids of providers, represented by $(s_1, \ldots, s_n)$ |
| $\Theta$ | set of profiles of the types of the providers $= \Theta_1 \times \Theta_2 \times \ldots \Theta_n$ |
| $\theta_{-i}$ | a profile of types without provider $i$, $= (\theta_1, \ldots, \theta_{i-1}, \theta_{i+1}, \ldots, \theta_n)$ |
| $s_{-i}$ | a profile of bids without provider $i$, $= (s_1, \ldots, s_{i-1}, s_{i+1}, \ldots, s_n)$ |
| $\Theta_{-i}$ | set of profiles of the types without provider $i$, $= \Theta_1 \times \ldots \times \Theta_{i-1} \times \Theta_{i+1} \times \ldots \times \Theta_n$ |
| $f(.)$ | social choice function |
| $k(.)$ | allocation rule of $f(.)$ |
| $k(\theta)$ | allocation vector when type profile is $\theta$ |
| $k_i(\theta)$ | allocation to resource provider $i$ when type profile is $\theta$ |
| $t_i(.)$ | payment to a provider $i$ |
| $t(\theta)$ | payment vector when type profile is $\theta$ |
| $t_i(\theta)$ | payment to resource provider $i$ when type profile is $\theta$ |
| $f(\theta)$ | outcome of the SCF for type profile $\theta$; $f(\theta) = (k(\theta), t(\theta))$ |
| $u_i(f(\theta), \theta_i)$ | utility of provider $i$ for outcome $f(\theta)$ and type $\theta_i$ |
| $v_i(k(\theta), \theta_i)$ | valuation of resource provider $i$ when allocation function is $k(.)$ and type is $\theta_i$ |
| $U_i(\theta_i \| f)$ | expected utility of provider $i$ when the provider's type is $\theta_i$ and the social choice functions is $f$ |
| $\phi_i(c_i, q_i)$ | joint probability density of provider $i$ corresponding to a task execution cost $c_i$ and capacity $q_i$ |
| $\Phi_i(c_i, q_i)$ | Cumulative distribution function corresponding to $\phi_i$ |

**Table 4.1** Notation used in the chapter

$$u_i((k(\theta), t(\theta)), \theta_i) = v_i(k(\theta), \theta_i) + t_i(\theta)$$

where $v_i(k(\theta), \theta_i)$ is the value that resource provider $i$ attaches to the allocation $k(\theta)$ when his type is $\theta_i$.

In this chapter, we seek to design a mechanism that would ensure truth elicitation from all the resource providers. Once true values are obtained, the allocation scheme would be an appropriate optimization algorithm that minimizes the cost to the grid user.

## 4.3 The G-DSIC Mechanism

In the case of dominant strategy implementation, recall that truth revelation becomes a best response strategy for each resource provider irrespective of what the other resource providers reveal. That is, we require that

$$u_i\left(f\left(\theta_i, \theta_{-i}\right), \theta_i\right) \geq u_i(f(s_i, \theta_{-i}), \theta_i)$$

$$\forall i \in N, \forall \theta_i \in \Theta_i, \forall \theta_{-i} \in \Theta_{-i}, \forall s_i \in \Theta_i \tag{4.1}$$

where the significance of the symbols used is explained in Table 4.1.

The *G-DSIC* mechanism implements the following social choice function truthfully in dominant strategies.

$$f(\theta) = (k_1(\theta), \ldots, k_n(\theta), t_1(\theta), \ldots, t_n(\theta)) \; \forall \theta \in \Theta.$$

As per our setting, the resource providers submit their bids, and we have to decide the allocation and payments based on the bids. We use the Clarke mechanism introduced in Section 2.15 to design an auction that achieves truth elicitation in dominant strategies.

Before we begin discussing the mechanism itself, we would like to define the notion of a *critical resource provider*. This notion is required because the structure of the proposed G-DSIC mechanism demands that the allocation problem be solvable with each resource provider excluded from the set of bidders.

**Definition 4.1.** *A critical resource provider in a resource procurement scenario is a resource provider without whose presence the job cannot be completed by any subset of the remaining resource providers and whose presence in some subset of winning resource providers makes the job executable. That is, the sum of the capacities of all the other resource providers is less than the number of tasks m.*

The above definition is equivalent to saying that a resource provider $i$ is critical if and only if,

$$\sum_{j \in N} q_j \geq m \text{ and } \sum_{j \in N, j \neq i} q_j < m.$$

The structure of the proposed *G-DSIC* mechanism requires that there be no critical resource provider in the system.

### 4.3.1 An Illustrative Example

Consider a grid user with a job that can be split into 10 tasks, and consider three resource providers who meet the requirements of the user, with values of $(c_i, q_i)$ being $RP_1$: (8,6), $RP_2$: (7,4), and $RP_3$: (9,7). Also let the values of $c_i$ be uniformly distributed in the interval $[5, 10]$ and the values of $q_i$ be uniformly distributed in the interval $[3, 8]$. Note that this assignment of values to the resource providers satisfies the condition that no critical resource provider exists in the system. The optimal choice of resource providers, given these values, is 3 units to $RP_2$, 6 units to $RP_1$, and 1 unit to $RP_3$. But $RP_1$, $RP_2$, and $RP_3$, being rational agents, would try to maximize their profit from the process and would report higher values of $c_i$. Now, the grid user, in order to minimize the total cost of execution, needs the true values of the types of the resource providers. To achieve this, the grid user has to pay the providers an appropriate information rent in addition to the payment he is required to make for getting the tasks executed. This is achieved by the *G-DSIC* mechanism as explained below.

As an incentive to encourage truthful bidding, the grid user agrees to pay to each of the resource providers the marginal value they contribute to the grid user. As per the Clarke mechanism, this marginal contribution of a given resource provider is computed as the difference in the total payment required to be made by the grid user with and without the resource provider. In our example,

$k = (6, 4, 0)$
$t_1 = 6 \times 8 +$ cost to user without $R_1 -$ cost to user with $R_1$
$\quad = 48 + (28 + 54) - (28 + 48) = 48 + 6 = 54$
$t_2 = 28 + (48 + 36) - (48 + 28) = 28 + 8 = 36$
$t_3 = 0 + (48 + 28) - (48 + 28) = 0.$

We reiterate that the absence of a critical resource provider in the mechanism is the key to obtaining the above solution. This is an important limitation of the proposed *G-DSIC* mechanism.

### 4.3.2 The G-DSIC Mechanism: Allocation and Payment Rules

The strength of the *G-DSIC* mechanism lies in the payment scheme. The allocation scheme is based on the types announced by the resource providers, which are assumed to be true. This assumption, of course, is justified by the payment scheme, which induces truth revelation. In this scheme, resource providers who do not execute any tasks do not receive any payment. Once the bids are received, they are sorted in increasing order of $\hat{c}_i$ value, and then progressively we allocate $\hat{q}_i$ tasks to the resource providers in that same order, removing from the sorted list providers to whom the allocation has just been made, until all tasks are over or we are left with $m'$ tasks with $m' <$ current value of $\hat{q}_i$. In the latter case, we allocate $m'$ tasks to the current provider. More formally, if we denote by $[i]$ the resource provider with the $i^{th}$ lowest cost valuation, and by $[\bar{i}]$ the resource provider who satisfies the property:

$$\sum_{j=[1]}^{[\bar{i}]-1} q_{[j]} < m \text{ and } \sum_{j=[1]}^{[\bar{i}]} q_{[j]} \geq m.$$

We can describe the allocation as

$$k_{[i]}^* = \begin{cases} \hat{q}_{[i]}, & [i] < [\bar{i}] \\ m - \sum_{j=[1]}^{[\bar{i}]-1} \hat{q}_{[j]}, & [i] = [\bar{i}] \\ 0, & \text{otherwise.} \end{cases} \tag{4.2}$$

Let us denote by $k^{-i}$ the optimal allocation of $m$ tasks that result when resource provider $i$ is not present in the system. The payment to each provider is characterized as follows.

**Payment function**

If there are no critical resource providers in the auction, then the payment to each resource provider is given by

$t_i(s) = k_i(s)c_i +$ optimal cost without provider $i$ $-$ optimal cost with provider $i$.

More formally,

$$t_i(s) = k_i^*(s)c_i + \sum_{j \in N} k_j^{-i}(s)c_j - \sum_{j \in N} k_j^*(s)c_j.$$

## 4.3.3 Properties of the G-DSIC Mechanism

In this section, we prove a few important properties of the *G-DSIC* mechanism. We show both *incentive compatibility* and *individual rationality*.

**Observation 1: Incentive Compatibility**

G-DSIC is a dominant strategy incentive compatible mechanism. This can be shown as a consequence of the fact that the G-DSIC mechanism uses the allocation rule and payment rule of the Clarke mechanism.

**Observation 2: Individual Rationality**

Recall from Section 2.10 that if a mechanism is individual rational, then agents participate in the mechanism voluntarily. That is, the resource providers get nonnegative utility by participating in the mechanism. Recall that there are three types of individual rationality depending on the stage at which the resource providers decide for or against participating in the mechanism. They are *ex-ante individual rationality*, *interim individual rationality*, and *ex-post individual rationality*. The notion of ex-ante individual rationality applies in a situation where the players decide whether or not to participate in the mechanism before knowing their types. Interim individual rationality is appropriate when the players decide whether or not to participate in the mechanism after knowing their types. Ex-post individual rationality is an appropriate notion to consider when the players decide whether or not to participate in the mechanism after announcing their types. In the current scenario, interim individual rationality is the most appropriate notion to consider since the resource providers already know their actual types before announcing their bids. It may also be recalled from Section 2.10 that ex-post individual rationality is the strongest form of individual rationality, and it implies interim individual rationality, which in turn implies ex-ante individual rationality.

Since interim individual rationality is most appropriate in the current context, we now investigate this property. In fact, below, we show that the G-DSIC mechanism is interim individually rational. Let $f(\theta) = (k(\theta), t(\theta))$ be the social choice function. It is sufficient to show that

$$U_i(\theta_i|f) \geq 0, \qquad \forall\, \theta_i \in \Theta_i, \ \forall\, i \in N$$

where $U_i(\theta|f)$ is the interim expected utility of resource provider $i \in N$ given by

$$U_i(\theta_i|f) = E_{\theta_{-i}}[u_i((k(\theta), t(\theta)), \theta_i)].$$

We now define the expected utility in the following way. We first define the valuation function of a resource provider as his cost to execute the allocated tasks. Now, the utility of a resource provider $i$ is the sum of his valuation function and the payment received. If a resource provider is not allocated any tasks for execution, he receives no payment and he does not lose anything either. Define $opt_{i-}$ as the optimal cost of procurement without provider $i$ and $opt_{i+}$ as the optimal cost with provider $i$. Utility of a resource provider $i$ is thus,

$$\begin{aligned} u_i((k(\theta), t(\theta)), \theta_i)) &= v_i(k(\theta), \theta_i) + t_i(\theta) \\ &= -k_i(\theta)c_i + k_i(\theta)c_i + opt_{i-} - opt_{i+}. \end{aligned}$$

If resource provider $i$ is not among the winners, $v_i(k(s), \theta_i)$ is 0 and so is $t_i(s)$. But if provider $i$ is among the winners, the overall cost reduces with him present in the system. Thus optimal cost without provider $i$ minus the optimal cost with provider $i$ can never be negative. This is obviously true for every $i \in N$.

### 4.3.4 The G-DSIC Algorithm

Algorithm 1 describes an algorithm for implementation of the *G-DSIC* mechanism for grid resource procurement. It is easy to compute the computational complexity of the *G-DSIC* mechanism. The limiting computation in the algorithm for the *G-DSIC* mechanism is the payment calculation step. This involves performing $\mathbf{O}(n \log n)$ operations for each of the $n$ resource providers. This seems to lead to an overall computational complexity of $\mathbf{O}(n^2 \log n)$. However, since it is enough to do the sorting step exactly once, the worst case computational complexity is actually $\mathbf{O}(n^2)$.

---

**Algorithm 1: G-DSIC Algorithm**

1. The user sends out information about the job on hand through the Grid Information System.
2. The resource providers respond with bids in the prescribed format $(\hat{c}_i, \hat{q}_i)$.
3. The user then sorts the bids in increasing order of the $c_i$ values.
4. *WHILE* some tasks are still unallocated *DO*
    *IF* number of remaining tasks $\geq q_{[1]}$
        Allocate $q_{[1]}$ units to resource provider $[1]$.
        Remove the first bid (lowest remaining bid) from the list
    *ELSE*
        Allocate all remaining tasks to resource provider $[1]$.
    end *WHILE*
5. Having obtained the allocation function $y$, the user computes the payment to be made to provider $i$ as
$t_i(.) = k_i(.)\hat{c}_i$ + optimal cost without provider $i$ − optimal cost with provider $i$.

---

### 4.3.5 Limitations of G-DSIC

The G-DSIC scheme is based upon the VCG mechanism, so it inherits both the advantages and disadvantages of the VCG mechanism. Here we list two important limitations of the G-DSIC scheme.

- **L1**: There should be no critical resource provider. This makes the mechanism unusable at certain times even if there are enough resources with the set of resource providers to complete the job. For example, consider a grid user who has a job with $m = 10$. Suppose there are 3 resource providers each of whose capacity valuation is 4. Now in this case each of them is a critical resource provider and the job cannot be completed by the other two alone. In this case, we cannot apply the *G-DSIC* mechanism though the job can be completed by the three of them together. This limitation can, however, be overcome by designing a Groves mechanism with appropriate $h_i$ functions. In our case, we can use a single function $h$ for each of the resource providers to take care of this limitation.

- **L2**: VCG mechanisms, on which the *G-DSIC* protocol is based, are *not budget balanced* in general (See Section 2.14). Lack of budget balance is a well known limitation of VCG mechanisms. It also results in quite high payment for the grid user.

## 4.4 The G-BIC Mechanism

As we have seen in Section 4.3, there are two important drawbacks of the *G-DSIC* mechanism that limit its applicability. Also, the cost of procurement tends to be high when using the *G-DSIC* mechanism. As a solution to these problems, we design a *Bayesian incentive compatible* mechanism (*G-BIC*) for the grid resource procurement problem.

The basic model for this section is the same as the one discussed in Section 4.3. Here, we seek to achieve incentive compatibility in Bayesian Nash Equilibria.

### *4.4.1 The G-BIC Mechanism: Allocation and Payment Rules*

The dAGVA mechanism, as any other mechanism with quasi-linear utilities, consists of an allocation function and a payment function. In this section, we design a grid resource procurement mechanism based on the dAGVA mechanism 2.17. This Bayesian incentive compatible mechanism implements the social choice function (SCF) $f(s) = (k(s), t_1(s), \ldots, t_n(s))$, $\forall s \in \Theta$ corresponding to the job allocation problem in computational grids. The allocation rule in the *G-BIC* mechanism is the same as the one for *G-DSIC*.

Thus, following the same allocation as in Section 4.3, once the bids are received, they are sorted in increasing order of $c_i$ values, and then progressively we allocate $q_i$ tasks to the providers in that same order, removing from the sorted list resource providers to whom the allocation has just been made, until all tasks are over or we are left with $m'$ with $m' <$ current value of $q_i$. In the latter case, we allocate $m'$ tasks to the current provider.

Suppose all resource providers who receive one or more tasks to be executed are classified as winners belonging to set $W$, and the rest are in the set $L$. As explained in the allocation scheme, all except one resource provider in the set $W$ receive $q_i$ tasks while the remaining provider in $W$ receives $m'$ tasks with $m' \leq q_i$. The valuation function $v_i(k(s), \theta_i)$ of provider $i$ is the cost of completing the tasks given to him at per unit cost $c_i$. Now the valuation function is different for each resource provider, depending on his type profile and the number of tasks he is allocated.

$$v_i(k^*(s), \theta_i) = -c_i q_i \quad \text{if } i \in W, k^*{}_i(s) = q_i \tag{4.3}$$

$$= -c_i m' \quad \text{if } i \in W, k^*{}_i(s) = m' \tag{4.4}$$

$$= 0 \qquad \text{if } i \notin W. \tag{4.5}$$

Using the dAGVA expression for computing the payments, we get

$$t_i(s) = E_{\theta_{-i}} \left[ \sum_{p \neq i} v_p(k(s), s_p) \right]$$
$$- \left( \frac{1}{n-1} \right) \sum_{j \neq i} E_{\theta_{-j}} \left[ \sum_{p \neq j} v_p(k(s), s_p) \right].$$

Now replacing the value of $v_i(k(s), \theta_i)$ from (4.3), (4.4) and (4.5), we get $\forall i \in N$, $\forall s \in \Theta$,

$$t_i(s) = \left( \frac{1}{n-1} \right) \sum_{j \neq i} E_{s_{-j}} \left[ \sum_{p \in W, p \neq j} c_p k_p(s) \right] - E_{s_{-i}} \left[ \sum_{p \in W, p \neq i} c_p k_p(s) \right]. \tag{4.6}$$

### 4.4.2 Properties of the G-BIC Mechanism

The *G-BIC* mechanism overcomes both the limitations of the *G-DSIC* mechanism that we had discussed earlier.

**Observation 1**

Since the payments to the resource providers are computed in one go, the allocation need not be computed multiple times. This also eliminates the need that there be no critical resource providers in the system. Thus the *G-BIC* mechanism can be applied to a wider range of problems when compared to the *G-DSIC* mechanism.

**Observation 2**

The *G-BIC* mechanism achieves budget balance. This can be easily seen as follows: Let $\tau_i(\theta_i) = E_{s_{-i}} \left[ \sum_{j \neq i} v_j(k(\theta_i, s_{-i}), \theta_j) \right]$. So, the payment to the resource providers according to our scheme is

$$t_i(s) = \tau_i(\theta_i) - \left( \frac{1}{n-1} \right) \sum_{j \neq i} \tau_j(\theta_j).$$

Now,

$$\sum_i t_i(s) = \sum_i \tau_i(\theta_i) - \left( \frac{1}{n-1} \right) \sum_i \sum_{j \neq i} \tau_j(\theta_j) = \sum_i \tau_i(\theta_i) - \left( \frac{1}{n-1} \right) \sum_i (n-1) \tau_i(\theta_i)$$

$$= 0.$$

This completes the proof of the observation.


**Observation 3**

It can be noted that all resource providers end up paying a positive amount to the grid user before the auction while the winners receive a positive amount as payment after the allocated tasks are completed. This positive amount paid to the grid user before the auction can be viewed as a participation fee. It is mainly due to this that the *G-BIC* mechanism loses individual rationality, which is one of the drawbacks of the scheme.


**Observation 4**

The *G-BIC* mechanism is very attractive to the grid user as it is budget balanced and hence he need not have to pay for using any of the services. But it might not be feasible as it does not satisfy individual rationality conditions. In a scenario where the grid user can enforce participation through other means, *G-BIC* would be the most preferred mechanism.


## 4.4.3 The G-BIC Algorithm

Algorithm 2 describes an algorithm that shows different steps in the *G-BIC* protocol.

---
**Algorithm 2: G-BIC Algorithm**
1. The user sends out information about the job on hand through the Grid Information System (GIS).
2. The resource providers respond with bids in the prescribed format $(c_i, q_i)$.
3. The user then sorts the bids in increasing order of the $c_i$ values.
4. *WHILE* some tasks are still unallocated *DO*
      *IF* number of remaining tasks $\geq q_{[1]}$
            Allocate $q_{[1]}$ units to resource provider [1].
            Remove the first bid (lowest remaining bid) from the list
      *ELSE*
            Allocate all remaining tasks to resource provider [1].
   end *WHILE*
5. Having obtained the allocation function $y$, the user calculates the payment to be made to provider $i$ as given in Equation 4.6.

---

In order to find the complexity of the *G-BIC* algorithm, one should note that in finding the actual allocation, the first step is the sorting of bids, which takes $\mathbf{O}(nlogn)$ time. But, when the actual payments are calculated, we are computing the expected externality imposed by a resource provider's valuation on the others. This takes **O**

($n$) time for each resource provider and therefore takes total $\mathbf{O}\,(n^2)$ time. So, the overall complexity of the algorithm is also $\mathbf{O}\,(n^2)$.


## 4.5  G-OPT: An Optimal Auction Mechanism

We have shown in the previous section that the *G-BIC* mechanism cannot guarantee *individual rationality*. In real applications of interest, the resource providers are free to withdraw from the mechanism and due to this, the mechanisms that we design should satisfy *individual rationality* constraints in addition to *incentive compatibility*, in order to ensure that the resource providers voluntarily participate in the system.

We had already shown that the *G-DSIC* mechanism is a *individually rational* mechanism. But the *G-DSIC* mechanism results in higher costs to the grid user. We would like to design a mechanism that achieves both *incentive compatibility* and *individual rationality* and minimizes the cost to the grid user compared to the *G-DSIC* mechanism. As a grid user, one would not be concerned about the efficiency of an instance of an allocation of tasks to the resource providers. That is to say, the grid user, himself being a rational agent in the system, will not be averse to choosing an allocation that gives less utility to some of the resource providers, if, in the process, the grid user can ensure that he minimizes his cost. So, we would like to design a mechanism, which, while slightly sacrificing efficiency, can ensure that the grid user minimizes his cost of procurement. For this setting, optimal auction mechanisms seem to be the logical choice for implementing the goals of the grid user. The work in this section is adapted from the work in [14] and [20].


### 4.5.1  Preliminaries

Here we recall the model that was introduced in Section 4.3. We would be using a similar model with some additional assumptions to enable us to design the required mechanism. Table 4.1 describes the notation. In this section, we use the following additional notation:

$$T_i(\hat{c}_i, \hat{q}_i) = \mathbf{E}_{\theta_{-i}}[t_i((\hat{c}_i, \hat{q}_i), \theta_{-i})]$$

$$K_i(\hat{c}_i, \hat{q}_i) = \mathbf{E}_{\theta_{-i}}[k_i((\hat{c}_i, \hat{q}_i), \theta_{-i})].$$


#### 4.5.1.1  Some Definitions

We first define some notions that would be of use to us in the development of the *G-OPT* mechanism. The offered expected surplus gives us a measure of the surplus that is on offer to a resource provider under a certain mechanism. That is, it characterizes the expected value of the profit to a resource provider, were he to participate in the mechanism.

**Definition 4.2. Offered Expected Surplus:** For a resource procurement mechanism $(k,t)$, we define

$$\rho_i(\hat{c}_i, \hat{q}_i) = T_i(\hat{c}_i, \hat{q}_i) - \hat{c}_i K_i((\hat{c}_i, \hat{q}_i)).$$

The offered expected surplus when the resource provider $i$ bids $(\hat{c}_i, \hat{q}_i)$ is a measure of the expected transfer payment. The expected surplus $\pi_i(\hat{c}_i, \hat{q}_i)$ of resource provider $i$ when the bid is $(\hat{c}_i, \hat{q}_i)$ is given by

$$\pi_i(\hat{c}_i, \hat{q}_i) = T_i(\hat{c}_i, \hat{q}_i) - c_i K_i((\hat{c}_i, \hat{q}_i)) = \rho_i(\hat{c}_i, \hat{q}_i) + (\hat{c}_i - c_i) K_i((\hat{c}_i, \hat{q}_i)).$$

The true surplus $\pi_i$ is equal to the offered surplus only if the mechanism is incentive compatible.

Following the idea from [20] and [14], we use a specific function called the *virtual cost function*, which is then used to rank the resource providers. The use of this virtual cost function in place of the actual cost bids, increases the expected profit of the grid user possibly at the cost of efficiency.

**Definition 4.3. Virtual Cost:** We define virtual cost as

$$H_i(c_i, q_i) = c_i + \frac{\Phi_i(c_i|q_i)}{\phi_i(c_i|q_i)}$$

where $\phi_i(.)$ is the conditional probability density function of cost of task execution of resource provider $i$ conditioned on his capacity, and $\Phi_i(.)$ is the corresponding conditional distribution function. This virtual cost parameter is similar to the virtual value parameter used in [20], except that here we allow for the cost and the quantity values to be correlated.

#### 4.5.1.2 Two Assumptions

The following assumptions are required. While these assumptions are not so strong as to limit the application of the results presented, they simplify the model enough to derive certain results that would not otherwise be achievable.

- **Assumption 1:** For all $i = 1, 2, \ldots, n$, the joint density function $\phi_i(c_i, q_i)$ is completely defined for all $c_i \in [\underline{c}, \bar{c}]$ and $q_i \in [\underline{q}, \bar{q}]$, and the conditional density function $\phi_i(c_i|q_i)$ has full support. This means that $\phi_i(c_i|q_i)$ will not be 0 in the interval of interest.
- **Assumption 2:** For all $1 = 1, 2, \ldots, n$, the virtual cost function $H_i(c_i, q_i)$ defined in Definition 4.3 is nondecreasing in $c_i$ and nonincreasing in $q_i$. It is to be noted that this assumption imposes a condition on the relation between the cost and quantity of each resource provider. This is similar to the regularity assumption in [20]. It holds especially when the cost and the quantity are independent of each other.

### 4.5.1.3 The G-OPT Mechanism: Formulation

As auction designers, we are looking for the allocation function $k$ and the transfer function $t$ that minimize the expected payment to be made by the grid user. Now the total expected profit that we are trying to maximize for the user is

$$\pi(k,t) = \mathbf{E}_s[R(q) - \Sigma_{i=1}^n t_i(s)]$$

subject to:

1. *Feasibility*: : $k_i(s) \le q_i \ \forall i = 1, \ldots, n$ and $s \in \Theta$.
2. *Individual Rationality* : The expected interim payoff for each bidder is nonnegative. That is,
   $\pi_i(\theta_i) = \mathbf{E}_{\theta_{-i}}[t_i(s) - c_i k_i(s)] \ge 0$.
3. *Bayesian Incentive Compatibility*: The bidders must be induced by the mechanism to truthfully bid their valuations; in particular, we require that truth revelation be optimal for each resource provider, provided the other resource providers bid truthfully. That is,
   $E_{\theta_{-i}}[t_i(\theta_i, \theta_{-i}) - c_i k_i(\theta_i, \theta_{-i})] \ge E_{\theta_{-i}}[t_i(s_i, \theta_{-i}) - c_i k_i(s_i, \theta_{-i})], \forall i \in N, \forall \theta_i \in \Theta_i, \forall s_i \in \Theta_i$.

We receive a cost minimizing mechanism for the grid user as solution to this optimization problem. Any mechanism that satisfies the constraints given above and achieves a minimum cost for the grid user is said to be an optimal auction on the lines of Myerson [20].

The problem of designing an optimal mechanism for the model we had discussed in Section 4.2 is different from the classic optimal auction presented in [20] in the sense that the demand is not of unit quantity. The grid user has a finite demand and allocates possibly multiple tasks to each resource provider. It is this multidimensional type of the resource providers that make this an interesting and nontrivial problem. We not only have to consider the cost but also maximum quantity as private information. This is nontrivial in the sense that the resource providers are not of unit capacity type or uncapacitated. This two-dimensional type profile of the resource providers makes the application of some traditional optimizing schemes unsuitable. This is illustrated in the example below.

### 4.5.1.4 An Illustrative Example

We consider the same situation that we had discussed in Section 4.3. The grid user has a job that can be split into 5 tasks. Suppose that there are three resource providers with $(c_i, q_i)$ values of $RP_1 : (1,5), RP_2 : (1,1)$, and $RP_3 : (5,5)$. Now assume that the user were to conduct the classic $K^{th}$ price auction (the $K$ here is not to be confused with the $K(.)$ used earlier), where the marginal payment to a resource provider is equal to the cost of the first losing resource provider. The user then calls for bids from the resource provider for the job he has on hand with details of the job. Assume that all the three resource providers meet the requirements of the user and participate in the bidding process.

The auction may ensure that it is optimal for the resource providers to bid their true costs but does not deter them from possibly altering their quantity bids. To see this, consider that all resource providers bid truthfully both the cost and the quantity bids. In this case, resource provider $RP_1$ gets all the five tasks allocated to him, and he receives as marginal payment the cost value of the first losing resource provider, $RP_2$, which is also 1 unit. So the total payment made to resource provider $RP_1$ is 5 units, which leaves him with a profit of 0 units.

But instead of bidding his true quantity valuation of 5, if he had bid his quantity value as 4, then 4 units would have been allocated to resource provider $RP_1$, and 1 unit would have been allocated to resource provider $RP_2$, and both the winning resource providers, $RP_1$ and $RP_2$, would have been made a marginal payment equal to the value of the cost value of the losing resource provider, $RP_3$, which is 5 units. Thus resource provider $RP_1$ would get 20 units as payment, which leaves him with a profit of 16 units. Thus it is quite evident that such uniform price mechanisms are not applicable to the case where both cost and quantity are private information. The intuitive explanation for this could be that by underbidding their capacity values, the resource providers create a fictitious shortage of resources in the system thereby forcing the grid user to pay overboard for use of the virtually limited resources.

## 4.5.2 Designing the G-OPT Mechanism

Before designing the *G-OPT* mechanism, we offer some insights into the problem itself by presenting some simplified models which have been obtained by relaxing certain conditions in the original model. These relaxations present us with a holistic view of the problem without some of the restrictions that are present in the original formulation we have discussed.

### 4.5.2.1 First-Fit Solution

In the first relaxation, we assume that the grid user has complete information about all the resource providers present in the bidding process. That is to say, the grid user has accurate knowledge of the cost and quantity valuations of all the bidders in the system. It can also be viewed as a process in which the bidders will always have to reveal their true types without the need for incentives to be provided by the grid user. This makes the problem easy to analyze. The problem can be viewed as the grid user being faced with a single bidder with a price curve and infinite capacity. This can be easily solved as explained in Section 4.3. The grid user can assign tasks at full capacity to some resource providers in ascending order of their cost valuations, and at most one resource provider may have to be allocated tasks at less than his full capacity. The set of resource providers who are allocated some tasks all have cost valuations less than or equal to the set of resource providers who do not receive an allocation. The payment made to each provider is of course exactly his cost valuation. The payment is needed as we need to ensure individual rationality (IR). This gives us an indication on the structure of the optimal allocation scheme.

### 4.5.2.2 Common Execution Costs Assumption

We have seen that if both cost and capacity are common knowledge, the optimal procurement scheme is, in fact, a simple allocation problem. Making the problem one step more complicated, if we are to assume that only the capacities of the resource providers are private information while each resource provider's marginal execution cost is common knowledge, then our problem reduces to a single dimensional type case. For the sake of convenience, let us denote $t_i(.)$ as $t_i(\hat{q})$ and $k_i(.)$ as $k_i(\hat{q})$. Resource provider $i$ bids $\hat{q}_i$ when he observes $q_i$ as his capacity type. If the user were to set the transfer payment $t_i(\hat{q}) = c_i k_i(\hat{q})$ where $c_i$ is accurately known to the user, then the resource provider's surplus is zero. Since the allocation scheme $k_i(\hat{q}_i, q_{-i})$ is weakly increasing in $\hat{q}_i$ for all $q_{-i}$, it becomes a weakly dominant strategy for the resource provider to bid his true capacity information. Thus the same solution for the first-best case works for this case also. This proves that the grid user does not need to make any extra payment for obtaining the capacities of the resource providers once the marginal costs are known. This implies that the grid user can safely ignore the incentive compatibility constraints when allocating the tasks as long as he has complete information about the resource providers' marginal costs. Though this is not the generalized optimal scheme, it is not an uncommon situation by itself in the grid scenario. After repeatedly bidding in many auctions, the resource providers' marginal costs may become fairly common knowledge, whereas, depending on their own current load of jobs at a particular time and the amount of tasks executing from other users, their capacities will continue to be private information. In this case, it is evident that no extra payment is to be made by the grid user in order to extract this information.

### 4.5.2.3 Characterizing the Optimal Solution

We are now ready to characterize the optimal procurement scheme for the grid resource procurement problem. We provide certain characterizations of all incentive compatible and individually rational mechanisms. From this set of the mechanisms, the one that minimizes the procurement cost for the grid user is the optimal mechanism that we are looking for. We now present an important lemma that characterizes the set of all incentive compatible and individually rational mechanisms.

**Lemma 4.1.** *For the resource procurement problem under discussion, a feasible allocation rule $k(.)$ is Bayesian incentive compatible and individually rational if and only if the expected allocation $K_i(c_i, q_i)$ is nonincreasing in cost valuation $c_i$, and the offered surplus $\rho_i(\hat{c}_i, \hat{q}_i)$ is of the form*

$$\rho_i(\hat{c}_i, \hat{q}_i) = \rho_i(\bar{c}, \hat{q}) + \int_{\hat{c}_i}^{\bar{c}} K_i(x, \hat{q}_i)dx. \tag{4.7}$$

*Also $\rho_i(\hat{c}_i, \hat{q}_i)$ must be nonnegative and nondecreasing in $\hat{q}_i$ for all $\hat{c}_i \in [\underline{c}, \bar{c}]$.*

**Proof:** The proof here follows arguments similar to those put forward by Kumar and Iyengar [14]. Suppose that $K_i(c_i, q_i)$ is nonincreasing in cost valuation $c_i$ for all $q_i$. Also suppose that the offered expected surplus is of the form

$$\rho_i(\hat{c}_i, \hat{q}_i) = \rho_i(\bar{c}, \hat{q}) + \int_{\hat{c}_i}^{\bar{c}} K_i(x, \hat{q}_i)dx$$

such that $\rho_i(\hat{c}_i, \hat{q}_i)$ is nonnegative and nondecreasing in $\hat{q}_i$ for all $\hat{c}_i \in [\underline{c}, \bar{c}]$. It can be easily seen that the resource providers do not benefit from ever overbidding their capacity valuation because they may end up not being able to complete the tasks allocated to them, which may result in a penalty for them. Similarly, the resource providers, being rational, will not bid below their cost valuations as they will incur a loss by doing so. So, we can safely assume $\hat{q}_i \leq q_i$. This is a crucial observation for the rest of the proof.

For the $\rho_i(\hat{c}_i, \hat{q}_i)$ function that we have chosen, the expected surplus of the provider, $\pi_i(\hat{c}_i, \hat{q}_i)$, is,

$$= \rho_i(\hat{c}_i, \hat{q}_i) + (\hat{c}_i - c_i)K_i((\hat{c}_i, \hat{q}_i))$$
$$= \rho_i(\bar{c}_i, \hat{q}_i) + \int_{\hat{c}_i}^{\bar{c}} K_i((x, \hat{q}_i))dx + (\hat{c}_i - c_i)K_i((\hat{c}_i, \hat{q}_i))$$
$$= \rho_i(\bar{c}_i, \hat{q}_i) + \int_{\hat{c}_i}^{c_i} K_i((x, \hat{q}_i))dx + \int_{c_i}^{\bar{c}} K_i((x, \hat{q}_i))dx + (\hat{c}_i - c_i)K_i((\hat{c}_i, \hat{q}_i))$$
$$= \rho_i(\bar{c}_i, \hat{q}_i) + \int_{c_i}^{\bar{c}} K_i((x, \hat{q}_i))dx + \int_{\hat{c}_i}^{c_i} K_i((x, \hat{q}_i))dx + (\hat{c}_i - c_i)K_i((\hat{c}_i, \hat{q}_i)).$$

Now, since $K$ can never be negative and $\hat{c}_i \geq c_i$, we obtain,

$$\pi_i(\hat{c}_i, \hat{q}_i) \leq \rho_i(\bar{c}_i, \hat{q}_i) + \int_{\hat{c}_i}^{c_i} K_i(x, \hat{q}_i)dx.$$

As per our assumption, $\rho_i(\hat{c}_i, \hat{q}_i)$ is nonnegative and nondecreasing in $\hat{q}_i$ for all $\hat{c}_i \in [\underline{c}, \bar{c}]$. Replacing $\hat{q}_i$ by $q_i$ and using the fact that $\hat{q}_i \leq q_i$

$$\pi_i(\hat{c}_i, \hat{q}_i) \leq \rho_i(\bar{c}_i, q_i) + \int_{\hat{c}_i}^{c_i} K_i(x, \hat{q}_i)dx$$
$$= \rho_i(c_i, q_i) = \pi_i(c_i, q_i).$$

This proves the *if* part of the lemma. To prove the *only if* part, we pick an arbitrary incentive compatible mechanism $(k(s), t(s))$. Now the expected surplus of provider $i$ is given by

$$\pi_i(c_i, q_i) = \max\{T_i(\hat{c}_i, \hat{q}_i) - c_i K_i(\hat{c}_i, \hat{q}_i)\}. \tag{4.8}$$

As we noted before, the capacity will never be overbid, that is, $\hat{q}_i \leq q_i$. It follows that for fixed $q_i$ the surplus $\pi_i(c_i, q_i)$ is convex in the cost bid $c_i$. This means that $\pi_i(c_i, q_i)$ is absolutely continuous in cost and differentiable almost everywhere in $c$. Since we assumed $k(s)$ is incentive compatible, truthful bidding is a solution to equation (4.8). Thus,

$$c_i \in \text{argmax}\{T_i(\hat{c}_i, q_i) - c_i K_i(\hat{c}_i, q_i)\}. \tag{4.9}$$

Since $\pi_i(c_i, q_i)$ is convex in $c_i$, equation (4.9) implies

$$\frac{\partial \pi_i(c_i, q_i)}{\partial c_i} = -K_i(c_i, q_i).$$

This means that $K_i(c_i, q_i)$ is nonincreasing in $c_i$. This proves the "only if" part of the lemma.

We have shown that for all $\hat{c}_i \in [\underline{c}, \bar{c}]$ and all $\hat{q}_i \in [\underline{q}, \bar{q}]$, $\pi_i(\hat{c}_i, \hat{q}_i) \leq \pi_i(c_i, q_i)$. This proves that the mechanism is Bayesian incentive compatible under the conditions mentioned in the lemma. Also since it is Bayesian incentive compatible, $\pi_i(c_i, q_i) = \rho_i(c_i, q_i)$, and since according to the conditions $\rho_i(c_i, q_i)$ is nonnegative, so is $\pi_i(c_i, q_i)$. This means that the mechanism that satisfies the conditions specified in Lemma 4.1 is individually rational as well.

### 4.5.3 The G-OPT Mechanism: Allocation and Payment Rules

In this section, we present the allocation rule and the payment function for the *G-OPT* mechanism. The first thing to note before we proceed with the design of the mechanism is that in order to minimize the cost of procurement of resources, the mechanism should be such that the additional cost paid by the user as informational rent must be minimized as much as possible.

Here we present an allocation rule and a payment function that we claim is an optimal mechanism for the grid resource procurement problem. Before we proceed, recall the definition of the virtual cost function

$$H_i(c_i, q_i) = c_i + \frac{\Phi_i(c_i|q_i)}{\phi_i(c_i|q_i)}.$$

We would be using the virtual cost function rather than the actual costs in the optimal mechanism. We would be using these $H_i(c_i, q_i)$ values to compute the assignment vector. Thus, using a method similar to the *first-fit solution* case except for the fact that we use $H_i$ values rather than $c_i$ values, we rank the resource providers on the basis of their $H_i(c_i, q_i)$ values and keep allocating tasks to them at their full capacity in increasing order of $H_i$ values until all the tasks are allocated. Let $[i]$ denote the resource provider with $i^{th}$ lowest $H_i(c_i, q_i)$ value and $[\bar{i}]$ be the resource provider such that $[\bar{i}]$ satisfies,

$$\sum_{j=[1]}^{[\bar{i}]-1} q_{[j]} < m \text{ and } \sum_{j=[1]}^{[\bar{i}]} q_{[j]} \geq m.$$

Then the allocation function is given by

$$k_{[i]} = \begin{cases} \hat{q}_{[i]}, & [i] < [\bar{i}] \\ m - \sum_{j=[1]}^{[i]-1} \hat{q}_{[j]}, & [i] < [\bar{i}] \\ 0, & \text{otherwise.} \end{cases} \quad (4.10)$$

With this allocation function, it is easy to see that $K_i(c_i, q_i)$ is nonincreasing in $c_i$ and nondecreasing in $q_i$. Given this allocation rule, we present the payment function that would make this an optimal auction for the resource procurement problem:

$$t_i(s) = c_i k_i(s) + \int_{c_i}^{\bar{c}} K_i((x, \hat{q}_i)) dx. \quad (4.11)$$

### 4.5.4 The G-OPT Algorithm

Algorithm 3 describes an algorithm for implementation of the *G-OPT* mechanism for grid resource procurement.

---

**Algorithm 3: G-OPT Algorithm**

1. The user sends out information about the job on hand through the Grid Information System (GIS).
2. The resource providers respond with bids in the prescribed format $(c_i, q_i)$.
3. The user computes the $H_i(c_i, q_i)$ values from the bids.
4. The user then sorts the bids in increasing order of the $H_i$ values.
4. *WHILE* some tasks are still unallocated *DO*
        *IF* number of remaining tasks $\geq q_{[1]}$
                Allocate $q_{[1]}$ units to resource provider [1].
                Remove the first bid (lowest remaining bid) from the list
        *ELSE*
                Allocate all remaining tasks to resource provider [1].
    end *WHILE*
5. Having obtained the allocation $y$, the user computes the payment to be made to provider $i$ as
$t_i(s) = c_i k_i(s) + \int_{c_i}^{\bar{c}} K_i((x, \hat{q}_i)) dx.$

---

The analysis of this algorithm is very similar to the analysis of the *G-BIC* algorithm. On similar lines it can be shown that the overall complexity of this algorithm is also $\mathbf{O}(n^2)$.

### 4.5.5 Properties of the G-OPT Mechanism

We now present some of the properties of the *G-OPT* mechanism that we have designed.

**Lemma 4.2.** *The allocation mechanism defined by (4.10) has the following properties:*

1. $K_i(c_i, q_i)$ *is nonincreasing in $c_i$ for all fixed $q_i \in [\underline{q}, \bar{q}]$.*
2. $K_i(c_i, q_i)$ *is nondecreasing in $q_i$ for all fixed $c_i \in [\underline{c}, \bar{c}]$.*

**Proof**: From the design of the allocation scheme, it can be seen that $k_i(c_i, q_i)$ is nonincreasing in $H_i(c_i, q_i)$. Now from assumption 2, we know that $H_i(c_i, q_i)$ is nondecreasing in $c_i$. Hence $k_i(c_i, q_i)$ is nonincreasing in $c_i$. Now, taking expectation over $\theta_{-i}$, we get the result that $K_i(c_i, q_i)$ is nonincreasing in $c_i$. This proves the first part of the lemma. The second part is easy to see. For a fixed cost value, $c_i$, the value of $k_i(c_i, q_i)$ is nondecreasing with increase in the value of $q_i$. Hence by taking expectation over $\theta_{-i}$, we obtain the second part of the lemma.

Q.E.D.

**Theorem 4.1.** *The* G-OPT *mechanism with allocation function, k defined by (4.10) and the payment function t as defined by (4.11), is* Bayesian incentive compatible, interim individually rational *and cost minimizing.*

**Proof**: First note that the offered surplus to a resource provider is given by

$$\rho_i(\hat{c}_i, \hat{q}_i) = \int_{\hat{c}_i}^{\bar{c}} K_i((x, \hat{q}_i)) dx.$$

Thus, according to this scheme, $\rho_i(\bar{c}, \hat{q}_i) = 0$. That is to say, the informational rent paid by the user to resource providers whose cost valuation is at its highest is zero. Also note that the payment made to a resource provider whose expected allocation is zero is also zero. Thus the mechanism minimizes these causes of loss of revenue to the minimum possible.

Now $K_i(c_i, q_i) = E_{\theta_{-i}}[k((c_i, q_i), \theta_{-i})]$. Then according to Lemma 4.2, $K_i$ is nonincreasing in cost valuation $c_i$ and nondecreasing in capacity valuation $q_i$. Also, the offered surplus is of the form specified by Lemma 4.1. Hence, it can be concluded that the *G-OPT* is indeed both Bayesian incentive compatible and individually rational. This completes the proof of the theorem.

**Observation 6**

The optimal mechanism reduces to the *G-DSIC* mechanism under the following assumptions.

1. The bidders are symmetric.

2. The joint distribution functions of the bidders are regular.

This means that for the symmetric bidders case that we have considered, the optimal auction mechanism would yield the same payment to the grid user as the *G-DSIC* mechanism. In order to then realize the value of this optimal auction, we have to move to a more general setting of asymmetric bidders, where the values of $\bar{c}, \underline{c}, \bar{q}, \underline{q}$ and the joint distribution functions $\Phi_i$ are in general different for different resource providers. In fact, this is the exact setting in our numerical experimentations.

## 4.6 Current Art and Future Perspective

We set out to develop resource procurement mechanisms with strong economic properties for the procurement of resources for running a parameter sweep job on a computational grid. We now briefly review the relevant work in the literature and provide a future perspective.

### 4.6.1 Current Art: The Field is Young but Rich

The reader has a plethora of interesting work in this field to assimilate and understand the need for economic mechanism design for the grid environment. It has remained an interesting idea to study the behavior of the resource providers in the presence of economic incentives that would make the resource providers voluntarily participate in the grid and perform to their maximum capacity to fulfill the overall goals of the grid and increase its efficiency. A few researchers [4], [23], [26] have studied both market-based and auction-based schemes for achieving the economic incentives. But the introduction of economic incentives tends to induce rational individuals to alter their bids or prices in order to increase their revenue. So studying the efficacy of auction mechanisms for procurement without a game theoretic analysis of the resource providers' actions will not provide a good insight into the actual working of these mechanisms. In such a setting where we have to ensure truth-elicitation, game theory and mechanism design [21, 19] become immediately applicable. There are many recent research initiatives that look into the question of looking at these game theoretic issues in grid resource allocation through the glass of mechanism design theory; examples of these include:

- analyzing loss in efficiency due to the presence of rational agents [15];
- modeling the load balancing problem as a game and developing a mechanism for it [11]; and
- modeling the resource allocation problem as a cooperative game and analyzing the dynamics involved [10].

In [24], the authors consider a classical centralized load balancing problem and give a game-theoretic flavor to it. They consider the problem of rational and self-interested clients who have to choose a server from a list of permissible servers. They model this scenario as an atomic congestion game. They derive tight bounds on the price of anarchy of the problem. The price of anarchy is a measure of the loss in overall efficiency due to the presence of selfish agents compared to the efficiency in a centralized system.

In [11], the authors investigate the problem of designing mechanisms for resource allocation problem in grids. They consider the problem of incentive compatibility, but the model they consider is severely limited because of their usage of a single dimensional type.

A slightly different approach to the grid resource management problem is using cooperative games. In [10], the authors formulate the static load balancing problem as a cooperative game between the resource providers and the grid user. Utilizing the Nash bargaining solution, the authors derive an optimal allocation scheme.

In [12], the authors consider design of online algorithms for scheduling of reusable resources. They provide bounds on the performance of these online mechanisms and characterize the class of incentive compatible mechanisms. They also provide a randomized algorithm that is shown to be better than any of the existing mechanisms.

In [1], a case is made for the application of game theoretic techniques in distributed systems. Specifically, the work studies application of cooperative games in distributed systems. The emphasis of the authors was completely on information secrecy rather than cost as a factor. This is a different approach to a problem in distributed computing where possession of information is a desired outcome for the participating entities

In [6] and [5], the authors explore the use of strategy-proof mechanisms to schedule divisible loads on bus-networked and linear distributed system, respectively. They study the use of incentives and punishments to processors as a method for truth elicitation and verify that the proposed algorithm is indeed strategy-proof.

## 4.6.2 Avenues for Further Exploration

The famous adage "Known is a drop, Unknown is an ocean" is very pertinent to the field of game theoretic modeling and analysis of resource procurement in grids. The field is all the more exciting because of this unique marriage of systems engineering, infrastructure, algorithms, parallel programming, economics and social aspects. Some of the interesting research directions are:

- Finding strong bounds for mechanisms in terms of the cost of procurement.
- Relaxing the assumptions on which the optimal scheme is based and developing a mechanism that will carry forward the same nice properties of the *G-OPT* mechanism.
- Considering other job types apart from parameter sweep and developing robust mechanisms for resource procurement for these jobs.

- Generalizing the results obtained to all forms of grid resource procurement. That is, apart from just computational grids, we can consider general grids and develop mechanisms for procurement of network bandwidth, storage space, etc.
- Allowing for resource providers to present cost curves instead of a single cost valuation. These types of auctions are known as volume discount auctions. A particular version of such an application is dealt with in the paper by Gautam et al [9].
- Considering the use of combinatorial auctions and combinatorial exchanges as a model for studying the problem of grid resource management.
- Circumventing the requirement of absence of critical resource providers in the *G-DSIC* mechanism using a Groves' payment rule that is more general than the Clarke's payment rule.

# References

1. I. Abraham, D. Dolev, R. Gonen, J. Halpern  Distributed computing meets game theory: robust mechanisms for rational secret sharing and multiparty computation. In *Proceedings of the Twenty-Fifth Annual ACM SIGACT-SIGOPS Symposium on Principles of Distributed Computing (PODC 2006)*, Denver, Colorado, USA, 2006.
2. L. Ausubel and P. Crampton.  Demand reduction and inefficiency in multi-unit auctions. Technical Report Mimeo, University of Maryland, College Park, MD, 1995.
3. F. Branco. Multiple unit auctions of an indivisible good. *Economic Theory*, 8:77–101, 1996.
4. R. Buyya, D. Abramson, J. Giddy, and H. Stockinger. Economic models for resource management and scheduling in Grid computing. *Concurrency and Computation: Practice and Experience*, 14(13):1507–1542, 2002.
5. T.E. Carroll and D. Grosu. A Strategy-Proof mechanism for scheduling divisible loads in linear networks. In *Proceedings of the 21st IEEE International Parallel and Distributed Processing Symposium (IPDPS 2007)*, March 2007, Long Beach, California, USA.
6. T.E. Carroll and D. Grosu. Strategy-proof mechanisms for scheduling divisible loads in bus-networked distributed systems. *IEEE Transactions on Parallel and Distributed Systems*, 19(8):1124–1135, 2008.
7. I. Foster, C. Kesselman, and S. Tuecke. The anatomy of the grid: Enabling scalable virtual organizations. *International Journal of Supercomputer Applications*, 15(3):200–222, 2001.
8. D. Garg, Y. Narahari and S. Gujar. Foundations of mechanism design: A tutorial — Part 1 and 2. In Sadhana — Indian Academy Proceedings in Engineering Sciences, Volume 33, Part 2, pages 83–130 and pages 131–174, April 2008.
9. R. Gautam, N. Hemachandra, Y. Narahari, V.H. Prakash. Optimal auctions for multi-unit procurement with volume discount bids. *The 9th IEEE International Conference on E-Commerce Technology and The 4th IEEE International Conference on Enterprise Computing, E-Commerce and E-Services (CEC-EEE 2007)*, pages 21–28, 2007.
10. D. Grosu, A.T. Chronopoulos, and M. Leung. Load balancing in distributed systems: An approach using cooperative games. In *Proceedings of IEEE International Parallel and Distributed Processing Symposium (IPDPS02)*, 2002.
11. D. Grosu and A.T. Chronopoulos. Algorithmic mechanism design for load balancing in distributed systems. In *Proceedings of IEEE International Parallel and Distributed Processing Symposium (IPDPS04)*, 2004.
12. M.T. Hajiaghayi, R.D. Kleinberg, D.C. Parkes, and M. Mahdian. Online auctions with reusable goods. In *Proceedings of the ACM conference of Electronic Commerce (ACM EC)*, 2005.

13. V. Hastagiri Prakash. A mechanism design approach to resource procurement in computational grids with rational resource providers. Master of Science Dissertation, Department of Computer Science and Automation, Indian Institute of Science. Bangalore, India, 2006.

14. A. Kumar and G. Iyengar. Optimal procurement auctions for divisible goods with capacitated suppliers. *Review of Economic Design*, 12(2):129–154, 2008.

15. Y. Kwok, S. Song, and K. Hwang. Selfish grid computing: Game theoretic modeling and performance results. In *Proceedings of IEEE/ACM International Symposium on Cluster Computing and the Grid (CCGRID 2003)*, pages 1143–1150, 2003.

16. A. Malakhov and R.V. Vohra. Single and multi-dimensional optimal auctions — A network approach. Technical report, http://www.kellogg.northwestern.edu/faculty/vohra/htm/res.htm, 2005.

17. A. Malakhov and R.V. Vohra. An optimal auction for capacitated bidders — A network perspective. Technical report, http://www.kellogg.northwestern.edu/faculty/vohra/htm/res.htm, 2005.

18. A.M. Manelli and D.R. Vincent. Multidimensional mechanism design: Revenue maximization and the multiple-good monopoly. *Journal of Economic Theory*, 137(1):153–185, 2007.

19. A. Mas-Colell, M.D. Whinston, and J.R. Green. *Micro-economic Theory*. Oxford University Press, New York, 1995.

20. R.B. Myerson. Optimal auction design. *Mathematics of Operations Research*, 6(1):58–73, 1981.

21. R.B. Myerson. *Game theory: Analysis of conflict*. Harvard University Press, Cambridge, Massachusetts, 1997.

22. J. Nabrzyski, J.M. Schopf, and J. Weglarz. *Grid resource management: State of the art and future trends*. Kluwer Academic Publishers, Norwell, Massachusetts, 2003.

23. N. Nisan, S. London, O. Regev, and N. Camiel. The popcorn market-online markets for computational resources. In *Proceedings of First International conference on Information and Computational Economies*, 1998.

24. S. Suri, C.D. Toth, and Y. Zhou. Selfish load balancing and atomic congestion games. In *Proceedings of Sixteenth Annual ACM Symposium on Parallelism in Algorithms and Architectures (SPAA)*, pages 188–195, 2004.

25. R. Wilson. Auctions of shares. *Quarterly Journal of Economics*, 94:675–689, 1979.

26. R. Wolski , J.S. Plank, J. Brevik, and T. Bryan, G-commerce: Market formulations controlling resource allocation on the computational grid. In *Proceedings of International Parallel and Distributed Processing Symposium (IPDPS)*, 2001.

# Chapter 5
# Incentive Compatible Broadcast Protocols for Ad hoc Networks

In this chapter, we consider a class of wireless networks, namely ad hoc wireless networks. We bring out certain issues that naturally lead to a rational (or selfish) behavior of the wireless nodes. When the nodes behave rationally, they may not follow the prescribed protocols faithfully. One way in which to stimulate cooperation by the rational nodes is to offer them appropriate incentives through mechanism design approaches. In this chapter, we undertake a study of this approach in the specific context of implementing a robust broadcast protocol in ad hoc networks. We call the problem that we address the *incentive compatible broadcast (ICB)* problem. We offer two different solution approaches to the ICB problem using mechanism design. In the first approach, we develop a dominant strategy incentive compatible protocol based on the Clarke mechanism, and we call it the *Dominant Strategy Incentive Compatible Broadcast (DSIC-B)* protocol. Though the DSIC-B protocol has strong economic properties, it has certain practical limitations. The second solution approach, which we call BIC-B (Bayesian incentive compatible broadcast), overcomes the limitations of the DSIC-B protocol and also has several attractive properties. This chapter is a detailed extension of the results presented in [32], [39].

## 5.1 Introduction to Ad Hoc Wireless Networks

The world has become increasingly mobile over the past few years. As a result, the traditional networking technologies have become inadequate to meet the challenges and demands posed by emerging new applications. If users must be connected to a network by physical cables, their movement is dramatically reduced. Wireless connectivity, however, poses no such restriction and allows a great deal more free movement on the part of the network user. As a result, wireless technologies are taking over [40].

*Wireless networks* are computer networks that use radio frequency channels as their physical medium for communication. Each node in the network broadcasts information that can be received by all nodes within its direct transmission range. Since nodes transmit and receive over the air, they need not be physically connected to any network. Hence, *mobility* is the most obvious advantage of wireless networking. Wireless network users can connect to existing networks and are then allowed

Y. Narahari et al., *Game Theoretic Problems in Network Economics and Mechanism Design Solutions*, Advanced Information and Knowledge Processing,
DOI: 10.1007/978-1-84800-938-7_5, © Springer-Verlag London Limited 2009

to roam freely. These networks have a great deal of *flexibility*, which can translate into rapid deployment.

The wireless communications industry has several segments such as cellular telephony, satellite-based communication networks, wireless LANs, and ad hoc wireless networks. An *ad hoc wireless network* is an autonomous system of nodes connected through wireless links. It does *not have any fixed infrastructure* such as base stations. The nodes in the network coordinate among themselves for communication. Hence, each node in the network, apart from being a source or destination, is also expected to route packets for other nodes in the network. Such networks find varied applications in real-life environments such as communication in battle fields, communication among rescue personnel in disaster affected areas, wireless sensor networks, and collaborative and distributive computing.

There are, however, several major issues and challenges that need to be considered when an ad hoc wireless network is to be designed. The major issues that affect the design, deployment, and performance of an ad hoc wireless system include medium access scheme, routing, multicast, broadcast, transport layer protocol, quality of service provisioning, self-organization, security, energy management, addressing and service discovery, and scalability. Here in this chapter our focus is on the broadcast. It is a variant of routing and multicast. Broadcast is a communication paradigm that allows us to send data packets from a source node to the rest of the nodes in the network.

Broadcast in ad hoc wireless networks is of critical importance in applications such as information diffusion, maintaining consistent global network information, etc. Broadcast is often necessary in some routing protocols, and these protocols rely on a simplistic form of broadcast called flooding, in which each node (or all nodes in a localized area) retransmits each received packet exactly one time. The main problems with flooding are that it typically causes unproductive and often harmful bandwidth congestion, as well as inefficient use of node resources. Broadcast is also more efficient than sending multiple copies of the same packet through unicast. It is important to use power-efficient broadcast algorithms for such networks since wireless devices are often powered by batteries only. Recently, a number of research groups have proposed more efficient broadcast techniques [26] with various goals such as minimizing the number of retransmissions, minimizing the total power used by all transmitting nodes, minimizing the overall delay of the broadcast, etc.

## 5.2 Ad Hoc Networks with Selfish Nodes

In the discussion above, it is implicitly assumed that nodes follow the prescribed protocol without any deviation, and they cooperate with one another in performing the network functions, for example packet forwarding. But, in typical applications of ad hoc wireless networks, nodes may be owned by rational and intelligent users. In general, if nodes are owned by autonomous users and their only objective is to maximize their individual goals, then such nodes are called *selfish* nodes. The behavior exhibited by a network of *selfish* nodes can be modeled in a natural way using *game theory*. Indeed, there are many recent efforts exploring the use of game theoretic models in the modeling and analysis of key problems in ad hoc networks, such as power control, routing, resource sharing, etc. In this chapter, we study how to model ad hoc networks in a game theoretic framework and explore the use of mechanism design to develop robust broadcast protocols.

## 5.2.1 Modeling Ad Hoc Wireless Networks as Games

As we know, in a strategic form game, there are three components: players, strategies, and utilities [42]. The players are independent decision makers where the utility or payoff depends on a player's own action and the other players' actions. Selfish nodes in an ad hoc network are also characterized by a similar feature since the utilities of the nodes depend on their resources such as battery power, bandwidth, CPU cycles, etc. and the packet forwarding policies of other nodes in the network. This similarity leads to a natural mapping between game theory components and elements of an ad hoc network. In this game, the nodes in the network become the players. The packet forwarding policies of the nodes constitute the strategy space of the nodes. Finally, the utilities of nodes will be determined by performance measures such as end-to-end delay, number of retransmissions, etc.

Game theory has been applied to the modeling of an ad hoc network at various levels of the protocol stack [35], [31]. It has been applied at (1) the physical layer level to analyze distributed power control [3] and waveform adaption; (2) at the data link layer level to analyze medium access control [27], [34] and the reservation of bandwidth [6]; and (3) at the network layer level to model the behavior of the packet forwarding strategies [16], [11], [10], [14]. Applications at the transport layer and above also exist, although less pervasive in the literature. A question of interest in all those cases is that of how to provide the appropriate incentives to get around the selfish behavior of the nodes. Selfishness is generally detrimental to overall network performance. Examples include: a node increasing its power without regard for interference it may cause on its neighbors, a node immediately retransmitting a frame in the case of a collision without going for the back-off phase, and a node refusing to forward the transit packets of the other nodes in the network.

Consider in particular the network layer to get an intuitive idea. The functions of the network layer include the establishment and updating of routes and the forwarding of packets along those routes. The establishment of multihop routes in an ad hoc network relies on the packet forwarding behavior of nodes for one another. However, a selfish node, in order to conserve its limited energy resources, could decide not to participate in the forwarding activity by switching off its interface. If all nodes decide to alter their behavior in this way, acting selfishly, this may lead to the collapse of the network. The recent works [36], [33], [10] in this direction develop game theoretic models to analyze the selfish behavior of the nodes in forwarding packets. Consider the network in Figure 5.1 to illustrate what is popularly known as the *Forwarder's Dilemma* [37], to understand how the selfish behavior of the nodes leads to inefficient situations. In the figure, node $p_1$ wishes to send packets to node $r_1$, and similarly node $p_2$ wants to send packets to node $r_2$. If $p_1$ forwards the packets of $p_2$ to $r_2$, it incurs the node $p_1$ a fixed cost $c$, which represents the energy and computation spent in the forwarding activity. By doing so, it enables the communication between $p_2$ and $r_2$, which gives $p_2$ some value. A similar reasoning applies to the forwarding action of player $p_2$, that is, node $p_2$ incurs a fixed cost $c$ to facilitate the communication between $p_1$ and $r_1$. The payoff to each node is the difference between its value and its incurred cost. The dilemma is the following: Each player is tempted to drop the packet that it should forward, as this would save some of its resources; but if the other player also reasons in an identical manner, then the packet that the first player wanted to be relayed would get dropped. Finally, none of the nodes would get any utility. However, they could do better by mutually relaying each other's packets.

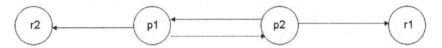

**Fig. 5.1** The network scenario in a packet forwarding context

Rational behavior of a node may suggest that forwarding the transit traffic is not a best response strategy, since the forwarding activity may consume significant resources. Such behavior may lead to suboptimal situations where nodes, through their actions, reach an undesirable steady state from a network point of view. Re-imbursing the transit costs of nodes is a solution to avoid such worst network-wide equilibrium situations. In the forward dilemma game, it is assumed that the transit cost of each node is known to every other node in the network. However in real world applications, such an assumption is not realistic since the transit cost of a node depends only on its consumed battery power, CPU cycles, bandwidth, etc. Hence it is reasonable to assume that the transit cost of a node is not known to other nodes. The transit cost of a node could be considered as its *private value* or *type*. If we want to reimburse the incurred cost of a node, first we need to elicit the true transit cost from that node, since the transit cost is its private value. Nodes may not reveal their true transit costs to maximize their utilities, since they are selfish and intelligent. So, we need to provide incentives to the selfish nodes to make them reveal the true transit costs. In the literature, two different ways of providing incentives have been explored. The first category of incentive schemes is called *credit based methods*. In this technique, the idea is to use various nonmonetary approaches including auditing, use of appropriate hardware, system-wide optimal point analysis [18], etc. Such methods, however, suffer from major problems. These methods look for a system wide optimum point that may not be individually optimal. Further, the use of hardware is not always possible in network settings. The second category of incentive schemes is based on mechanism design. In this chapter, our focus is on mechanism design based techniques in the specific context of broadcast.

## 5.2.2 The Incentive Compatible Broadcast (ICB) Problem

As already stated, broadcast is useful in ad hoc wireless networks in many situations, for example, route discovery, paging a particular host, or sending an alarm signal, etc. Successful broadcast requires appropriate forwarding of the packet(s) by individual wireless nodes. The nodes incur certain costs for forwarding packets since packet forwarding consumes the resources of the nodes. If the nodes are rational, we can expect the nodes to be selfish, and packet forwarding cannot be taken for granted. Reimbursing the forwarding costs or transit costs incurred by the nodes is a way to make them forward the packets. For this, we need to know the exact transit costs at the nodes; however, the nodes may not be willing to reveal the true transit costs. Broadcast protocols that induce revelation of true transit costs by the individual wireless nodes can be called *incentive compatible*, following mechanism design terminology. We can design an incentive compatible broadcast protocol by embedding an appropriate allocation rule and payment rule into the broadcast protocol. We shall refer to the problem of designing such robust broadcast protocols as the *incentive compatible broadcast (ICB)* problem.

## 5.2.3 Modeling the ICB Problem

An ad hoc wireless network can be represented by an undirected graph $G = (N, E)$, where $N = \{1, 2, ..., n\}$ is the set of $n$ wireless nodes and $E$ is the set of communication links between the nodes. There exists a communication link between two nodes $i$ and $j$, if a node $i$ is reachable from node $j$, and node $j$ is also reachable from node $i$. Thus we have an undirected graph representation. Assume that wireless nodes use directional antennas, and a single transmission by a node may be received by only a subset of nodes in its vicinity. Note that all nodes in the graph $G$ are connected. Assume that the nodes are owned by rational and intelligent individuals, and their objective is to maximize their individual goals. For this reason, they may not always participate loyally in key network functions, such as forwarding the packets, since such activity might consume the node's resources, say battery power, bandwidth, CPU cycles, etc. Let each node $i$ incur a cost $\theta_i$ for forwarding a packet. For simplicity, we assume that $\theta_i$ is independent of the neighbor from which the packet is received and the neighbor to which the packet is destined. We call $\theta_i$ the *private value* or *type* of node $i$. It is assumed that the types of the nodes are drawn from a common prior distribution $\varphi$. Using this common prior, individual belief functions $p_i$ are calculated. $p_i$ describes the belief of node $i$ about the types of the remaining nodes. We denote by $\Theta_i$ the set of types of node $i$, $i = 1, 2, ..., n$. Let $\Theta = \times_{i \in N} \Theta_i$

be the set of all type profiles of the nodes and $\theta = (\theta_1, \theta_2, \ldots, \theta_n)$ be a typical type profile of the nodes. Now $\forall i \in N$, $\theta_i$ can be represented as the weight of node $i$ in the graph $G$. This results in a *node weighted graph*. We first assume that the graph $G$ is bi-connected. This guarantees that the graph that remains after removing any node and its incident links from the graph is still connected. Thus, the bi-connectivity of the graph prevents monopoly on the network by the cut vertices. This restriction on the graph will be relaxed in the later part of the analysis. Following the standard principles of noncooperative game theory, assume that the wireless nodes do not collude with one another to improve their utility.

Consider the task of broadcast in such a setting. Let the source of the broadcast be node $s$. If not all remaining nodes are connected to node $s$, then a few nodes have to forward the broadcast packet to ensure that the packet reaches all the nodes in the network. For the sake of simplicity, call the nodes that forward the packets *routers*. As explained above, the routers incur costs. This means there is a need to look for a tree that spans all the nodes with node $s$ being the root of the tree. Such a tree could be called a *broadcast tree*. Let $X$ be the set of different possible (or allowable) outcomes, that is, the set of different possible broadcast trees. The selection of a particular broadcast tree depends on the type profile $\theta = (\theta_1, \theta_2, \ldots, \theta_n)$ of the nodes. This is captured by the social choice function $f(\theta) = (k(\theta), t_1(\theta), \ldots, t_n(\theta))$ $\forall \theta \in \Theta$, following the concepts of mechanism design. Here $k(.)$, and $t_i(.)$, $\forall i \in N$ are interpreted in the following way. $k(\theta)$ is the allocation rule that indicates which nodes in the network have to forward the packet(s), given the type profile $\theta$ of the nodes. That is, $k(\theta)$ selects one particular broadcast tree from the set $X$ based on $\theta$. $(t_1(\theta), \ldots, t_n(\theta))$ is the vector of payments received by the nodes, given the type profile $\theta$ of the nodes. For any $i \in N$, if $t_i(\theta) > 0$, then, the interpretation is that $i$ receives some positive amount, and if $t_i(\theta) < 0$, then the interpretation is that $i$ pays some positive amount.

Now the value that accrues to node $i$, if it forwards a packet, is captured using the valuation function, $v_i(.)$. That is, if the allocation rule is $k(.)$ and the type of the node $i$ is $\theta_i$, then $v_i(k(\theta), \theta_i) = -\theta_i$. Here the negative sign indicates that the resources (such as CPU cycles, bandwidth, etc.) of the node are consumed. If the node $i$ forwards a packet, it receives an amount $t_i(\theta)$. Then the utility, $u_i(\theta)$, of the node $i$ is defined as $u_i(\theta) = v_i(k(\theta), \theta_i) + t_i(\theta)$. This is the general structure of a utility function in quasilinear environments, introduced in Section 2.13.

Let us focus on allocation rule $k(.)$. The allocation rule is required to find a broadcast tree with a minimum amount of incurred costs to the routers. Unfortunately, finding such a tree with these requirements turns out to be a hard task from a computational viewpoint for the following reasons. Since an ad hoc wireless network can consist of heterogeneous nodes, the incurred costs to the nodes could vary across the nodes. In such a setting, we want to broadcast a packet over the network with minimum total amount of incurred cost. That is, we need to select a set, call it $R$, of routers in a way such that the sum of incurred costs to the routers is a minimum. Now consider a special case where the nodes are homogeneous. It turns out that the incurred costs to the nodes for forwarding the packets are the same since the nodes are homogeneous. In this context, the problem of finding a set of routers with minimum incurred cost reduces to the problem of finding a set $R$ with a minimum number of routers. It is well known that finding such a set $R$ is an $NP$-complete problem [5], [38]. The hardness of this problem can be proved by reducing an instance of multipart relaying to an instance of broadcast scheme from $s$ in $G$ with a minimum number of routers [38]. We note that the problem of finding the set $R$ of routers with minimum cost in the general case where the nodes in the ad hoc wireless network are heterogeneous will be much harder computationally.

An immediate alternate approach to get around this intriguing situation is to look for an approximate solution to construct a broadcast tree for which the set of routers incur minimal cost. Based on such a broadcast tree, a payment rule is developed that is built into the broadcast protocol.

## 5.3 Relevant Work on Incentive Compatible Protocols

In recent times, routing in the presence of selfish nodes has received significant attention driven by the need to design protocols, like routing protocols, multicast protocols, etc., for networks with selfish nodes. Several varieties of solution approaches have been proposed based on different techniques for the routing activity with selfish nodes. The following are a few important classes of solution methods.

### 5.3.1 Reputation-Based Solution Methods

The early research on stimulating cooperation in ad hoc wireless networks used a reputation mechanism [30]. This approach uses techniques such as auditing, use of appropriate hardware, and system-wide optimal point analysis to identify selfish nodes and isolate non-cooperative nodes from the network. The watchdog mechanism in Marti, Giuli, Lai, Baker [15], the *core* mechanism in Michiardi and Molva [19], the *confidant* mechanism in Buchegger and Boudec [2], etc., are a few examples of reputation mechanisms. These methods look for a system-wide optimum point that may not be individually optimal. Also, the use of hardware is not always feasible in network settings.

### 5.3.2 Non-Cooperative Game Theoretic Solution Methods

Srinivasan, Nuggehalli, and Chiasserini [10] modeled the routing situation using the repeated prisoner's dilemma problem. According to evolutionary game theory, an effective strategy in this kind of setting is the so-called TIT-FOR-TAT strategy. In Altman, Kherani, Michiardi, and Molva [36], each node is assumed to forward packets with some probability which is independent of the source. These models do not take the dynamics of the network into consideration. A game theoretic model was introduced by Urpi, Bonuccelli, and Giordano [25], based on static Bayesian games [42], to model forwarding behavior of selfish nodes in ad hoc wireless networks. Although this model properly formulates the game the nodes are playing, it does not capture nonsimultaneous decision making. In addition, the strategies in this framework are not dependent on past behavior. To get around these problems, Nurmi [21] modeled routing in ad hoc wireless networks with selfish nodes as a dynamic Bayesian game. This model is rich in the sense that it allows non-simultaneous decision making, incorporating history information into the decision making process.

## 5.3.3 Mechanism Design Based Solution Methods

Another important approach to designing incentive mechanisms is based on mechanism design. Feigenbaum, Papadimitriou, Sami, and Shenker [20] and Hershberger and Suri [7] developed an incentive mechanism to address the truthful low cost routing (unicast) problem. The model consists of $n$ nodes where each node represents an autonomous system. They assume that each node $k$ incurs a transit cost $c_k$ for forwarding one transit packet. For any pair of nodes $i$ and $j$ of the network, $T_{i,j}$ is the total amount of traffic originating from $i$ and destined for node $j$. The payments to nodes are computed using the Vickrey–Clarke–Groves (VCG) payment rules. Feigenbaum, Papadimitriou, Sami, and Shenker [20] and Hershberger and Suri [7] presented a distributed method such that each node $i$ can compute its payment $p_{i,j}^l$ to node $l$ for carrying the transit traffic from node $i$ to node $j$ if node $l$ is on the least cost path from $i$ to $j$.

A similar type of model is presented by Anderegg and Eidenbenz [16] for stimulating cooperation among selfish nodes in ad hoc networks using an incentive scheme. This model generalizes one aspect of the model in [20] by associating several costs to each node, one per each neighbor, instead of just one. This leads to a model based on *edge weighted graph* representation of the ad hoc network. The VCG mechanism is used to compute a power efficient path, where each node determines the power level required to transit/forward the packets. A node alone cannot determine its required power level because it needs feedback information in the form of packets from its neighbors. As the nodes are selfish and noncooperative, this feedback information may allow a node to cheat its neighbors in order to raise its own payoff. The authors of [16] did not address this issue. Eidenbenz, Santi, and Resta [11] modified the model in Anderegg and Eidenbenz [16] by using the VCG mechanism to compute the payments to the nodes, but the sender is charged the total declared cost of a second least cost path, that is the least cost path with all nodes in the cost efficient path removed. This requires the existence of at least two node disjoint paths between the sender and the receiver. Wang and Li [28] proposed strategy-proof mechanisms for the truthful unicast problem. They also presented an algorithm for fast computation of payments to nodes and a distributed algorithm for payment computation. Another piece of work towards designing incentive compatible routing protocols in ad hoc networks is presented in [8]. A new metric, which is called maximum expected social welfare, is defined in [9] to efficiently address the routing problem in ad hoc networks. In this new metric, they integrate the cost and stability of nodes in a unified model to evaluate the optimality of routes.

Zhong, Li, Liu, and Yang [13] used a two-stage routing protocol to model the routing problem in ad hoc wireless networks with selfish nodes. They integrated a novel cryptographic technique into the VCG mechanism to solve the link cost dependence problem. Zhong, Chen, and Yang [29] proposed a system called Sprite, which combines incentive methods and cryptography techniques to implement a group cheat-proof ad hoc routing system. Lu, Li, and Wu [12] embed an incentive-compatible, efficient, and individual rational payment scheme into the routing protocol in ad hoc networks that consist of selfish nodes. Unlike traditional routing protocols in ad hoc networks, which only elicit cost information from selfish nodes, this model motivates selfish nodes to report truthfully both their stability and cost information.

In all the above solution approaches for the incentive compatible unicast problem, all the intermediate nodes on the path between the source node and the destination node are compensated for forwarding the packet. Let us apply this technique of service reimbursement to the *ICB* problem to see the consequences. Consider a portion of the network as shown in Figure 5.2. Let node 1 be the source of broadcast. Nodes 2, 3, 4, and 5 are the intended destinations of the broadcast packet. Node 5 needs to reimburse the nodes 2, 3, and 4 since these nodes are intermediate nodes on the path between node 1 and node 5, and they are required to forward the broadcast packet. For similar reasons, node 4 needs to reimburse nodes 2 and 3; and node 3 needs to reimburse node 2. On the whole, node 2 receives payments from nodes 3, 4, and 5 for forwarding a packet once. Similarly, node 3 receives payments from node 4 and node 5. Hence we end up with high values of payments leading to an inefficient solution to the *ICB* problem. It is important to observe, in the context of broadcast, that all the nodes in the network are intended destinations, and there is no notion of intermediate nodes as in case of unicast.

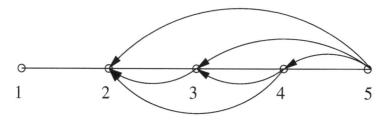

**Fig. 5.2** Applying incentive compatible unicast solution techniques to the ICB problem

Wang and Li [17], [14] have proposed strategy-proof mechanisms for the truthful multicast problem in ad hoc wireless networks with selfish nodes. The authors, in this model, assume that the nodes in the multicast set forward the packets for free among themselves. However, this assumption may not be credible in real world applications, especially when the nodes are rational. In our work, we assume that nodes, being rational, do not necessarily forward the broadcast packets for free. This assumption clearly captures the real world more accurately.

To summarize, several solutions have been proposed for the incentive compatible unicast and multicast problems in the literature. If these solution approaches are used for the incentive compatible broadcast problem, they could lead to inefficient solutions. These provide a strong motivation to look for more efficient incentive compatible protocols for the *ICB* problem.

## 5.4 A Dominant Strategy Incentive Compatible Broadcast Protocol

Here we design a dominant strategy incentive compatible broadcast (DSIC-B) protocol for the ICB problem. Towards this end, we consider the network model described in Section 5.2.3. Let $\theta = (\theta_1, \ldots, \theta_n)$ be the announced cost profile of the nodes. The goal is to design a DSIC mechanism that implements social choice function $f(\theta) = (k(\theta), t_1(\theta), \ldots, t_n(\theta))$, $\forall \theta \in \Theta$. To progress in this direction, first a broadcast tree needs to be constructed corresponding to the underlying graph $G = (N, E)$ of the ad hoc wireless network under consideration.

### 5.4.1 Construction of a Broadcast Tree

As mentioned in the previous section, the construction of an optimal broadcast tree is a hard problem. For this reason, an approximate method is used here for the required situation. First we define a few relevant concepts. A path between two nodes is a sequence of nodes such that any two consequent nodes in the sequence are connected. The cost of a path is the sum of transit costs of nodes in the path except the first and the last nodes. A *least cost path* between two nodes is a path with minimum cost. Based on the announced cost vector $\theta = (\theta_1, \theta_2, \ldots, \theta_n)$, all the least cost paths $LCP(\theta, s, i)$ are computed, $\forall i \in N \setminus \{s\}$, by breaking the ties appropriately. We represent a least cost path using indicator functions $I(.)$, which we define in the following. $I_i(\theta, s, j) = 1$, if a node $i$ is on the least cost path between $s$ and $j$, and $I_i(\theta, s, j) = 0$, otherwise. So, we can represent the least cost path $LCP(\theta, s, j)$ between $s$ and $j$ as $\{I_i(\theta, s, j)\}_{i \in N}$. Here we note that $I_s(\theta, s, j) = 0$ and $I_j(\theta, s, j) = 0$. Finally merge the least cost paths $LCP(\theta, s, i)$, $\forall i \in N \setminus \{s\}$ to form a broadcast tree. Let such a broadcast tree be called *Source Rooted Broadcast Tree (SRBT)*. All the internal nodes of the *SRBT* are packet forwarders, and they form the set $R$ of routers. Note that a node is called an internal node in a rooted tree if it is neither the root nor a leaf. More formally, the allocation function $k(.)$ gives a vector that specifies the set of routers corresponding to the source of the broadcast task. The following is an example that illustrates how an *SRBT* is constructed for a better understanding.

## 5.4.2 DSIC-B: The Idea

Here a broad outline of the working of the proposed approach for broadcast is presented with an illustrative example. Consider the underlying graph of an ad hoc wireless network with selfish nodes as shown in Figure 5.3.

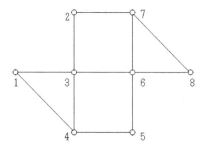

**Fig. 5.3** Underlying graph of an ad hoc wireless network

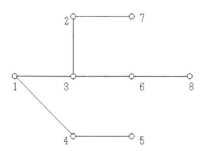

**Fig. 5.4** Broadcast tree rooted at node 1 for the graph in Figure 5.3

Let *node* 1 be the source of broadcast, and the corresponding broadcast tree is shown in Figure 5.4. Similarly when *node* 2 is the source of broadcast, the corresponding broadcast tree is shown in Figure 5.5. Recall that the nodes that forward the packets are called *routers*. For the broadcast tree in Figure 5.4, the set of routers is $\{2,3,4,6\}$. Similarly for the broadcast tree in Figure 5.5, the set of routers is $\{3,4,7\}$. For the reasons already explained above, it is difficult to get an optimal

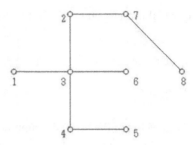

**Fig. 5.5** Broadcast tree rooted at node 2 for the graph in Figure 5.3

broadcast tree. Hence an approximate method of constructing such a broadcast tree is described below.

Assume, for computational simplicity, that the transit cost of each node in the network is the same, and let its value be $c$. This means the cost profile is $\theta = (c, \ldots, c)$. Assume that node 1 is the source of the broadcast. We first compute the least cost paths to all nodes from node 1. The least cost path from node 1 to node 8 is represented by $LCP(\theta, 1, 8) = (1, 3, 6, 8)$. We determine all such least cost paths and join all least cost paths to form a tree rooted at node 1. In this way, we get the *SRBT* shown in Figure 5.4 corresponding to the graph in Figure 5.3. Once the broadcast tree is constructed, the payments for the nodes are determined using the payment scheme. The incentives provided are such that the nodes are not worse off by forwarding the packets.

Now consider the situation at node 6. When node 6 receives the packet along $LCP(\theta, 1, 6)$, according to our approach, node 6 pays an amount, call it $p_{1,6}^3$, to node 3. Here node 3 is the immediate predecessor of node 6 in the *SRBT*. The value of $p_{1,6}^3$ is computed using the Clarke payment rule (refer to the chapters) in the following way.

$$
\begin{aligned}
p_{1,6}^3 = {}& \text{(cost of node 3 to forward a packet)} \\
& + \text{(cost of } LCP(\theta, 1, 6) \text{ in the network without node 3)} \\
& - \text{(cost of } LCP(\theta, 1, 6) \text{ in the network with node 3)} \\
= {}& c + 3c - 2c \\
= {}& 2c.
\end{aligned}
$$

Similarly when node 6 forwards the packet, node 8 receives and pays an amount, $p_{1,8}^6$, to node 6 because node 6 is the immediate predecessor of node 8 in the *SRBT*. Node 8 need not pay to the other intermediate node 3, on $LCP(\theta, 1, 8)$, because node 3 is the immediate predecessor for node 6 in the *SRBT*, and node 6 has already paid to node 3 before node 8 receives the packet. The reason is that node 6 is also an intended recipient of the packet in a broadcast activity. From this example, the main idea behind the mechanism can be summarized in the following way. When a node receives a packet along the path in the given *SRBT*, it pays only to its immediate predecessor node in the tree. This payment mechanism is more formally defined in the following way.

### 5.4.3 DSIC-B: Payment Rule

Now based on the allocation rule $k(\theta)$ corresponding to the announced type profile $\theta$, we define two quantities. We first define $F_i(k(\theta))$, $\forall i \in N \backslash \{s\}$, which indicates the set of nodes that are immediate successors for node $i$ in the $SRBT$. Since the source node $s$ is not a router and for the sake of defining of the payment rule, it is convenient to assume that $F_s(k(\theta)) = \emptyset$. We then define the valuation function, $v_i(k(\theta), \theta_i)$, of node $i$ as its cost to forward a transit packet. From the allocation rule $k(.)$ in our SCF $f(.)$, we get, $\forall \theta \in \Theta$, $\forall i \in N$,

$$v_i(k(\theta), \theta_i) = -\theta_i, \quad \text{if } F_i(k(\theta)) \neq \emptyset \tag{5.1}$$
$$= 0, \quad \text{if } F_i(k(\theta)) = \emptyset. \tag{5.2}$$

Using the payment rule, the nodes that do not forward packet(s) should receive zero payment. For the nodes that forward the packets, the payments are defined in the following way:

$$t_i(\theta) = \begin{cases} 0, & \text{if } i \text{ is a leaf,} \\ \sum_{j \in F_i(k(\theta))} \, p^i_{s,j}(\theta), & \text{otherwise} \end{cases} \tag{5.3}$$

where,

$$p^i_{s,j}(\theta) = -\{ \sum_{m \in N, m \neq i} I_m(\theta, s, j)\theta_m \} + \{ \sum_{m \in N, m \neq i} I_m(\theta_{-i}, s, j)\theta_m \}. \tag{5.4}$$

Recall that $\sum_{m \in N} I_m(\theta, s, j)\theta_m$ is the least cost path from $s$ to $j$ and node $i$ present on that path. $\sum_{m \in N, m \neq i} I_m(\theta_{-i}, s, j)\theta_m$ is the least cost path from $s$ to $j$ when node $i$ does not present in the network. For each $i, j \in N$, $p^i_{s,j}(\theta)$ represents the per-packet price paid to node $i$ from node $j$ when the source of the packet is $s$. However, using expression (5.4), it is easy to see that the payment received by a nonrouter node (i.e., a node that is not forwarding packets) is zero. Observe that each non-router node $j$ is occurring as the leaf of $SRBT$. For example, for the SRBT in Figure 5.4, the payments received by routers 2, 3, 4 and 6 are given in the following way. $t_2(\theta) = p^2_{1,7}(\theta)$; $t_3(\theta) = p^3_{1,2}(\theta) + p^3_{1,6}(\theta)$; $t_4(\theta) = p^4_{1,5}(\theta)$; and $t_6(\theta) = p^6_{1,8}(\theta)$. The remaining nodes 1, 5, 7, and 8 receive *zero* payments since they are leaf nodes. A discussion regarding the computation of the quantity $p^i_{s,j}(\theta)$, $\forall i, j \in N$ is given in the following sections.

## 5.4.4 Properties of DSIC-B Mechanism

Here a few key properties of the *DSIC-B* mechanism are discussed.

**Lemma 5.1.** *The* DSIC-B *mechanism is strategy-proof.*

*Proof.* We use the notion of *maximal in the range* [4] to test the effectiveness of allocation rule $k(.)$ since finding an optimal broadcast tree is NP-hard. Following this notion, it is required to define the set, say $\chi$, of allowable broadcast trees that are constructed by combining the paths from all the remaining nodes to the source node of broadcast. Here the set $\chi$ is called the range of $k(.)$. By the construction of SRBT, it clear that SRBT belongs to set $\chi$. Moreover, SRBT minimizes the cost of internal nodes (i.e., routers) among the other broadcast trees in $\chi$ since it is constructed by merging the *shortest* paths from all the remaining nodes to the source node of broadcast. Hence $k(.)$ is maximal in its range, and thus the *DSIC-B* mechanism is strategy-proof [4].                                                    *Q.E.D.*

We now investigate the *individual rationality (IR)* of the *DSIC-B* mechanism. Individual rationality implies that the nodes participate in the mechanism voluntarily [41]. That is, nodes get nonnegative utility by participating in the mechanism. In our context, the *DSIC-B* mechanism provides payments to the nodes for forwarding the broadcast packets and also provides incentives to the nodes to participate truthfully in the mechanism.

Recall from Chapter 2 (Section 2.10) that there are three versions of individual rationality. They are *ex-ante individual rationality*, *interim individual rationality*, and *ex-post individual rationality*. In ex-ante individual rationality, nodes are expected to decide whether or not to participate in the mechanism even before knowing their types. In interim individual rationality, nodes would decide whether or not to participate in the mechanism after learning their types but before playing out their actions. In ex-post individual rationality, nodes are allowed to decide whether or not to drop out of the mechanism after announcing their types and playing out their actions. In the current setting, nodes already know their types. That is, nodes know their own transit costs to forward the packets. So, interim individual rationality (IIR) is more appropriate.

**Lemma 5.2.** *The* DSIC-B *mechanism is interim individually rational (IIR) .*

*Proof.* Recall from the model that the types of the nodes have been assumed to be statistically independent. We have to show that

$$U_i(\theta_i) \geq 0, \qquad \forall \, \theta_i \in \Theta_i, \, \forall \, i \in R$$

where $R$ is the set of routers and $U_i(\theta_i)$ is the utility of a router $i \in R$, given $\theta_i$. We now define the expected utility in the following way.

The utility of a node $i$ in the network is the sum of its valuation and the payment received. Obviously, for each leaf node in the *SRBT*, the utility is zero since the leaf node does not forward packets, and it also does not receive any payments. For this reason, it is sufficient to show the utility of router nodes is nonnegative. The utility of each router $i \in R$ is:

$$U_i(\theta_i) = v_i(k(\theta), \theta_i) + t_i(\theta_i)$$
$$= -\theta_i + \sum_{j \in F_i(k(\theta))} \left[ -\{\sum_{m \in N, m \neq i} I_m(\theta, s, j)\theta_m\} + \{\sum_{m \in N \, m \neq i} I_m(\theta_{-i}, s, j)\theta_m\} \right].$$

Note that $\theta = (\theta_i, \theta_{-i})$ is the announced cost profile, where $\theta_i$ is the announcement of node $i$ and the profile $\theta_{-i}$ is the announcement of remaining nodes. Since node $i$ is a router, it is an immediate predecessor for at least one $j \in N \setminus \{s\}$.

Therefore $\sum_{j \in F_i(k(\theta))} [I_i(\theta, s, j)\theta_i] \geq \theta_i$. So, we get

$$U_i(\theta) \geq -\theta_i - \{\sum_{m \in N \, , m \neq i} I_m(\theta, s, j)\theta_m\} + \{\sum_{m \in N \, m \neq i} I_m(\theta_{-i}, s, j)\theta_m\}$$
$$= -\{\sum_{m \in N} I_m(\theta, s, j)\theta_m\} + \{\sum_{m \in N \, m \neq i} I_m(\theta_{-i}, s, j)\theta_m\}.$$

Recall that $\{\sum_{m \in N} I_m(\theta, s, j)\theta_m\}$ is the value of the least cost path from $s$ to $j$ in the presence of node $i$ in the network and $\{\sum_{m \in N \, m \neq i} I_m(\theta_{-i}, s, j)\theta_m\}$ is the value of the least cost path from $s$ to $j$ in the absence of node $i$ in the network. Hence $\{\sum_{m \in N, \, m \neq i} I_m(\theta_{-i}, s, j)\theta_m\} \geq \{\sum_{m \in N} I_m(\theta, s, j)\theta_m\}$. That is,

$$U_i(\theta_i) \geq 0.$$

This is true for any $i \in R$. This completes the proof of the lemma.

### 5.4.5 DISC-B: Protocol Implementation

Here an approach is given how to implement the *DSIC-B* mechanism. Since ad hoc wireless networks are distributed in nature, the approach that implements the mechanism should also be distributed. Based on the given type profile $\theta$ of the nodes, we first compute all the least cost paths between the nodes and the per-packet VCG payments to the nodes, using the distributed scheme [20]. Each node $j$ determines its packet forwarder, say node $i$ depending upon the source node $s$, by sending a control message $SEND\_CTRL\_MSG(s, j, p_{s,j}^i)$ to node $i$. This control message conveys a significant amount of information to node $i$: (a) node $i$ needs to forward the broadcast packet to node $j$ when node $s$ is the source of broadcast, and (b) node $i$ would receive a payment $p_{s,j}^i$ when forwarding a packet. The following is a high level algorithm of the implementation procedure.

---

**Algorithm 1**

1. Use the distributed scheme [20] to find the least cost paths and the per-packet payment values $p_{s,j}^i$, $\forall s, i, j \in N$. If there are multiple least cost paths from a node to the other node, then the node chooses any one least cost path arbitrarily.

2. Each node $j \in N$ does the following:
       *FOR* $s = 1, 2, ..., n$ and $s \neq j$ *DO*
           (i) *IF* node $i$ is the immediate predecessor node for $j$ on the least cost
               path from $s$ to $j$ and $i \neq s$, *THEN*, it sends a control message
               $SEND\_CTRL\_MSG(s, j, p_{s,j}^i)$ to $i$.
       end *FOR*

---

Now, any node $i$ processes the control message, $SEND\_CTRL\_MSG()$, after receiving from any of its neighbors by making use of *Algorithm 2*.

---

**Algorithm 2**
On receiving the control message $SEND\_CTRL\_MSG(s, j, p)$, DO

*FOR source* $= 1, 2, \ldots, n$ *DO*
       *IF* there is a record in $INT\_TAB_i$ with node $s$ as *source THEN*
          (i) Append the pair $(j,p)$ to the *(Node,Price)* entry in the
             record with node $s$ as *source* in $INT\_TAB_i$.
       *ELSE*
          (ii) create a record with node $s$ as *source* in $INT\_TAB_i$
          (iii) Insert the pair $(j,p)$ into the *(Node,Price)* entry in the
             newly created record with node $s$ as *source*.
end *FOR*

---

Node $i$ creates an internal table, which is called $INT\_TAB_i$, to store all this processed information for future use. This table has two fields. They are: *source* and *(Node,Price)*. For example, the construction of the internal table for node 7 of the graph shown in Figure 5.3 is given in Table 5.1.

| Source | (Node,Price) |
|--------|--------------|
| 1 | — |
| 2 | (8,2c) |
| 3 | — |
| 4 | — |
| 5 | — |
| 6 | — |
| 8 | (2,2c) |

**Table 5.1** Internal table contents of node 7 for the graph in Figure 5.3

For computational simplicity assume that the transit costs of the nodes are the same with value equal to $c$. This table conveys the following significant information. If any of the nodes $1, 3, 4, 5, 6$ is the source of broadcast, then node 7 does not need to forward the packets. In those situations, it means that node 7 is a leaf in the corresponding *SRBT*. On the other hand, if node 2 is the source of broadcast, then node 7 forwards the packet to node 8, and it gets the payment $2c$ from that node. In this situation, node 7 is the internal node in the corresponding *SRBT*. Similarly, when node 8 is the source of broadcast, then node 7 forwards the packet to node 2 and it gets payment $2c$ from that node.

The *DSIC-B* mechanism can be built into the broadcast protocol, and by doing this, cooperation is achieved among the selfish wireless nodes, and deviation from the prescribed broadcast protocol is avoided. We call such a broadcast protocol the *DSIC-B* protocol.

*If a node receives a broadcast packet, it checks its INT\_TAB record corresponding to the source of broadcast. If the (Node,Price) field in that table entry is empty, then the node does not forward the packet. Otherwise, it forwards the packet.*

To illustrate this protocol, consider the graph in Figure 5.3. Let node 1 be the source of the broadcast and suppose the corresponding *SRBT* is as shown in Figure 5.4. If the broadcast packet reaches node 7, then it checks its internal table $INT\_TAB_7$, shown in Table 5.1. The entry against node 1 is empty, and hence node 7 does not forward the packet(s).

## 5.4.6 DSIC-B: Performance Evaluation

To evaluate the performance of the *DSIC-B* protocol, there is a need to have a benchmark mechanism. To this end, we make the following observations. Since the *DSIC-B* protocol is distributed in nature, it may be difficult to achieve or implement more efficient packet forwarding strategies to broadcast packet(s). On the other hand, if there is an abstract central authority, it can prescribe more efficient packet forwarding strategies to the selfish nodes than the strategies that are implemented in a distributed setting. Let the mechanism that is used by the central authority in such a setting be called the *OPT* mechanism. We provide the details of the OPT mechanism in the following.

### 5.4.6.1 A Centralized Mechanism (OPT)

Consider the network model described in Section 5.2.3 together with the assumption of an abstract central authority. Let $\theta = (\theta_1, \dots, \theta_n)$ be the announced cost profile of the nodes. The goal is to design the *OPT* mechanism that implements the social choice function $f'(\theta) = (k'(\theta), t'_1(\theta), \dots, t'_n(\theta))$, $\forall \theta \in \Theta$ to address the *ICB* problem in a node weighted ad hoc wireless network. Here $k'(.)$, and $t'_i(.)$, $\forall i \in N$ are interpreted in the following way. $k'(\theta)$ is the allocation rule that represents which nodes in the network have to forward the packet, given the type profile, $\theta$, of the nodes. $(t'_1(\theta), \dots, t'_n(\theta))$ is the vector of payments received by the nodes, given the type profile, $\theta$, of the nodes.

To progress in this direction, first an *SRBT* needs to be constructed corresponding to the underlying graph $G = (N, E)$ of the ad hoc wireless network under consideration.

### 5.4.6.2 Construction of Broadcast Tree

Here is a method for cost efficient construction of a broadcast tree:

1. Based on the announced transit cost vector (or profile) $\theta = (\theta_1, \theta_2, \dots, \theta_n)$ of the nodes, all the least cost paths $LCP(\theta, i, j)$ are computed, $\forall i \in N$ and $\forall j \in N$ where node $i$ and node $j$ are not adjacent in $G$.

2. Construct a new weighted undirected graph $G^*$ with all the $n$ nodes in the set $N$. If nodes $i$ and $j$ are not adjacent in $G$, then introduce an edge from node $i$ to node $j$ in the new graph $G^*$ with the weight of the edge equal to the cost of the $LCP(\theta, i, j)$. Finally consider the edges that are incident on node $s$ and include all of them in new graph $G^*$ with weight 0. Clearly, for each edge in $G^*$, there is an equivalent least cost path in $G$. Note that, $\forall x \in N - \{s\}$, $\forall y \in N - \{s\}$ and $x \neq y$. The edge from node $x$ to node $y$ is not included in the new graph $G^*$ with 0 weight to ensure that nodes do not forward the broadcast packet(s) for free to their neighbors.

3. Now compute the minimum spanning tree of the new graph $G^*$. Call it $T$.

4. Finally the remaining task is to unmark and mark the edges of the graph $G$ appropriately to build the broadcast tree. Initially unmark all the edges in $G$. For each edge in the spanning tree $T$, there is a corresponding least cost path in $G$, and mark the edges in $G$ on that least cost path. Finally retain all the marked edges, delete all the unmarked edges from $G$, and form a broadcast tree. This broadcast tree is the desired *(SRBT)*.

All the internal nodes of the *SRBT* are packet forwarders, and they form the set $R$ of routers. Note that a node is called an internal node in a rooted tree if it is neither the root nor a leaf. This is how the allocation rule finds the set of routers. More formally, the allocation function $k'(.)$ gives a vector that specifies the set of routers corresponding to the source of the broadcast task.

### 5.4.6.3 Comparison of *DSIC-B* and *OPT*

In this section, the performance of *DSIC-B* is compared with that of the *OPT* using simulation experiments. Consider a randomly generated graph of an ad hoc wireless network with the number of nodes $n = 5$, 10, 15, 20, 25, 30, 35, 40. According to the network model described above, the graph is node weighted where these weights are the transit costs of the nodes. We assume that the weights are chosen independently and uniformly from a range $[1, 50]$. After the nodes announce their types, first compute the least cost paths to all the nodes from the source node and then construct an *SRBT*. Using the *SRBT*, determine the set of routers. Then compute the payments to the nodes using the payment rule of the appropriate incentive mechanism. In all our simulation experiments, the results for the performance metrics are averages taken over 100 random instances. Note that we consider node $s$ as the root of the *SRBT*. Two performance metrics are considered here. The first metric is *total overpayment ratio (TOR)*. It is defined as:

$$TOR = \frac{\sum_{i \in N} t_i}{\sum_{j \in N} \{I_m(\theta, s, j)\theta_m\}_{m \in N}}. \tag{5.5}$$

Recall that $\{I_m(\theta, s, j)\theta_m\}_{m \in N}$ is the cost of the least cost path to node $j$ from the source node $s$. Thus TOR is the ratio of the sum of the payments in the network to the sum of the cost of all the least cost paths to the nodes from the source node $s$.

Figure 5.6 shows the comparison of TOR of *DSIC-B* and *OPT* mechanisms. It is clear from the figure that the TOR is decreasing as the number of nodes increases in the network using both the mechanisms. Hence we can conclude that *DSIB-B* protocol works well when compared to that of the centralized scheme.

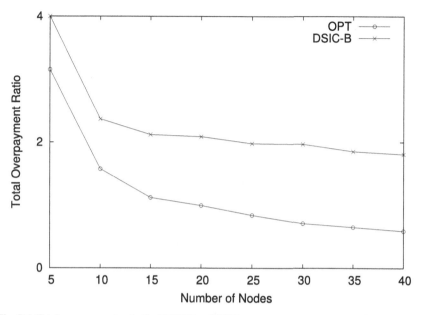

**Fig. 5.6** Total overpayment ratio for DSIC-B and OPT

The second performance metric that is considered is the *worst overpayment ratio (WOR)*. Let *overpayment ratio* be defined as the ratio of payment made by a node to the cost of its least cost path from the source node $s$. Now, define WOR as the maximum over the overpayment ratios of all the nodes in the network. That is,

$$\text{WOR} = \max_{i \in N} \frac{\text{payment made by node } i}{\{I_m(\theta, s, i)\theta_m\}_{m \in N}}$$

where $\{I_m(\theta, s, i)\theta_m\}_{m \in N}$ is the cost of the least cost path to node $i$ from the source node $s$. The comparison of WOR of *DSIC-B* and *OPT* is shown in Figure 5.7. It is clear, from the figure, that WOR of both the schemes is the same. That is, using the proposed distributed scheme, the maximum overpayment by a node in the network is the same as that of the optimal centralized scheme, *OPT*.

Hence from the graphs shown in Figure 5.6 and Figure 5.7, we can conclude that the performance of the proposed *DSIC-B* protocol compares adequately with that of the optimal centralized payment scheme, *OPT*, based protocol.

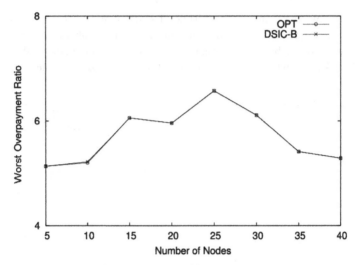

**Fig. 5.7** Worst overpayment ratio for DSIC-B and OPT

### 5.4.7 Groves Mechanism Based Broadcast Protocol

An important limitation of the DSIC-B mechanism is that it cannot be applied if
the underlying graph is biconnected. This is a direct consequence of the use of a
Clarke mechanism based payment rule. We now examine one way of overcoming
this limitation by invoking the more general Groves payment rule. If we recall the
payment rule of the Groves mechanism from Section 2.14, there is a function $h_i(.)$
that is an arbitrary function of $\theta_{-i}$. Note that this function is independent of $\theta_i$.
We can define this function according to our requirements of the ICB problem in
order to design a Groves mechanism-based payment rule. Of course, there could be
several ways of defining this function. In this context, we propose to define $h_i(.)$
as the sum of transit costs of all routers except $i$. This immediately implies that the
underlying graph of the ad hoc network need not be biconnected. We elaborate on
this payment rule below.

#### 5.4.7.1 Payment Rule

First, we note that we use the same allocation rule $k(.)$ as before. Based on the
allocation rule $k(\theta)$ corresponding to the announced type profile $\theta$, here we define
three quantities. Two of these three quantities are the same as in the case of DSIC-
B protocol. However, for the sake of completeness, we again define the required
quantities. We first define $F_i(k(\theta))$, $\forall i \in N \setminus \{s\}$, which indicates the set of nodes

that are immediate successors for node $i$ in the *SRBT*. Since the source node $s$ is not a router and for the sake of defining of the payment rule, it is convenient to assume that $F_s(k(\theta)) = \emptyset$. We then define $Z$ as the set of internal nodes (i.e., routers) of the corresponding SRBT of the ad hoc network. Finally we define the valuation function, $v_i(k(\theta), \theta_i)$, of node $i$ as its cost to forward a transit packet. From the allocation rule $k(.)$ in our SCF $f(.)$, we get $\forall \theta \in \Theta, \forall i \in N$,

$$v_i(k(\theta), \theta_i) = -\theta_i, \quad \text{if } F_i(k(\theta)) \neq \emptyset \tag{5.6}$$
$$= 0, \quad \text{if } F_i(k(\theta)) = \emptyset. \tag{5.7}$$

Using the payment rule, the nodes that do not forward packet(s) should receive zero payment. For the nodes that forward the packets, the payments are defined in the following way.

$$t_i(\theta) = \begin{cases} 0, & \text{if } i \text{ is a leaf,} \\ \sum_{j \in F_i(k(\theta))} p^i_{s,j}(\theta), & \text{otherwise} \end{cases} \tag{5.8}$$

where

$$p^i_{s,j}(\theta) = -\{ \sum_{m \in N, m \neq i} I_m(\theta, s, j)\theta_m \} + \{ \sum_{m \in Z, m \neq i} \theta_m \}. \tag{5.9}$$

Recall that $\sum_{m \in N} I_m(\theta, s, j)\theta_m$ is the least cost path from $s$ to $j$ and node $i$ present on the that path. $\sum_{m \in Z, m \neq i} \theta_m$ is the sum of announced costs of the internal nodes (or routers) except node $i$ in the corresponding SRBT of the ad hoc network. For each $i, j \in N$, $p^i_{s,j}(\theta)$ represents the per-packet price paid to node $i$ from node $j$ when the source of the packet is $s$.

## 5.5 A Bayesian Incentive Compatible Broadcast (BIC-B) Protocol

We first motivate the reasons for exploring a Bayesian model.

- The use of DSIC-B protocol requires the underlying graph to be biconnected while the Bayesian model would work for non-biconnected graphs also.
- Though the Groves payment rule discussed in Section 5.4.7 works for non-biconnected graphs, the payments of VCG mechanisms in general are known to be much higher when compared with the Bayesian models (This is natural because the dominant strategy incentive compatibility property guaranteed by the VCG model is a much stronger property and hence entails higher incentives to be paid). This is well known in the mechanism design literature.

- In the case of the VCG model, budget balance is difficult to achieve, except under very special settings. Budget balance is a desirable property because it ensures that the protocol does not require any external source of funding and therefore is self-sustaining. In the case of the Bayesian model, budget balance can be easily achieved.
- In the case of broadcast, the context is such that there is one source node, and the rest of the nodes are receivers of the broadcast packet. Since the nodes are rational, they may need to make payments to receive the broadcast packet(s). It makes sense to expect all the nodes that receive packets to make the same payments. This is because the source node has no way of distinguishing among the remaining nodes (which are all receivers). We will be showing that the payments as computed by the Bayesian model will be identical for all the receiving nodes. Thus the Bayesian model captures the real world in a natural way.

However, there are two issues with the use of the Bayesian approach. First, Bayesian incentive compatibility is a much weaker form of incentive compatibility. Second, we may lose out on individual rationality. We have however derived a sufficient condition under which individual rationality is also guaranteed by our Bayesian model.

## 5.5.1 The BIC-B Protocol

Assume that we are given the *SRBT* corresponding to the graph under consideration. We design, based on the given *SRBT*, the following payment scheme that determines the payments $(t_i(\theta))_{i \in N}$ to the individual nodes for a broadcast. In the *SRBT*, all the internal nodes forward the broadcast packet. We call such packet forwarding nodes *routers*, and represent the set of routers by $R$. Note that each outcome of the SCF $f(.)$ has an allocation rule and a payment rule. The allocation function remains the same for DSIC-B protocol and BIC-B protocol. For completeness, we recall that the allocation rule is defined in the following way: $\forall \theta \in \Theta, \forall i \in N,$

$$k_i(\theta) = 1, \quad \text{if } i \in R$$
$$= 0, \quad \text{if } i \notin R.$$

The valuation function, $v_i(k(\theta), \theta_i)$, of node $i$ is its cost to forward a transit packet. From the allocation rule $k(.)$ in our SCF $f(.)$, we get, $\forall \theta \in \Theta, \forall i \in N,$

$$v_i(k(\theta), \theta_i) = -\theta_i, \quad \text{if } i \in R \tag{5.10}$$

$$= 0, \quad \text{if } i \notin R. \tag{5.11}$$

The broadcast packets from the source node travel through the paths specified in the *SRBT*. To compensate the incurred cost of the routers in the network, we need to determine payments to the nodes. We follow the payment rule of the classical *dAGVA mechanism* or *expected externality mechanism* (Section 2.17) to compute the payments to the nodes in our scheme. Using the payment rule of the dAGVA mechanism $\forall i \in N$, $\forall \theta \in \Theta$, we get

$$t_i(\theta) = E_{\theta_{-i}}\left[\sum_{l \neq i} v_l(k(\theta), \theta_l)\right] - \left(\frac{1}{n-1}\right)\sum_{j \neq i} E_{\theta_{-j}}\left[\sum_{l \neq j} v_l(k(\theta), \theta_l)\right].$$

From (5.10), (5.11) we get, $\forall i \in N$, $\forall \theta \in \Theta$,

$$t_i(\theta) = \left(\frac{1}{n-1}\right)\sum_{j \neq i} E_{\theta_{-j}}\left[\sum_{l \in R, l \neq j} \theta_l\right] - E_{\theta_{-i}}\left[\sum_{l \in R, l \neq i} \theta_l\right] \qquad (5.12)$$

where $E_{\theta_{-i}}\left[\sum_{l \in R, l \neq i} \theta_l\right]$ is interpreted as the total expected value to node $i$ that would be generated by all the remaining nodes in the absence of node $i$. This completes the characterization of the payment rule of the *BIC-B* mechanism. Note that this mechanism with the above payment rule is incentive compatible. This directly follows from the incentive compatibility property of the dAGVA mechanism (Section 2.17).

In the following, we first present an illustrative example to understand the details of the proposed payment scheme, and we then investigate the properties of the mechanism.

### 5.5.2 BIC-B: An Example

Now we provide an example to illustrate the payment scheme in *BIC-B* protocol. Let us consider the graph in Figure 5.8, which is *not biconnected*. We recall that the types of nodes are their incurred transit costs. We consider the type sets of nodes as $\Theta_1 = \{10, 11\}$, $\Theta_2 = \{15, 16\}$, $\Theta_3 = \{12, 13\}$, $\Theta_4 = \{7, 8\}$. Now let us assume that the belief probability functions of the players are independent discrete uniform distributions with equal probabilities for all types. Let $\theta = (10, 15, 13, 8)$ be the announced cost profile.

**Fig. 5.8**  Illustrative example 1

In this example, the *SRBT* is also the same as the original graph. Now, the allocation rule is $k(\theta) = (0, 1, 1, 0)$. We note that $N = \{1, 2, 3, 4\}$ and $R = \{2, 3\}$. The valuation functions of nodes are $v_1(k(\theta)) = 0$, $v_2(k(\theta)) = -15$, $v_3(k(\theta)) = -13$, $v_4(k(\theta)) = 0$. Now we compute the payments using the payment rule (5.12) of *BIC-B* protocol. The payment computation for node 1 is as follows.

$$t_1(\theta) = \left(\tfrac{1}{4-1}\right) \sum_{j \neq 1} E_{\theta_{-j}} \left[ \sum_{l \in R, l \neq j} \theta_l \right] - E_{\theta_{-1}} \left[ \sum_{l \in R, l \neq 1} \theta_l \right]$$

$$= \left(\tfrac{1}{3}\right) E_{\theta_{-2}} \left[ \sum_{l \in R, l \neq 2} \theta_l \right] + \left(\tfrac{1}{3}\right) E_{\theta_{-3}} \left[ \sum_{l \in R, l \neq 3} \theta_l \right] + \left(\tfrac{1}{3}\right) E_{\theta_{-4}} \left[ \sum_{l \in R, l \neq 4} \theta_l \right]$$
$$\quad - E_{\theta_{-1}} \left[ \sum_{l \in R, l \neq 1} \theta_l \right]$$

$$= \left(\tfrac{1}{3}\right) \left[ E_{\theta_{-2}} [\theta_3] + E_{\theta_{-3}} [\theta_2] + E_{\theta_{-4}} [\theta_2 + \theta_3] \right] - E_{\theta_{-1}} [\theta_2 + \theta_3]$$

$$= \left(\tfrac{1}{3}\right) \left[ E_{\theta_{-2}} [\theta_3] + E_{\theta_{-3}} [\theta_2] \right] + \left(\tfrac{1}{3}\right) \left[ E_{\theta_{-4}} [\theta_2] + E_{\theta_{-4}} [\theta_3] \right]$$
$$\quad - \left[ E_{\theta_{-1}} [\theta_2] + E_{\theta_{-1}} [\theta_3] \right]$$

(since types are statistically independent)

$$= \left(\tfrac{1}{3}\right) [12.5 + 15.5 + 15.5 + 12.5] - [15.5 + 12.5]$$

$$= -9.33.$$

In a similar fashion, we can compute the payments to the remaining nodes also. The payments to nodes are: $t_1(\theta) = -9.33$, $t_2(\theta) = 11.33$, $t_3(\theta) = 7.33$, $t_4(\theta) = -9.33$. In this example, node 1 and node 4 do not forward the packets. Observe that they pay the same amount, namely 9.33. Since node 2 and node 3 are routers, they receive amounts 11.33 and 7.33 respectively. Further observe that the sum of the payments is zero, i.e., $\sum_{i=1}^{4} t_i(\theta) = 0$. This implies that there is no need for any external budget to be injected into the network to sustain the running of the protocol.

For this particular example, we cannot apply the DSIC-B protocol since the graph in Figure 5.8 is not biconnected. Hence we can see that BIC-B protocol is applicable to any type of underlying graph of an ad hoc wireless network.

### 5.5.3 BIC-B: Some Properties

#### 5.5.3.1 Budget Balance

Assume that $\xi_i(\theta_i) = E_{\theta_{-i}} \left[ \sum_{l \in R, l \neq i} \theta_l \right]$, $\forall i \in N$. To show budget balance, we need to show that the sum of the payments received and the payments made by the nodes in the network is zero, i.e., $\sum_i t_i(\theta) = 0$, $\forall \theta \in \Theta$. Let us assume that $\xi_i(\theta_i) = E_{\theta_{-i}} \left[ \sum_{l \in R, l \neq i} \theta_l \right]$, $\forall \theta \in \Theta$, $\forall i \in N$. Then, from (5.12), we get $\forall \theta \in \Theta$

$$t_i(\theta) = \xi_i(\theta_i) - \left(\tfrac{1}{n-1}\right) \sum_{j \neq i} \xi_j(\theta_j)$$

then,

$$\sum_{i=1}^{n} t_i(\theta) = \sum_{i=1}^{n} \xi_i(\theta_i) - \left(\tfrac{1}{n-1}\right) \sum_{i=1}^{n} \sum_{j \neq i} \xi_j(\theta_j)$$
$$= \sum_{i=1}^{n} \xi_i(\theta_i) - \left(\tfrac{1}{n-1}\right) \sum_{i=1}^{n} (n-1) \xi_j(\theta_j)$$
$$= \sum_{i=1}^{n} \xi_i(\theta_i) - \sum_{i=1}^{n} \xi_j(\theta_j)$$
$$= 0.$$

It can be noted that each node $i$ distributes $\xi_i$ equally among the remaining $(n-1)$ nodes.

### 5.5.3.2 Payments by Nonrouter Nodes

Recall that $N$ is the set of nodes, and $R$ represents the set of routers in the network. We now state *Lemma 5.3*, which is useful in proving *Lemma 5.4* and *Theorem 5.1*.

**Lemma 5.3.** *For any* $i \in R$ *and for any* $j \notin R$, *we have*

$$E_{\theta_{-i}}\left[\sum_{l \in R, l \neq i} \theta_l\right] = E_{\theta_{-j}}\left[\sum_{l \in R, l \neq i} \theta_l\right].$$

**Proof**: Note that the types of the nodes are statistically independent according to the current network model. For this reason, it does not matter even though we take the expectation, in the above expression, with respect to $\theta_i$ or $\theta_j$, where $i \in R$ and $j \notin R$. Let the nodes in the set $R$ be indexed by the set $\{1, 2, ..., r\}$, where $r = |R|$. Now for any $j \notin R$ and $i \in R$,

$E_{\theta_{-j}}\left[\sum_{l \in R, l \neq i} \theta_l\right]$

$= \int ... \int \left[\sum_{l \in R, l \neq i} \theta_l\right] q(x_1)...q(x_{j-1})q(x_{j+1})...q(x_n)d(x_1)...d(x_{j-1})d(x_{j+1})...d(x_n)$

$= \int ... \int \left[\sum_{l \in R, l \neq i} \theta_l\right] q(x_1)...q(x_r) \quad d(x_1)...d(x_r)$

(since types are statistically independent)

$= \int ... \int \left[\sum_{l \in R, l \neq i} \theta_l\right] q(x_1)...q(x_{i-1})q(x_{i+1})...q(x_r)d(x_1)...d(x_{i-1})d(x_{i+1})...d(x_r)$

(since $\left[\sum_{l \in R, l \neq i} \theta_l\right]$ does not include $\theta_i$)

$= \int ... \int \left[\sum_{l \in R, l \neq i} \theta_l\right] q(x_1)...q(x_{i-1})q(x_{i+1})...q(x_n)d(x_1)...d(x_{i-1})d(x_{i+1})...d(x_n)$

(since types are statistically independent)

$= E_{\theta_{-i}}\left[\sum_{l \in R, l \neq i} \theta_l\right].$

*Q.E.D.*

**Lemma 5.4.** *For the* BIC-B *protocol,* $t_i(\theta) < 0$, $\forall i \notin R$, $\forall \theta \in \Theta$. *That is, the nodes other than the routers will pay a positive amount of money for receiving the packet(s).*

**Proof**: If $R$ is empty, then the source node can reach all the remaining nodes in the network within a single hop. In that case, the payments will be 0. We are not interested in such a trivial situation. So we assume that,

$$|R| > 0. \tag{5.13}$$

Let us assume that, $\forall j \notin R$,

$$\mathfrak{I}_j = E_{\theta_{-j}}\left[\sum_{l \in R, l \neq j} \theta_l\right]$$
$$= E_{\theta_{-j}}\left[\sum_{l \in R} \theta_l\right] \quad \text{(since } j \notin R\text{)}.$$

Since the types of nodes are statistically independent, the values of $\mathfrak{I}_j$, $\forall j \notin R$ are all the same. We represent this value with $\mathfrak{I}$. That is,

$$\mathfrak{I} = \mathfrak{I}_j, \qquad \forall j \notin R. \tag{5.14}$$

Now, let us assume that $\Upsilon_i = E_{\theta_{-i}}\left[\sum_{l \in R, l \neq i} \theta_l\right]$, $\forall i \in R$. Then

$$\begin{aligned}
\Upsilon_i &= E_{\theta_{-i}}\left[\sum_{l\in R, l\neq i}\theta_l\right]\\
&= E_{\theta_{-j}}\left[\sum_{l\in R, l\neq i}\theta_l\right]\\
&\quad\text{(consequence of Lemma 5.3)}\\
&< E_{\theta_{-j}}\left[\sum_{l\in R}\theta_l\right]\qquad\text{(since }i\in R\text{)}\\
&= \mathfrak{S}\qquad\text{(from equation (5.14))}.
\end{aligned}$$

So, we can conclude that

$$\Upsilon_i < \mathfrak{S}, \qquad \forall i\in R. \tag{5.15}$$

From the payment rule (5.12) of the *BIC-B* protocol, we have $\forall i\notin R$,

$$\begin{aligned}
t_i(\theta) &= \left(\tfrac{1}{n-1}\right)\sum_{j\neq i,\, j\in N} E_{\theta_{-j}}\left[\sum_{l\in R, l\neq j}\theta_l\right] - E_{\theta_{-i}}\left[\sum_{l\in R, l\neq i}\theta_l\right]\\
&= \left(\tfrac{1}{n-1}\right)\sum_{j\in R} E_{\theta_{-j}}\left[\sum_{l\in R, l\neq j}\theta_l\right] + \left(\tfrac{1}{n-1}\right)\sum_{j\neq i,\, j\notin R} E_{\theta_{-j}}\left[\sum_{l\in R, l\neq j}\theta_l\right]\\
&\quad - E_{\theta_{-i}}\left[\sum_{l\in R, l\neq i}\theta_l\right]\\
&= \left(\tfrac{1}{n-1}\right)\sum_{j\in R}\Upsilon_j + \left(\left(\tfrac{1}{n-1}\right)\sum_{j\neq i,\, j\notin R}\mathfrak{S}\right) - \mathfrak{S}\\
&\quad\text{(since from equation (5.14))}\\
&= \left(\tfrac{1}{n-1}\right)\sum_{j\in R}\Upsilon_j + \left(\tfrac{|N|-|R|-1}{n-1} - 1\right)\mathfrak{S}\\
&= \left(\tfrac{1}{n-1}\right)\sum_{j\in R}\Upsilon_j - \left(\tfrac{1}{n-1}\right)\sum_{j\in R}\mathfrak{S}\\
&= \left(\tfrac{1}{n-1}\right)\sum_{j\in R}(\Upsilon_j - \mathfrak{S})\\
&< 0,\qquad\text{(from (5.15))}.
\end{aligned}$$

According to the standard interpretation, $t_i(\theta) < 0$ means that node $i$ needs to pay the specified amount. This completes the proof of the lemma. (*Q.E.D.*)

**Observation 1:** From the proof of Lemma 5.4, we know that $\forall i\notin R$,

$$t_i(\theta) = \left(\tfrac{1}{n-1}\right)\sum_{j\in R}(\Upsilon_j - \mathfrak{S}).$$

Note that the right hand side of the above expression is independent from $i$. Hence, using the *BIC-B* protocol, $t_i(.)$, $\forall i\notin R$ are all the same. The immediate implication is that the payments made by the nodes other than the routers are the same.

### 5.5.3.3 Optimality of the *BIC-B* Payments

Here we prove the optimality of the payments prescribed by the *BIC-B* protocol. We note that an appropriate allocation rule is employed to determine the SRBT of the underlying graph of the ad hoc wireless network for the broadcast task. We define *cost of SRBT* as the sum of the forwarding costs of the router nodes. We first make the following observation.

**Observation 2:** Consider two type profiles $\theta$ and $\theta'$. Assume that these two type profiles are different only with respect to the type of node $i$. Using the BIC-B mechanism, when the type profiles are $\theta$ and $\theta'$ the payments to node $i$ are $t_i(\theta)$ and $t_i(\theta')$, respectively. Note that if the corresponding SRBT is the same for both the type profiles $\theta$ and $\theta'$, then the set of routers is the same and hence the payments $t_i(\theta)$ and $t_i(\theta')$ respectively to node $i$ are the same. This is because in payment rule (5.12) of the BIC-B mechanism, the quantities involved only look for expected values of the types of the routers.

**Theorem 5.1.** *For a given SRBT of the underlying graph G of the ad hoc wireless network, the payment to any node using the BIC-B mechanism is minimum among all other Bayesian incentive compatible mechanisms based on that SRBT.*

### 5.5.3.4  Individual Rationality of the *BIC-B* Protocol

We now investigate the *individual rationality (IR)* of the *BIC-B* protocol. In particular, we investigate the ex-post individual rationality, which is the strongest among the three notions of individual rationality (see Section 2.10). In the following theorem, we obtain a necessary and sufficient condition for the ex-post individual rationality of the BIC-B protocol. Let $\hat{\theta}_i$ be the announced cost of a node $i$ and $\theta_i$ be the actual cost of that node.

**Theorem 5.2.** *The BIC-B protocol is ex-post individual rational if and only if*

$$\hat{\theta}_i \leq \left(\tfrac{n}{n-1}\right)E[\theta_i], \quad \forall i \in R$$

*where the $\hat{\theta}_i$ is the announced cost of the node i.*

**Proof:** We show that the utility of each router is nonnegative after participating in the mechanism if and only if the specified condition holds. This is nothing but proving the ex-post individual rationality of the nodes.

Let $f$ be the mechanism for the ICB problem. Now for ex-post individual rationality to hold for the router nodes,

$$u_i(f(\hat{\theta}), \hat{\theta}_i) \geq 0, \forall \hat{\theta} \in \Theta, \quad \forall i \in R,$$

where $\hat{\theta}$ is the vector of announcements of the costs (or types) of the nodes. We also call $\hat{\theta}$ the announced cost (or type) profile of the nodes.

Now we characterize the utility of each node $i \in R$. All nodes are the intended recipients of the packet(s) in a broadcast in the network. We know, from Observation 1, that the payment made by a nonrouter node is equivalent to $t_m(\hat{\theta})$ for any $m \notin R$. Since the routers are also intended recipients, they also need to pay this amount. But actually, the routers do not pay this amount, and hence it is credited to their utility. Now we have, $\forall \hat{\theta} \in \Theta$,

$$u_i(f(\hat{\theta}), \hat{\theta}_i) = v_i(k(\hat{\theta}_i, \hat{\theta}_{-i})) - t_m(\hat{\theta}) + t_i(\hat{\theta}). \tag{5.16}$$

We have $t_m(\hat{\theta}) < 0$ for any $m \notin R$, from Lemma 5.4. Hence this term appears with a negative sign in the expression (5.16). Now substituting the expression (5.12) in (5.16) and rearranging the terms, we get $\forall \hat{\theta} \in \Theta$,

$$u_i(f(\hat{\theta}), \hat{\theta}_i) = -\hat{\theta}_i + E_{\theta_{-m}}\left[\textstyle\sum_{l \in R} \theta_l\right] - E_{\theta_{-i}}\left[\textstyle\sum_{l \in R, l \neq i} \theta_l\right]$$
$$+ \left(\tfrac{1}{n-1}\right) \textstyle\sum_{j \neq i} E_{\theta_{-j}}\left[\textstyle\sum_{l \in R, l \neq j} \theta_l\right] - \left(\tfrac{1}{n-1}\right) \textstyle\sum_{j \neq m} E_{\theta_{-j}}\left[\textstyle\sum_{l \in R, l \neq j} \theta_l\right].$$

By canceling out appropriate the terms, we get $\forall \hat{\theta} \in \Theta$,

$$u_i(f(\hat{\theta}), \hat{\theta}_i) = -\hat{\theta}_i + E_{\theta_{-m}}\left[\sum_{l \in R} \theta_l\right] - E_{\theta_{-i}}\left[\sum_{l \in R, l \neq i} \theta_l\right] + \left(\frac{1}{n-1}\right) E_{\theta_{-m}}\left[\sum_{l \in R} \theta_l\right]$$
$$- \left(\frac{1}{n-1}\right) E_{\theta_{-i}}\left[\sum_{l \in R, l \neq i} \theta_l\right]$$
$$= -\hat{\theta}_i + E_{\theta_{-m}}\left[\sum_{l \in R} \theta_l\right] - E_{\theta_{-m}}\left[\sum_{l \in R, l \neq i} \theta_l\right] + \left(\frac{1}{n-1}\right) E_{\theta_{-m}}\left[\sum_{l \in R} \theta_l\right]$$
$$- \left(\frac{1}{n-1}\right) E_{\theta_{-m}}\left[\sum_{l \in R, l \neq i} \theta_l\right]$$

(consequence of Lemma 5.3)

$$= -\hat{\theta}_i + E_{\theta_{-m}}\left[\sum_{l \in R} \theta_l\right] - E_{\theta_{-m}}\left[\sum_{l \in R} \theta_l\right] + E_{\theta_{-m}}[\theta_i]$$
$$+ \left(\frac{1}{n-1}\right) E_{\theta_{-m}}\left[\sum_{l \in R} \theta_l\right] - \left(\frac{1}{n-1}\right) E_{\theta_{-m}}\left[\sum_{l \in R} \theta_l\right] + \left(\frac{1}{n-1}\right) E_{\theta_{-m}}[\theta_i]$$

(by expanding the Expectation terms)

$$= -\hat{\theta}_i + E[\theta_i] + \left(\frac{1}{n-1}\right) E[\theta_i]$$
$$= \frac{nE[\theta_i] - (n-1)\hat{\theta}_i}{n-1}.$$

For the *BIC-B* protocol to be ex post individually rational,

$$u_i(f(\hat{\theta}), \hat{\theta}_i) \geq 0, \quad \forall \hat{\theta} \in \Theta, \quad \forall i \in R.$$

From the above characterization of utility function, we get

$$\frac{nE[\theta_i] - (n-1)\hat{\theta}_i}{n-1} \geq 0, \quad \forall i \in R.$$

This implies,

$$\hat{\theta}_i \leq \left(\frac{n}{n-1}\right) E[\theta_i], \quad \forall i \in R.$$

Furthermore, it is easy to see that each of above arguments are reversible and hence the specified condition is necessary and sufficient for the *BIC-B* protocol to satisfy ex-post individual rationality.

$$Q.E.D.$$

## 5.5.4 BIC-B Protocol: An Implementation

The protocol implementation is motivated by the emerging technology of hybrid ad hoc wireless networks, where there are base stations that provide fixed infrastructure for many network related functions. We propose that additional functionality (which we call mediation functionality) as described below be incorporated into each base station. Let us call this part of the base station *mediator*. The mediator could be a part of any other network infrastructure also.

The mediator elicits the types from all the nodes, computes the allocation and payments of the nodes, and announces the outcome. We assume that all the nodes can communicate with the mediator.

After receiving the messages from the mediator regarding the payments, each node constructs an internal table as shown in Table 5.2. Each row of the table corresponds to a node in the network. Each row contains three fields of information: (a) *Source ID*, which specifies the source node ID from which the packet is originating, (b) *Node List*, which specifies the set of nodes to which the packet needs to be forwarded, and (c) *Payment*, which specifies the payment to be received or paid. Table 5.2 is the structure of internal table of a node.

| Source ID | Node List | Payment |
|-----------|-----------|---------|
|           |           |         |
|           |           |         |

**Table 5.2**  Structure of internal table of a node

In view of the above, the BIC-B protocol can be implemented as follows.

*BIC-B Protocol:* If a node receives a broadcast packet, it checks its internal table entry corresponding to the source ID of the broadcast. Then it forwards the packet to the set of nodes specified in the *Node List* field of the entry and receives the payment as mentioned in the *Payment* field of the entry. On the other hand, if the *Node List* field is empty, then the node does not forward the packet to any node and makes a payment as mentioned in the *Payment* field of the entry.

In the above protocol, all the payment information by the node is communicated directly to the mediator, which takes care of all the book keeping.

## 5.5.5  BIC-B Protocol: Performance Evaluation

In this section, we show the efficacy of the proposed *BIC-B* protocol for the *ICB* problem. In our simulation experiments, we compare the performance of the *BIC-B* protocol with that of the *Dominant Strategy Incentive Compatible Broadcast (DSIC-B)* protocol [23].

### 5.5.5.1  Simulation Model

The *DSIC-B* protocol is based on *dominant strategy equilibrium* of the underlying game, and the *BIC-B* protocol is based on the *Bayesian Nash equilibrium* of the underlying game. Since every dominant strategy equilibrium is also a Bayesian Nash equilibrium, but not vice-versa, we first find a dominant strategy equilibrium of the underlying game and compute the payments to the nodes using the *DSIC-B* and the *BIC-B* protocols [23].

We work with a randomly generated graph of an ad hoc wireless network with the number of nodes $n = 5, 10, 15, 20, 25, 30, 35, 40$. According to our network model presented in Section 5.2.3, the graph is node weighted where these weights are the transit costs of the nodes chosen independently and uniformly from a range $[1, 50]$. After the nodes announce their types, we first compute the least cost paths to all the nodes from the source node and then construct an *SRBT*. Using the *SRBT*, we can decide the set of routers. This fixes the allocation rule. Then we compute payments to the nodes using the payment rule of the appropriate broadcast protocol. In all our simulation experiments, the results for the performance metrics are averages taken over 100 random instances.

### 5.5.5.2 Simulation Results

We consider two performance metrics. The first metric is *average payment to routers (APR)*. This specifies the payment on an average to each router for forwarding the transit packets. The graph in Figure 5.9 shows the comparison of the *BIC-B* protocol and the *DSIC-B* protocol using APR. In the figure, the lower curve corresponds to the *BIC-B* protocol. It is clear from the figure that the *BIC-B* protocol performs better than the *DSIC-B* protocol. This means that the system wide payments made by the nodes to forward a broadcast packet are less using the *BIC-B* protocol.

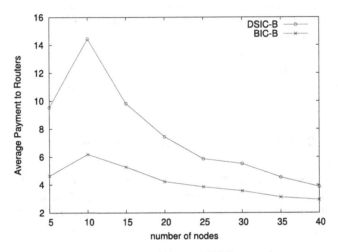

**Fig. 5.9**  Average payment to routers in the *DSIC-B* and *BIC-B* protocols

The second performance metric, we consider, is the *worst overpayment ratio (WOR)*. We first define *overpayment ratio* as the ratio of payment made by a node to its least cost path value from the source node $s$. Now, we can define WOR as the maximum over the overpayment ratios of all the nodes in the network. That is,

$$\text{WOR} = \max_{i \in N} \frac{\text{payment made by node } i}{\text{cost of path from source of broadcast to node } i}$$

where *cost of path from source of broadcast to node i* is the sum of forwarding costs of the nodes that lie on the path from the source of broadcast node to the node $i$. Ideally we expect this ratio to be 1. We compare the WOR of *BIC-B* protocol and the *DSIC-B* protocol in Figure 5.10. We note that the lower curve corresponds to the *BIC-B* protocol in the figure. From the graph, it is easy to see that the worst overpayment ratio in the network is higher using the *DSIC-B* protocol than the *BIC-B* protocol. WOR conveys the following significant information. When a node receives a packet from a router, then clearly the payment made by the receiver node to the router is higher than the value of its least cost path, since it has to give incentives to the router to make it reveal the true incurred cost. If we take a ratio of the payment to the value of least cost path, from Figure 5.10, this ratio is less than 2 times over all the nodes using the *BIC-B* protocol, and it is higher than 5 times over all the nodes using the *DSIC-B* protocol. This shows that nodes end up with very high payments using *DSIC-B* protocol when compared with the value of the corresponding least cost path.

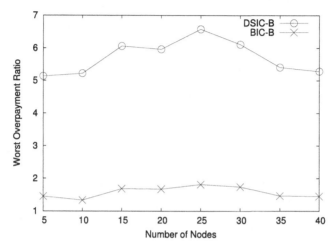

**Fig. 5.10**  Worst overpayment ratios for the *DSIC-B* and *BIC-B* protocols

## 5.6 DSIC-B Protocol Versus BIC-B Protocol: A Discussion

We now compare and contrast the two solution approaches for the *ICB* problem.

- **Underlying Network Structure:** The *DSIC-B* protocol requires the underlying graph of the ad hoc network to be biconnected. If the number of nodes in the network increases, sustaining such a condition may be difficult. On the other hand, the *BIC-B* protocol works for any topology of the ad hoc wireless network as long as all the nodes are connected. This provides better fault-tolerance to the *BIC-B* protocol.
- **Game Theoretic Properties:** The *DSIC-B* protocol is individually rational but not budget balanced. However they can be made budget balanced provided there is a single node whose type is common knowledge to the entire network. This may be possible with hybrid ad hoc wireless networks where that specific agent can be a base station. The *BIC-B* protocol is budget balanced, which is clear

from Lemma 5, and is ex-post individually rational under the specified condition in Theorem 5.2.

- **Computational Complexity:** In the *DSIC-B* protocol, it is required to compute an efficient allocation $(n+1)$ times to determine the payments to the nodes. On the other hand, using the *BIC-B* protocol, it is sufficient to determine an efficient allocation exactly once in order to compute the payments.
- **Strategic Complexity:** In the *DSIC-B* protocol, any node can choose its best strategy irrespective of behavior of the other nodes. On the other hand, in the *BIC-B* protocol, selection of the best strategy by a node is dependent on the other nodes.

## 5.7 Conclusions and Future Work

We considered the incentive compatible broadcast (ICB) problem in ad hoc wireless networks with selfish nodes. We presented two elegant solutions to this problem. In the first solution, we proposed a dominant strategy incentive compatible protocol and also proposed a distributed algorithm to implement this protocol. However this solution has certain limitations practically. For example, it requires the underlying graph of the ad hoc wireless network to be biconnected. The second solution method avoids these limitations. We proposed an incentive-based broadcast protocol that satisfies Bayesian incentive compatibility and minimizes the incentive budgets required by the individual nodes. The proposed protocol, *BIC-B*, also satisfies budget balance. We also derived a necessary and sufficient condition for the ex-post individual rationality of the BIC-B protocol. We showed that the *BIC-B* protocol exhibits superior performance when compared to a dominant strategy incentive compatible solution to the problem.

In terms of mechanism design, it would be interesting to explore optimal broadcast mechanisms along the lines of the mechanisms that were explored in Chapters 3 and 4. These are mechanisms that would minimize the incentive budgets subject to Bayesian incentive compatibility and individual rationality.

An important problem that could be explored is design of Bayesian incentive compatible protocols for unicast and multicast problems. Existing game theoretic approaches to unicast and multicast are all based on VCG mechanisms and have the usual limitations associated with the use of VCG mechanisms.

Also, it is important to address certain practical issues that arise in the implementation of these mechanisms as part of standard protocols. For example, the payment computation is performed in a centralized way in the BIC-B protocol. It would be interesting to design a distributed algorithm for this problem that could help deploy the BIC-B protocol in the real world.

# References

1. V. Conitzer and T. Sandholm. Failures of the VCG mechanism in combinatorial auctions and exchanges. In *Proceedings of 5th International Joint Conference on Autonomous Agents and Multiagent Systems (AAMAS)*, pages 521–528, 2006.
2. S. Buchegger and J.-Y.L. Boudec. Performance analysis of the CONFIDANT protocol: Co-operation of nodes – fairness in dynamic ad hoc networks. In *Proceedings of the 3rd ACM International Symposium on Mobile Ad Hoc Networking and Computing*, pages 157–164, 2002.
3. Q. Chen and Z. Niu. A game-theoretical power and rate control for wireless ad hoc networks with step-up price. *IEICE Transactions on Communications*, E88-B(9):3515–3523, 2005.
4. N. Nisan and A. Ronen. Computationally feasable VCG mechanisms. In *Proceedings of the 2nd ACM Conference on Electronic Commerce (ACM EC)*, pages 242–252, 2000.
5. S. Guha and S. Khuller. Improved methods for approximating node weighted steiner trees and connected dominating sets. In *Proceedings of 4th European Symposium on Algorithms (ESA)*, pages 179–193, 1996.
6. Z. Fang and B. Bensaou. Fair bandwidth sharing algorithms based on game theory frame-works for wireless ad-hoc networks. In *Proceedings of the 23rd IEEE Conference on Computer and Communications (INFOCOM)*, pages 1284–1295, 2004.
7. J. Hershberge and S. Suri. Vickrey prices and shortest paths: What is an edge worth? In *Proceedings of the 42th IEEE Symposium on Foundations of Computer Science (FOCS)*, pages 252–259, 2001.
8. M. Lu, F. Li, and J. Wu. Incentive compatible cost and stability-based routing in ad hoc networks. In *Proceedings of the 12th International Conference on Parallel and Distributed Systems (ICPADS)*, pages 495–500, 2006.
9. M. Lu and J. Wu. Social welfare based routing in ad hoc networks. In *Proceedings of International Conference on Parallel Processing (ICPP)*, pages 211–218, 2006.
10. V. Srinivasan, P. Nuggehalli, F. Chiasserini, and R.R. Rao. Cooperation in wireless ad hoc networks. In *Proceedings of the 22nd IEEE Conference on Computer and Communications (INFOCOM)*, pages 808–817, 2003.
11. S. Eidenbenz, P. Santi, and G. Resta. COMMIT: A sendercentric truthful and energy-efficient routing protocol for ad hoc networks. In *Proceedings of the 5th International Workshop on Wireless, Mobile, and Ad Hoc Networks (WMAN)* in conjunction with *19th IEEE International Parallel and Distributed Processing Symposium (IPDPS)*, page 239, 2005.
12. M. Lu, F. Li, and J. Wu. Incentive compatible cost- and stability-based routing in ad hoc networks. In *Proceedings of the 12th International Conference on Parallel and Distributed Systems (ICPADS)*, pages 495–500, 2006.
13. S. Zhong, L.E. Li, Y. Liu, and Y.G. Yang. On designing incentive-compatible routing and forwarding protocols in wireless ad-hoc networks: an integrated approach using game theo-retical and cryptographic techniques. In *Proceedings of the 11th ACM annual international conference on Mobile computing and networking (MOBICOM)*, pages 117–131, 2005.
14. W. Wang and X.Y. Li. Low-cost truthful multicast in selfish and rational wireless ad hoc networks. In *Proceeding of the 1st IEEE International Conference on Mobile Ad-hoc and Sensor Systems (MASS)*, pages 534–536, 2004.
15. S. Marti, T.J. Giuli, K. Lai, and M. Baker. Mitigating routing misbehaviour in mobile ad hoc networks. In *Proceedings of the 6th Annual ACM International Conference on Mobile Computing and Networking (MOBICOM)*, pages 255–265, 2000.
16. L. Anderegg and S. Eidenbenz. Ad hoc-VCG: A truthful and cost-efficient routing protocol for mobile ad hoc networks with selfish agents. In *Proceedings of the 9th Annual ACM International Conference on Mobile Computing and Networking (MOBICOM)*, pages 245–259, 2003.
17. W. Wang, X.Y. Li, and Y. Wang. Truthful multicast in selfish wireless networks. In *Proceedings of the 10th ACM annual international conference on Mobile computing and networking (MOBICOM)*, pages 245–259, 2004.

18. N.B. Salem, L. Butty, J.P. Hubaux, and M.A. Jakobsson. Charging and rewarding scheme for packet forwarding in multi-hop cellular networks. In *Proceedings of the 3rd ACM international symposium on Mobile ad hoc networking and computing (MOBIHOC)*, pages 13–24, 2003.

19. P. Michiardi and R. Molva. Core: a collaborative reputation mechanism to enforce node co-operation in mobile ad hoc networks. In *Proceedings of the 6th Joint Working Conference on Communications and Multimedia Security: Advanced Communications and Multimedia Security*, pages 107–121, 2002.

20. J. Feigenbaum, C.H. Papadimitriou, R. Sami, and S. Shenker. A BGP-based mechanism for lowest-cost routing. In *Proceedings of the 21st ACM Symposium on Principles Of Distributed Computing (PODC)*, pages 173–182, 2002.

21. P. Nurmi. Modelling routing in wireless ad hoc networks with dynamic Bayesian games. In *Proceedings of 1st Annual IEEE Communications Society Conference on Sensor and Ad Hoc Communications and Networks (SECON)*, pages 63–70, 2004.

22. S. Dobzinski and N. Nisan. Limitations of VCG-based mmchanisms. In *Proceedings of 39th ACM Symposium on Theory of Computing (STOC)*, pages 338–344, 2007.

23. N. Rama Suri. Design of incentive compatible protocols for wireless ad hoc networks: a game theoretic approch. In *Proceedings of the IEEE INFOCOM 2006 Student's Workshop*, pages 1–2, 2006.

24. N. Rama Suri and Y. Narahari. Broadcast in ad hoc wireless networks with selfish nodes: a Bayesian incentive compatibility approach. In *Proceedings of 2nd IEEE/Create-Net/ICST International Conference on COMmunication System softWAre and MiddlewaRE (COMSWARE)*, pages 1–9, 2007.

25. A. Urpi, M. Bonuccelli, and S. Giordano. Modeling cooperation in mobile ad hoc networks: a formal description of selfishness. In *Proceedings of the 1st Workshop on Modeling and Optimiztion in Mobile, Ad Hoc and Wireless Networks*, 2003.

26. B. Williams and T. Camp. Comparison of broadcasting techniques for mobile ad hoc networks. In *Proceedings of the ACM International Symposium on Mobile Ad Hoc Networking and Computing (MOBIHOC)*, pages 194–205, 2002.

27. E. Altman, R. El-Azouzi, and T. Jimenez. Slotted Aloha as a stochastic game with partial information. In *Proceedings of Modeling and Optimization in Mobile, Ad Hoc and Wireless Networks (WIOPT)*, pages 701–713, 2003.

28. W. Wang and X.Y. Li. Truthful low-cost unicast in selfish wireless networks. In *Proceedings of the 4th International Workshop on Algorithms for Wireless, Mobile, Ad Hoc and Sensor Networks (WMAN)* in conjunction with of IPDPS, pages 219, 2004.

29. S. Zhong, J. Chen, and Y. Yang. Sprite: A simple, cheatproof, credit-based system for mobile ad hoc networks. In *Proceedings of the 22nd IEEE Conference on Computer and Communications, (INFOCOM)*, pages 1987–1997, 2003.

30. P. Resnick, R. Zeckhauser, E. Friedman, and K. Kuwabara. Reputation systems: facilitating trust in internet interactions. *Communications of the ACM*, 43(12):45–48, 2000.

31. E. Altman, T. Boulogne, R.E. Azouzi, T. Jimenez, and L. Wynter. A survey on networking games in telecommunications. *Computers and Operations Research*, 33(2):286-311, 2006.

32. N. Rama Suri and Y. Narahari. Design of optimal Bayesian incentive compatible broadcast protocol for ad hoc networks with rational nodes. *IEEE Journal on Selected Areas of Communication (IEEE JSAC)*, Special Issue on Game Theory, 26(7):1138–1148, 2008.

33. M. Felegyhazi, L. Buttyan, and J.-P. Hubaux. Nash equilibria of packet forwarding strategies in wireless ad hoc networks. *IEEE Transactions on Mobile Computing*, 5(5):463–476, 2006.

34. P. Kyasanur and N.H. Vaidya. Selfish MAC layer misbehavior in wireless networks. *IEEE Transactions on Mobile Computing*, 4(5):502–516, 2004.

35. V. Srivastava, J. Neel, A.B. MacKenzie, R. Menon, L.A. DaSilva, J. Hicks, J.H. Reed, and R. Gilles. Using game theory to analyze wireless ad hoc networks. *IEEE Communications Surveys and Tutorials*, 7(4):46–56, 2005.

36. E. Altman, A.A. Kherani, P. Michiardi, and R. Molva. Non-cooperative forwarding in ad-hoc networks. Technical Report, INRIA, Sophin Antipolis, France, 2004.

37. M. Felegyhazi and J.P. Hubaux. Game theory in wireless networks: A tutorial. Technical Report, EPFL, Switzerland, 2006.
38. C. Guillaume and E. Fleury. NP-Completeness of ad hoc multicast routing problems. Technical Report, INRIA, France, 2005.
39. N. Rama Suri. Design of incentive compatible broadcast protocols for ad hoc wirless networks: A game theoretic approach, Master's Thesis, Department of Computer Science and Automation, Indian Institute of Science, Bangalore, 2006.
40. M.S. Gast. *802.11 Wireless Networks: The Definitive Guide*. Shroff Publishers, Mumbai, 2004.
41. A.S. Mas-Colell, M.D. Whinston, and J.R. Green. *Microeconomic Theory*. Oxford University Press, New York, 1995.
42. R.B. Myerson. *Game Theory: Analysis of Conflict*. Harvard University Press, Cambridge, Massachusetts, 1997.

# Chapter 6
# To Probe Further

In the last three chapters, we have explored mechanism design in the context of three representative network economics situations, after having studied the foundations and key results in mechanism design in Chapter 2. Clearly it is impossible to cover all aspects of mechanism design and all its application areas in any single monograph. In this chapter, we first provide a glimpse of what we perceive are some key topics in mechanism design that would be relevant for future work in this area. The list of topics mentioned here is by no means exhaustive. Next, we list a few application areas in network economics where mechanism design would be a valuable tool to use. Again this list is indicative and is by no means all inclusive.

## 6.1 Topics in Mechanism Design

In this section, we briefly touch upon several topics (in mechanism design) which we believe have important implications for network economics applications.

### 6.1.1 Combinatorial Auctions

Combinatorial auctions are those where the bidders can specify bids on combinations of items. Because of the nature of bids, the winner determination problem in combinatorial auctions turns out to be a hard problem. We introduced combinatorial auctions briefly when we discussed the generalized Vickrey auction in Section 2.16. These auctions have found widespread applications in numerous network economics situations. The edited volume by Cramton, Shoham, and Steinberg [1] is a comprehensive source of information on different aspects of combinatorial auctions. There are also many survey articles on combinatorial auctions. These include: de Vries and Vohra [2, 3], Pekec and Rothkopf [4], Narahari and Dayama [5], and Blumrosen and Nisan [6].

Y. Narahari et al., *Game Theoretic Problems in Network Economics and Mechanism Design Solutions*, Advanced Information and Knowledge Processing, DOI: 10.1007/978-1-84800-938-7_6, © Springer-Verlag London Limited 2009

## 6.1.2  Optimal Mechanisms

Optimal mechanisms are those that optimize a well defined performance metric (for example, revenue maximization, cost minimization, etc.) subject to incentive compatibility and individual rationality constraints. We have seen the classic Myerson optimal auction for a single indivisible item in Section 2.20. Optimal mechanisms represent a key current topic of research in mechanism design. The paper by Hartline and Karlin [7] is an excellent survey of results in this area. Optimal auctions with multidimensional types is recognized as a difficult problem to solve. We wish to draw attention to a few recent results in this topic:

- Malakhov and Vohra [8] have provided a novel interpretation to the optimal auctions problem with multidimensional types when the type sets are finite. They model this problem as a linear program that is an instance of the parametric shortest path problem on a lattice.
- Designing an optimal combinatorial auction is also generally recognized as a challenging problem. Recently, Ledyard [9] has developed an optimal combinatorial auction under the special case of single minded bidders (that is, bidders who are only interested in a single combination of items).
- An optimal auction for multi-unit auctions with volume discount bids has been designed by Gautam et al [10], extending the work of Kumar et al [11].

## 6.1.3  Efficient Auctions

Recall that Myerson's optimal auction achieves maximum revenue or minimum cost, as the case may be, subject to Bayesian incentive compatibility and interim individual rationality. In general, Myerson's auction is not allocatively efficient. Krishna and Perry [12] have argued in favor of an auction that will maximize the revenue (or minimize the cost as the case may be) subject to allocative efficiency, dominant strategy incentive compatibility, and interim individual rationality. The Green–Laffont theorem (Theorem 2.6) asserts that any DSIC and AE mechanism has to be necessarily a VCG mechanism. This means we have to look for a VCG mechanism that will maximize the revenue to the seller subject to IIR constraints. It has been shown by Krishna and Perry [12] that the classical Vickrey auction is an optimal and efficient auction for a single indivisible item. They have also shown that the Vickrey auction is an optimal one among VCG mechanisms for multi-unit auctions, when all the bidders have downward sloping demand curves. Further details on efficient auctions can be obtained from the book by Vijay Krishna [13].

## 6.1.4 Cost Sharing Mechanisms

In a cost sharing mechanism, there is a set of identical objects (goods or services) available, associated with a certain cost, and there are multiple players with private preferences who are interested in one object each. A cost sharing mechanism is a protocol that determines the players who are allocated the objects and the prices to be paid. The three major properties desired from a cost sharing mechanism are: (1) incentive compatibility, which guarantees that all players bid their true private values; (2) budget balance, which ensures that the mechanism recovers the incurred cost with the prices charged; and (3) efficiency, which ensures that the cost incurred and the value to the players are traded off in an optimal way. It is known that not all three of these properties can be simultaneously satisfied.

Moulin mechanisms represent a popular paradigm for cost sharing. Moulin has designed a Groves mechanism, which is not a Clarke mechanism but minimizes the budget imbalance [14]. The proposal of Moulin is to redistribute the surplus created in a way that the mechanism remains a Groves mechanism and achieves minimum budget imbalance. Being a Groves mechanism, Moulin mechanisms are allocatively efficient and dominant strategy incentive compatible. A survey of cost sharing in network economics applications appears in [15].

## 6.1.5 Iterative Mechanisms

In many situations, the decisions are naturally sequential rather than simultaneous. Two leading examples of well known mechanisms where decisions are made sequentially are the English auction and the Dutch auction. Such mechanisms are called iterative mechanisms. The computational and communication costs of the agents may be reduced by making the decisions sequentially. For example, in the case of combinatorial auctions, the total number of combinations is exponential, and the agents need to find out valuations for all the combinations. In such scenarios, rather than having a static combinatorial auction, iterative auctions could be used to reduce the cost of computing valuations and allocations. Linear programming duality has been used as a key tool in designing iterative mechanisms. The survey paper by Parkes [16] is an excellent source of material on iterative mechanisms. There is plenty of scope for further work in designing iterative mechanisms.

## 6.1.6 Dynamic Mechanisms

In the discussion so far, we have assumed that agents observe their private values and take actions depending upon their values and the rules of mechanism. The social planner has to take the decision exactly once based upon the actions by the agents. Such a mechanism can be categorized as a static implementation. Let us consider a dynamic scenario such as an air carrier that wishes to sell the tickets for the flights. The buyers are arriving at different times, and each buyer could have a unique demand. The valuation a buyer attaches to the airticket constitutes the buyer's private information. The air carrier's goal is to maximize revenue in such a dynamic environment.

To handle such scenarios, we need mechanisms that are *dynamic*. When a sequence of decisions is required to be made, a static mechanism cannot be employed if the parties receive information over time that should affect the decisions. For example, agents who are involved in a long-term relationship may need to make a sequence of trading and investment decisions in a changing environment. A procurement authority may wish to conduct a sequence of auctions, where bidders have serially correlated values or capacity constraints or learning-by-doing. In recent times, the design of mechanisms in a dynamic environment has emerged as an interesting area of research.

Athey and Segal [17] have considered a dynamic environment in which agents observe a sequence of private signals over a number of periods/time slots. The number of periods can either be finite or countably infinite. In each time slot, the agents report their private signals. Based on the reported signals, the public decision is made. The probability distribution over future signals may depend on both past signals and past decisions. In such an environment, Athey and Segal have designed a BIC mechanism that is also efficient and budget balanced.

Contracts to operate public facilities such as airports or to use natural resources such as forests, are renewed periodically. The current winner of the contract receives additional information about it, learns most about the value of the resource. This enables this current winner to revise its valuation. In such dynamic settings, Bapna and Weber [18] and Bergemann and Valimaki [19] have designed a Bayesian incentive compatible and efficient mechanism.

## Online Mechanisms

Online mechanisms constitute a category of dynamic mechanisms. In online mechanisms, the allocation and payment decisions need to be made dynamically on the lines of an online algorithm. The survey article by Parkes [20] is an excellent source of information on this topic.

## Optimal Dynamic Auctions

In a recent study, Vohra and Pai [21] have looked into a dynamic situation where a seller with multiple units of an item to sell faces bidders who arrive and depart over the course of a finite time horizon. The time at which a bidder arrives, their valuations, and their departure times are private information of the bidders. The study develops a revenue maximizing auction that is Bayesian incentive compatible.

### 6.1.7 Stochastic Mechanisms

Consider a scenario in which the decision maker has to solve a stochastic optimization problem, and the parameters for optimization are private to the agents. For example, consider the following situation discussed by Ieong et al [22]. There is a university book seller who sells course material. The seller has two options: (1) get it printed from a printing press, which costs less, but takes more time; (2) photocopy the material with less turnaround time but higher cost than the first option. If the seller has prior knowledge about how many students will be registering for the course, the seller can directly print the copies. The students are not sure, before the semester starts, whether or not they would be taking the course. However, each student may have a probability distribution that gives the probability of this student taking the course. If the students report their distributions to the book seller truthfully, the book seller can simply solve a two stage stochastic optimization problem. However, the students may not reveal this distribution function truthfully. In such a scenario, mechanism design can be used. The proposal by Ieong et al [22] is a two stage stochastic mechanism which is Bayesian incentive compatible. It is not known whether one can design a dominant strategy implementation of a stochastic mechanism.

### 6.1.8 Computationally Efficient Approximate Mechanisms

We have seen in Section 2.21 that mechanism design for realistic scenarios involves computationally hard problems to be solved. It is found that if nonoptimal solutions are used in mechanism design, then the mechanisms may no longer satisfy economic properties such as incentive compatibility and individual rationality. For example, Nisan and Ronen [23] show that all reasonable approximations or heuristics for combinatorial auctions yield non-truthful VCG mechanisms. There is a rich body of literature that studies how mechanisms could be designed that satisfy economic properties almost exactly despite using approximate algorithms or randomized algorithms in mechanism design. The use of these algorithms will ensure that the mechanisms are computationally efficient. Lavi [24] provides a neat review of the results in this area.

### 6.1.9 Mechanisms without Money

Recall the Gibbard–Satterthwaite impossibility theorem (Section 2.11) which asserts that only dictatorial social choice functions can be implemented in dominant strategies if the preferences are unrestricted. One way of circumventing this problem was to restrict the preferences, for example, quasilinear preferences. The use of quasilinear preferences involves monetary transfers. There are many situations where monetary transfers are to be avoided. Examples include political decisions, organ donations, etc. Schummer and Vohra [25] provide an overview of mechanism design in such a setting.

## 6.2 Key Application Areas

In this monograph, we have seen three applications of mechanism design to network economics problems. There are many other important applications; we now present very briefly a few selected current and emerging applications.

### 6.2.1 Procurement Auctions

Procurement is a ubiquitous activity in industrial organizations. Auctions have emerged as a leading technology within industries for procuring direct or indirect materials from a pool of qualified suppliers. There is now a rich body of literature on this topic including many published case studies. The major requirements of a procurement mechanism are cost minimization for the buyer, incentive compatibility, individual rationality, and allocative efficiency. Usually there would be several business constraints also to be satisfied in the procurement process (such as a lower bound on the number of suppliers, upper bound on the number of suppliers, etc.); the business constraints add another level of complexity to the design of procurement auctions. Taking into account multiple attributes such as quality, lead time, service costs, etc makes the design problem even more challenging. Chandrashekar and co-authors [26] provide a comprehensive survey of the literature in this topic. A survey of the current state-of-the-art also appears in Hohner et al [27].

### 6.2.2 Spectrum Auctions

Selling frequency spectrum to telecommunication companies through online auctions was first attempted in New Zealand (1989) and in England (1990) [28]. Later, auctions were used for selling spectrum rights in Australia in 1993. These auctions failed to generate much revenue due to flaws in auction design. In 1994, the Federal Communications Commission (FCC) in USA conducted landmark auctions for frequency spectrum in which major telecommunication firms (long distance, local, cellular telephone companies, and cable television companies) participated [28]. This auction went through an elaborate design exercise in which many celebrated auction theorists provided their technical advice. The total revenue generated by the auction was estimated as 10 billion US dollars.

Designing auction mechanisms for selling spectrum is quite a fascinating problem and is discussed in detail in [29, 30]. One of the popular mechanisms used is the simultaneous ascending auction, which is an iterative auction. This auction allows bidders to take advantage of allowing any information revealed during the successive rounds and provides flexibility to aggregate their items.

## 6.2.3 Supply Chain Formation

Consider a decentralized supply chain formation problem where the supply chain planner (SCP) is faced with the decision of choosing a partner or service provider at individual supply chain echelons so as to meet delivery quality targets at minimum cost. The information required by the SCP is distributed across the entire supply chain network. Moreover, the SCP does not always have access to accurate or truthful information. The primary reason for this is the fact that typical supply chain entities (such as echelon managers and service providers) are autonomous, rational, and intelligent, and consequently exhibit strategic behavior. As a result, these entities may not reveal their true cost information; in fact, they may provide false information in the best hope of maximizing their individual utility functions. Eliciting truthful information is key to forming an optimal supply chain but calls for payment of appropriate incentives to the rational entities.

There is a rich body of literature that deals with mechanism design approaches to supply chain formation and logistics network formation. Interesting variants of this problem are discussed in [31, 32, 33, 34].

## 6.2.4 Communication Networks

Pricing, incentive design, and protocol design in communication networks (both wired and wireless) with selfish nodes based on mechanism design techniques represents a major current area engaging the attention of a large number of groups in the world. Some of the literature has already been surveyed in Chapter 5 on ad hoc networks. The paper by Ozdaglar and Srikant [35] is a recent survey in this area.

## 6.2.5 Peer-to-Peer Networks

Babaioff, Chuang, and Feldman [36] present an interesting application of mechanism design to peer-to-peer networks. Peer-to-peer systems support a wide range of applications from file sharing to sophisticated distributed computation (like in computational grids). Strategic behavior is common in such networks. A wide variety of incentive based techniques are being explored to overcome the problems arising out of strategic behavior by peers. Mechanism design is one of the key techniques being used.

## 6.2.6 Prediction Markets

Prediction markets, which are also called as information markets, are markets that aggregate knowledge and opinions about the likelihood of future events. Immediate applications include financial forecasting, options trading, weather forecasts, etc. The design of prediction markets involves many interesting mechanism design issues. A survey is provided by Pennock and Sami [37].

## 6.2.7 Social Networks

Social networks have now become an extremely popular medium for communication and information dissemination. A social network essentially consists of rational and intelligent players who are required to perform tasks such as information transfer, responding to queries, providing recommendations, etc. The individual players in a social network could exhibit highly strategic behavior. Thus game theory and mechanism design provide a valuable tool in understanding the dynamics of a social network and to accomplish a well-defined function through a social network. Kleinberg [38] provides a glimpse of the issues in social network dynamics.

## 6.3 In Conclusion

There are several conferences, some of them recently launched, which report the applications of mechanism design to network economics problems. These include: ACM Conference on Electronic Commerce (ACM EC), International Workshop on Network and Internet Economics (WINE), International Symposium on Algorithmic Game Theory (SAGT), IEEE Conference on Electronic Commerce Technology (CEC), International Conference on Autonomous Agents and Multi-Agent Systems (AAMAS), International Conference on the World Wide Web (WWW), International Joint Conference on Artificial Intelligence (IJCAI), and International Conference of the American Association of Artificial Intelligence (AAAI). Occasionally, conferences such as the IEEE Conference on Foundations of Computer Science (FOCS), ACM Symposium on Theory of Computing (STOC), ACM-SIAM Symposium on Discrete Algorithms (SODA), ACM SIG Conference on Communications (SIGCOMM), and ACM Principles of Distributed Computing (PODC) also publish relevant work. These conferences and workshops constitute a valuable source to look up the current art in the area. Archival results appear in the journals, which have been listed as part of the references in the monograph.

# References

1. P. Cramton, Y. Shoham, and R. Steinberg. Introduction to combinatorial auctions. In P. Cramton, Y. Shoham, and R. Steinberg (eds.), *Combinatorial Auctions*, pages 1–14. The MIT Press, Cambridge, Massachusetts, 2006.
2. S. de Vries and R.V. Vohra. Combinatorial auctions: A survey. *INFORMS Journal of Computing*, 15(1):284–309, 2003.
3. S. de Vries and R.V. Vohra. Design of combinatorial auctions. In D. Simchi-Levi, S.D. Wu, and Z.J. Shen (eds.), *Handbook of Quantitative Supply Chain Analysis: Modeling in the E-Business Era*, pages 247–292. International Series in Operations Research and Management Science, Kluwer Academic Publishers, Norwell, MA, USA, 2005.
4. A. Pekec and M.H. Rothkopf. Combinatorial auction design. *Management Science*, 49:1485–1503, 2003.
5. Y. Narahari and P. Dayama. Combinatorial auctions for electronic business. *Sadhana - Indian Academy Proceedings in Engineering Sciences*, 30(2-3):179–212, 2005.
6. L. Blumrosen and N. Nisan. Combinatorial auctions. In N. Nisan, T. Roughgarden, E. Tardos, and V.V. Vazirani (eds.), *Algorithmic Game Theory*, pages 267–300. Cambridge University Press, New York, 2007.
7. J.D. Hartline and A.R. Karlin. Profit maximization in mechanism design. In N. Nisan, T. Roughgarden, E. Tardos, and V.V. Vazirani (eds.), *Algorithmic Game Theory*, pages 331–362. Cambridge University Press, New York, 2007.
8. A. Malakhov and R.V. Vohra. Single and multi-dimensional optimal auctions — a network perspective. Discussion paper, Kellogg School of Management, Northwestern University, 2004.
9. J.O. Ledyard. Optimal combinatoric auctions with single-minded bidders. In *Proceedings of the 8th ACM conference on Electronic commerce (ACM EC)*, pages 237–242, 2007.
10. R. Gautam, N. Hemachandra, Y. Narahari, H. Prakash, , D. Kulkarni, and J.D. Tew. An optimal mechanism for multi-unit procurement with volume discount bids. *International Journal of Operational Research (To appear)*, 2008.
11. A. Kumar and G. Iyengar. Optimal procurement auctions for divisible goods with capacitated suppliers. Technical report, Columbia University, New York, 2006.
12. V. Krishna and M. Perry. Efficient mechanism design. *SSRN eLibrary*, 1998.
13. V. Krishna. *Auction Theory*. Academic Press, San Diego, California, USA, 2002.
14. H. Moulin. Efficient, strategy-proof and almost budget-balanced assignment. Technical report, Northwestern University, Center for Mathematical Studies in Economics and Management Science, 2007.
15. K. Jain and M. Mahdian. Cost sharing. In N. Nisan, T. Roughgarden, E. Tardos, and V.V. Vazirani (eds.), *Algorithmic Game Theory*, pages 385–410. Cambridge University Press, New York, 2007.
16. D.C. Parkes. Iterative combinatorial auctions. In P. Cramton, Y. Shoham, and R. Steinberg (eds.), *Combinatorial Auctions*. The MIT Press, Cambridge, Massachusetts, 2005.
17. S. Athey and I. Segal. An efficient dynamic mechanism. Technical report, UCLA Department of Economics, 2007.
18. A. Bapna and T.A. Weber. Efficient Dynamic Allocation with Uncertain Valuations. *SSRN eLibrary*, 2005.
19. D. Bergemann and J. Valimaki. Efficient dynamic auctions. Cowles Foundation Discussion Paper 1584, Cowles Foundation, Yale University, New Haven, 2006.

20. D.C. Parkes. Online mechanisms. In N. Nisan, T. Roughgarden, E. Tardos, and V.V. Vazirani (eds.), *Algorithmic Game Theory*, pages 411–442. Cambridge University Press, New York, 2007.

21. R. Vohra and M. Pai. Optimal dynamic auctions. Technical report, Working Paper, Kellogg School of Management, Northwestern University, Illinois, 2008.

22. S. Ieong, A. Man-Cho So, and M. Sundararajan. Mechanism design for stochastic optimization problems. Technical report, Working Paper, Stanford University, Palo Alto, 2006.

23. N. Nisan and A. Ronen. Computationally feasible VCG mechanisms. *Journal of Artificial Intelligence Research*, 29:19–47, 2007.

24. R. Lavi. Computationally efficient approximation mechanisms. In N. Nisan, T. Roughgarden, E. Tardos, and V.V. Vazirani (eds.), *Algorithmic Game Theory*, pages 301–330. Cambridge University Press, New York, 2007.

25. J. Schummer and R.V. Vohra. Mechanism design without money. In N. Nisan, T. Roughgarden, E. Tardos, and V.V. Vazirani (eds.), *Algorithmic Game Theory*, pages 243–266. Cambridge University Press, New York, 2007.

26. T.S. Chandrashekar, Y. Narahari, C.H. Rosa, D. Kulkarni, J.D. Tew, and P. Dayama. Auction based mechanisms for electronic procurement. *IEEE Transactions on Automation Science and Engineering*, 4(3):297–321, 2006.

27. G. Hohner M. Bichler, A. Davenport and J. Kalagnanam. Industrial procurement auctions. In P. Cramton, Y. Shoham, and R. Steinberg (eds.), *Combinatorial Auctions*, pages 593–612. The MIT Press, Cambridge, Massachusetts, 2006.

28. P. Cramton. Spectrum auctions. In M. Cave, S. Mazumdar, and I. Vogelsang, (eds.), *Handbook of Telecommunications Economics*, pages 605-639. Amsterdam: Elsevier Science B.V., 2002.

29. P. Cramton. Simultaneous ascending auctions. In P. Cramton, Y. Shoham, and R. Steinberg (eds.), *Combinatorial Auctions*, pages 99–114. The MIT Press, Cambridge, Massachusetts, 2006.

30. L.M. Ausubel and P. Milgrom. Ascending proxy auctions. In P. Cramton, Y. Shoham, and R. Steinberg (eds.), *Combinatorial Auctions*, pages 79–98. The MIT Press, Cambridge, Massachusetts, 2006.

31. W.E. Walsh, M.P. Wellman, and F. Ygge. Combinatorial auctions for supply chain formation. In *2nd ACM Conference on Electronic Commerce (ACM EC)*, pages 260–269, 2000.

32. M. Babaioff and W.E. Walsh. Incentive compatible, budget balanced, yet highly efficient auctions for supply chain formation. *Decision Support Systems*, 39(1):123–149, 2004.

33. M. Babaioff and W.E. Walsh. Incentive compatible supply chain auctions. In B. Chaib-draa and J. Muller (eds.), *Multiagent Based Supply Chain Management*, Springer, Berlin, Heidelberg, 2006.

34. Y. Narahari, N. Hemachandra, N.K. Srivastava, D. Kulkarni, and J.D. Tew. A Bayesian incentive compatible mechanism for decentralized supply chain formation. *International Journal of Operational Research (To appear)*, 2008.

35. A. Ozdaglar and R. Srikant. Incentives and pricing in communication networks. In N. Nisan, T. Roughgarden, E. Tardos, and V.V. Vazirani (eds.), *Algorithmic Game Theory*, pages 571–592, Cambridge University Press, New York, 2007.

36. J. Chuang, M. Babaioff, and M. Feldman. Incentives in peer-to-peer systems. In N. Nisan, T. Roughgarden, E. Tardos, and V.V. Vazirani (eds.), *Algorithmic Game Theory*, pages 593–612, Cambridge University Press, New York, 2007.

37. D.M. Pennock and R. Sami. Computaional aspects of prediction markets. In N. Nisan, T. Roughgarden, E. Tardos, and V.V. Vazirani (eds.), *Algorithmic Game Theory*, pages 651–676, Cambridge University Press, New York, 2007.

38. J. Kleinberg. Cascading behavior in networks: Algorithmic and economic issues. In N. Nisan, T. Roughgarden, E. Tardos, and V.V. Vazirani (eds.), *Algorithmic Game Theory*, pages 613–632, Cambridge University Press, New York, 2007.

# Index